Equations of Evolution

Equations of Evolution

Hiroki Tanabe
Osaka University

Translated from Japanese by
N. Mugibayashi and H. Haneda
Kobe University

Pitman
London   San Francisco   Melbourne

PITMAN PUBLISHING LIMITED
39 Parker Street, London WC2B 5PB

FEARON PITMAN PUBLISHERS INC.
6 Davis Drive, Belmont, California 94002, USA

Associated Companies
Copp Clark Pitman, Toronto
Pitman Publishing New Zealand Ltd, Wellington
Pitman Publishing Pty Ltd, Melbourne

AMS Subject Classifications: (main) 47F05
                        (subsidiary) 35J30, 35K35, 35L45, 47D05, 47D05, 47H15, 49A20

British Library Cataloguing in Publication Data

Tanabe, Hiroki
   Equations of evolution. – (Monographs and studies
   in mathematics; 6).
   1. Equations
   I. Title II. Series
   512.9'42   QA211

ISBN 0-273-01137-5

Translated from HATTENHOTEISHIKI by Hiroki Tanabe
© Hiroki Tanabe 1975
First published in Japanese by IWANAMI SHOTEN, Publishers, Tokyo
First English edition published by Pitman Publishing Ltd. 1979
Translated by N. Mugibayashi and H. Haneda

© Hiroki Tanabe 1979

Typeset in Northern Ireland by The Universities Press (Belfast) Ltd.

Printed in Great Britain by Pitman Press, Bath.

Contents

Preface

The aim of this book is to give a systematic presentation of the theory of equations of evolution based mainly on the results from semigroups of linear operators. It was quickly recognized that the theory of semigroups known as Hille–Yosida's theory is capable of far-reaching applications to the initial and mixed problems of hyperbolic and parabolic equations with coefficients independent of the time variable. It was Kato [76] who first produced significant results concerning the semigroup-theoretic treatment of equations with coefficients that depend on the time variable. Around 1959, the application of the theory to parabolic equations was initiated. Since then the research in this field has become very active and an extensive literature has appeared. A number of interesting books on these subjects have already been published, such as Kreĭn's book [9] and others listed in the bibliography. I felt some hesitation in writing a similar book on the same subject, but I have dared to undertake it and have endeavoured to reduce the overlap with other published books as far as possible. In the study of equations of evolution there is another method involving the use of some representation theorems of linear functionals by quadratic forms. On this subject the reader is referred to Lions [11], Carroll [3], etc.

The contents of this book are as follows. Chapter 1 consists of fundamental facts from functional analysis which are used frequently in the subsequent chapters. Chapter 2 deals with dissipative operators and fractional powers of operators, both of which are closely related to linear semigroups. Chapter 3 is devoted to the theory of the semigroups of linear operators together with its application. Although the basic theory of linear semigroups is throughly described in many books, a complete account is given since it plays a fundamental role throughout the present volume. Chapters 4 and 5 develop the method of solving temporally inhomogeneous equations and apply the results to the initial and mixed problems of hyperbolic and parabolic equations. In Chapter 5 a brief account of the regularity and asymptotic behaviour of the solutions is given in addition to their existence and uniqueness. In some parts of the applications in Chapters 3 and 5, the theory by Agmon, Douglis, Nirenberg and Schechter on elliptic boundary-value problems becomes essential, and the results will be employed without verification since their proof is too lengthy to be included in this book. Chapter 6 consists of a discussion of a semilinear hyperbolic equation, which is important in physics, and a brief survey of the theory of monotone operators, which has become very active in recent years. Finally, in Chapter 7, a general idea of the optimal control of an evolution equation is given.

The prerequisite for reading this book is an elementary course on linear operators in Banach and Hilbert spaces and distributions.

The greater part of this book consists of Professor T. Kato's extensive work, and the author is very grateful to him for having constantly influenced the author in this field. The author also expresses has sincerest gratitude to Professor H. Fujita of the University of Tokyo for providing the opportunity to write this book. Mr K. Maruo of Osaka University read through the manuscript and gave valuable advice. Joint research with Messrs Y. Fujie, N. Yamada, M. Nasu and A. Yagi, in addition to Mr Maruo, was most helpful during the writing of this book. The author also expresses has sincerest thanks to all of these people. Finally, the author wishes to express his appreciation to the Iwanami Shoten for their most efficient handling of publication of this book.

March 1975   Hiroki Tanabe

Preface to the English Edition

It is a great honour and pleasure that an English edition of my book on equations of evolution has been published. The present translation follows the original Japanese edition without change except for a small supplement to the bibliography and the correction of some misprints and errors suggested to me by the translators and Mr J. Shirota, to whom I express my profound thanks. I also wish to express my sincerest gratitude to Professors N. Mugibayashi and H. Haneda of Kobe University for their excellent work of translation. Finally my thanks are extended to the Iwanami Shoten and Pitman Publishing Ltd for the publication of this English edition.

December 1977   Hiroki Tanabe

Notation

$\mathbb{R}$	field of real numbers
$\mathbb{C}$	field of complex numbers
$\mathbb{R}^n$	n-dimensional Euclidean space
$X \backslash Y$	set of all elements in X which do not belong to Y
int	interior
$\partial \Omega$	boundary of a set Ω
X^*	conjugate space of a normed space X. (In 2.2, 3.6, 5.4, 5.5 and Application 2 in 5.9 only, X^* stands for the set of all continuous antilinear functionals defined on X.)
T^*	adjoint operator of T
$f(u) = (f, u) = \overline{(u, f)}$	value of f at u when $f \in X^*$ and $u \in X$
$Y^\perp$	set of all elements which are orthogonal to a subset Y of a normed space or its conjugate space
$D(T)$	domain of an operator T
$R(T)$	range of an operator T
$G(T)$	graph of an operator T
$B(X, Y)$	set of all bounded linear operators from a normed space X into a normed space Y
$B(X) = B(X, X)$	
$\bar{B}(X, Y)$	set of all bounded antilinear operators from a normed space X into a normed space Y
$T \subset S$	extension of an operator T to an operator S
$\rho(T)$	resolvent set of T
$\sigma(T)$	spectrum of T
$\exp(tA)$	semigroup generated by A
$G(X, M, \beta)$	set of all generators of a semigroup on a Banach space X which satisfies $\|T(t)\| \leqslant Me^{\beta t}$

$G(X, \beta) = \bigcup_{M \geqslant 1} G(X, M, \beta)$

$G(X) = \bigcup_{-\infty < \beta < \infty} G(X, \beta)$ set of all linear semigroups on a Banach space X

$\rightarrow$ strongly or $\rightarrow$ strong convergence

$\rightarrow$ weakly or $\rightharpoonup$ weak convergence

$D^\alpha, |\alpha|$ $D^\alpha = (i^{-1}\partial/\partial x_1)^{\alpha_1} \cdots (i^{-1}\partial/\partial x_n)^{\alpha_n}$, $|\alpha| = \alpha_1 + \cdots + \alpha_n$ when $D = (i^{-1}\partial/\partial x_1, \ldots, i^{-1}\partial/\partial x_n)$, $\alpha = (\alpha_1, \ldots, \alpha_n)$

$W_p^m(\Omega)$ set of all functions whose derivatives up to degree m in distribution sense belong to $L^p(\Omega)$

$\| \ \|_{p,\Omega} = \| \ \|_p$ norm as an element of $L^p(\Omega)$

$\| \ \|_{m,p,\Omega} = \| \ \|_{m,p}$ norm as an element of $W_p^m(\Omega)$

$H_m(\Omega) = W_2^m(\Omega)$

$W_p^{m-1/p}(\partial\Omega)$ set of all restrictions of functions belonging to $W_p^m(\Omega)$ $(m > 0)$ to a boundary $\partial\Omega$

$[\ ]_{m-1/p,\partial\Omega} = [\ ]_{m-1/p}$ norm as an element of $W_p^{m-1/p}(\partial\Omega)$

$C^m(\Omega)$ set of all m-times continuously differentiable functions on Ω

$C_0^m(\Omega)$ set of all functions in $C^m(\Omega)$ with compact support

$B^m(\Omega)$ set of all functions whose derivatives up to degree m are continuous and bounded on Ω

$|\ |_{m,\Omega} = |\ |_m$ norm as an element of $B^m(\Omega)$

$\mathring{W}_p^m(\Omega)$ closure of $C_0^m(\Omega)$ in $W_p^m(\Omega)$

$\mathring{H}_m(\Omega) = \mathring{W}_2^m(\Omega)$

$C^m((a, b); X), C^m([a, b]; X)$ set of all m-times continuously differentiable functions from (a, b), $[a, b]$ into a Banach space X

$C((a, b); X) = C^0((a, b); X)$

$C([a, b]; X) = C^0([a, b]; X)$

$L^p(a, b; X), 1 \leqslant p < \infty$ set of all strongly measurable functions from (a, b) into a Banach space X whose pth power of norms are integrable

$L^\infty(a, b; X)$ set of all essentially bounded and strongly measurable functions from (a, b) into a Banach space X

${}^t(u_1, \ldots, u_N)$ transposed matrix of $(u_1, \ldots, u_N) = \begin{bmatrix} u_1 \\ \vdots \\ u_N \end{bmatrix}$

$|A|$ Lebesgue measure of $A \subset \mathbb{R}^n$

1

Preliminaries from functional analysis

In this chapter, basic facts from functional analysis are explained for later use. Most of the proofs will be omitted. *See* for the details Dunford and Schwartz [4], Yosida [18], Hille and Phillips [7], etc.

1.1 Banach spaces

Let $\mathbb{R}$ and $\mathbb{C}$ be real and complex number fields. A linear space over $\mathbb{R}$ or $\mathbb{C}$ is called a real or complex linear space, respectively. The Hahn–Banach theorem states that any linear functional given in a subspace of a real linear space X can be extended to the entire space X. If X is a normed space, the norm of an element of X is usually denoted by $\|\cdot\|$. The collection of all continuous linear functionals defined on the whole of X constitutes a space conjugate to X which is denoted by X^*. The space X^* is a Banach space with the norm $\|f\| = \sup_{\|u\| \le 1} |f(u)|$, where $f \in X^*$ and $u \in X$. The Hahn–Banach theorem immediately implies the following three theorems concerning the existence of continuous linear functionals on a normed space.

Theorem 1.1.1 *Let f_0 be a continuous linear functional defined on a subspace X_0 of a normed space X, then there exists an extension $f \in X^*$ of f_0 with the norm*

$$\|f\| = \sup_{u \in X_0, \|u\| \le 1} |f_0(u)|.$$

Theorem 1.1.2 *Given an arbitrary element u of a normed space X, there is an $f \in X^*$ satisfying $f(u) = \|u\|$ and $\|f\| = 1$.*

Theorem 1.1.3 *Let Y be a closed subspace of a normed space X and $u_0 \in X \backslash Y$, then there exists an $f_0 \in X^*$ such that $f_0(u_0) = 1$ and, moreover, $f_0(u) = 0$ for all $u \in Y$.*

Let us now denote by f_0 the element of X^* which, Theorem 1.1.2 assures us, exists for any element u of a normed space X. Then, on putting $f = \|u\| f_0$, the relation

$$f(u) = \|u\|^2 = \|f\|^2 \tag{1.1}$$

holds.

Definition 1.1.1 *Let X be a normed space and X^* its conjugate. The set of all $f \in X$ which satisfy (1.1) for every $u \in X$ is denoted by Fu. The F is called a* duality mapping *from X into X^*.*

The duality mapping is, in general, multivalued.

Remark 1.1.1 If $f(u) = 0$ for $u \in X$ and $f \in X^*$, then it is said that f and u are *orthogonal* to each other. Let M and N be non-void subsets of X and X^*, respectively. Denote by $M^\perp$ (respectively, $N^\perp$) the set of those elements of X^* (respectively, X) which are orthogonal to M (respectively, N). $(X^*)^\perp = \{0\}$ by Theorem 1.1.2. Furthermore, Theorem 1.1.3. says that $Y^{\perp\perp} = Y$ for each closed subspace Y of X.

For $u \in X$ and $f \in X^*$, we often write

$$f(u) = (u, f). \tag{1.2}$$

If X is a complex normed space, it is convenient to define the product of $f \in X^*$ and $\alpha \in \mathbb{C}$ by $(\alpha f)(u) = \bar{\alpha} f(u)$, because, by following the convention (1.2), we have a familiar expression $(u, \alpha f) = \bar{\alpha}(u, f)$, as in the Hilbert space description. Let u be an element of a normed space X and put $F(f) = \overline{f(u)}$ for each $f \in X^*$, then it is evident that $F \in X^{**}$ and $\|F\| \leq \|u\|$. But, by Theorem 1.1.2, there exists an $f \in X^*$ satisfying $f(u) = \|u\|$ and $\|f\| = 1$, so that we obtain $\|u\| = F(f) \leq \|F\|$ and, hence, $\|F\| = \|u\|$. Thus we can identify u and F and regard X as a subspace of X^{**}. In particular, if X can be regarded as being identical with X^{**}, it is said to be *reflexive*.

Let X and Y both be real or complex normed spaces. Also, let T be a linear mapping which carries each element of some subspace D of X into Y. D is called the *domain* of T and is denoted by $D(T)$. The image of D by the mapping T is called the *range* of T and is denoted by $R(T)$. A linear operator T is continuous if and only if

$$\|T\| \equiv \sup_{u \in D(T), \|u\| \leq 1} \|Tu\| < \infty.$$

In particular, if T is defined and continuous on the entire space X, it is called a *bounded operator* from X into Y, and the collection of all such operators is denoted by $B(X, Y)$. $B(X, Y)$ is a normed space. If Y is complete, then so is $B(X, Y)$ and the latter becomes a Banach space. $B(X, X)$ is simply written as $B(X)$. Let I be the identity mapping of X onto X and for $\lambda \in \mathbb{R}$ or $\mathbb{C}$ let us write simply λ instead of λI. A linear operator T has a continuous inverse if and only if there is a positive number α such that

$$\|Tu\| \geq \alpha \|u\| \tag{1.3}$$

holds for every $u \in D(T)$. For two real (or complex) normed spaces X and Y we agree to define $X \times Y = \{[u, v]: u \in X, v \in Y\}$. If a norm is introduced by $\|[u, v]\| = \|u\| + \|v\|$, then $X \times Y$ becomes a real (or complex) normed space. It is easy to see that $(X \times Y)^* = X^* \times Y^*$ and $([u, v], [f, g]) = (u, f) + (v, g)$ for $[u, v] \in X \times Y$ and $[f, g] \in X^* \times Y^*$, where we have used the convention (1.2). If T is a linear operator from X into Y, then the set $G(T) = \{[u, Tu]: u \in D(T)\}$ is called the *graph* of T. $G(T)$ is a subspace of $X \times Y$. When $D(T)$ is dense in X, the *adjoint operator* T^* of T is defined as follows:

$D(T^*) = \{g \in Y^*: g(Tu)$, as a functional of u, is continuous
in $D(T)$ under the norm of $X\}$.

If $g \in D(T^*)$, then $g(Tu) = (T^*g)(u)$ for every $u \in D(T)$.

When X and Y are both Hilbert spaces, it is customary to regard $X = X^*$ and $Y = Y^*$ by Riesz's theorem and to define T^* by

$D(T^*) = \{v \in Y: (Tu, v)$ is continuous in $D(T)$ with respect to $u\}$.

If $v \in D(T^*)$, then $(Tu, v) = (u, T^*v)$ for every $u \in T(D)$.

$T \in B(X, Y)$ implies $T^* \in B(Y^*, X^*)$ and the following relation holds:

$$\|T^*\| = \|T\|. \tag{1.4}$$

Let $V[u, v] = [-v, u]$ for each $[u, v] \in X \times Y$, then $V \in B(X \times Y, Y \times X)$ and the graph of T^* is represented as $G(T^*) = (VG(T))^{\perp}$. A linear operator T from X into Y is said to be *closed* if its graph is a closed subspace of $X \times Y$. An alternative definition of the closed operator T is that the relations $u_n \in D(T)$, $u_n \to u$, $Tu_n \to v$ imply the relations $u \in D(T)$ and $Tu = v$. The inverse of a closed operator, if it exists, is also closed. An adjoint operator is always closed. In the remainder of this section, we shall be concerned exclusively with linear operators.

Theorem 1.1.4 *Let X and Y both be reflexive Banach spaces, and T a closed operator from X into Y whose domain $D(T)$ is dense in X. Then $D(T^*)$ is also dense in Y and $T^{**} = T$.*

Proof In order to see the denseness of $D(T^*)$, it is sufficient by virtue of

Theorem 1.1.3 to ascertain that $f(v) = 0$, $v \in Y$, for every $f \in D(T^*)$ implies $v = 0$. Since $([v, 0], [f, Tf]) = 0$, we have $[v, 0] \in G(T^*)^\perp = (VG(T))^{\perp\perp}$, from which, by Remark 1.1.1, it follows that $[v, 0] \in VG(T)$ and, accordingly, $v = T0 = 0$. The relation $T^{**} = T$ is evident. $\square$

Consider two linear operators T and S from X into Y with domains $D(T) \subset D(S)$. If $Tu = Su$ for each $u \in D(T)$, then S is called an *extension* of T and we write $T \subset S$. A linear operator T is said to be *closable* if the operator itself is not closed but has a closed extension. A necessary and sufficient condition for T to be a closable operator is that the relations $u_n \in D(T)$, $u_n \to 0$, $Tu_n \to v$ imply that $v = 0$. If the condition is satisfied, $G(T)$ is a closure of the graph of a certain closed operator $\bar{T}$. The operator $\bar{T}$ is called the *smallest closed extension* of T. Henceforth, we will simply say linear operators, closed operators, bounded operators in X in place of those operators from X into X.

Let u_0 be an element of a normed space X. Take families of elements of X^*, $\{f_i\}$, $i = 1, \ldots, n$, $n = 1, 2, \ldots$ and put

$$U(u_0 ; f_1, \ldots, f_n, \varepsilon) = \{u \in X : |f_i(u - u_0)| < \varepsilon, i = 1, \ldots, n\}$$

for $\varepsilon > 0$. Since the totality of these sets, obtained by arbitrarily varying $f_1, \ldots, f_n$, n and ε, satisfies the axioms for a neighbourhood system, a topology is introduced in X by considering them as a neighbourhood of u_0. It is called the *weak topology*. The weak topology does satisfy Hausdorff's axiom of separation by Theorem 1.1.2, but not the first countability axiom. As a distinction from the weak topology, a topology defined by the norm of X is called the *strong topology*. The strong topology is stronger than the weak topology. When a sequence $\{u_n\}$ of points of X converges to u in the weak topology, we say that $\{u_n\}$ *converges weakly* to u and denote it by w-$\lim_{n \to \infty} u_n = u$ or $u_n \to u$ weakly. A sequence $\{u_n\}$ converges weakly to u if and only if $\lim_{n \to \infty} f(u_n) = f(u)$ for every $f \in X^*$. If u_n converges to u in the strong topology, namely $\|u_n - u\| \to 0$, we speak of the *strong convergence* of $\{u_n\}$ to u and denote it by s-$\lim_{n \to \infty} u_n = u$ or $u_n \to u$ strongly. Unless specified otherwise, the topology of X will mean the strong topology.

Let X be a normed space and f_0 an element of X^*. Take families of elements of X, $\{u_i\}$, $i = 1, \ldots, n$, $n = 1, 2, \ldots$ and put

$$U(f_0 ; u_1, \ldots u_n, \varepsilon) = \{f \in X^* : |(f - f_0)(u_i)| < \varepsilon, i = 1, \ldots, n\}$$

for $\varepsilon > 0$. The totality of these sets obtained by arbitrarily varying $u_1, \ldots, u_n$, n and ε satisfies the axioms for a neighbourhood system. A topology introduced in X^* by considering them as a neighbourhood of f_0 is called the w^* *topology*. The w^* topology does satisfy Hausdorff's axiom of separation, but not the first countability axiom. By regarding X^* as a

Banach space, we can also introduce the weak topology into it. The w^* topology of X^* is weaker than the weak topology, but they coincide if X is reflexive. A sequence of points $\{f_n\}$ converges to f in the w^* topology if and only if $f_n(u) \to f(u)$ for every $u \in X$.

If a normed space is not finite-dimensional, its bounded set is not relatively compact. Therefore the Weierstrass–Bolzano theorem is not applicable for the strong topology. However, we can replace it by the following theorems.

Theorem 1.1.5 *The unit ball of a space X^* conjugate to a Banach space X, i.e., the set $\{f \in X^* : \|f\| \leqslant 1\}$, is compact in the w^* topology.*

Since X is the conjugate space of X^* if X is reflexive, this theorem implies that the unit ball of X is compact in the weak topology or, in short, is weakly compact. The converse of this statement is also true, i.e., we have

Theorem 1.1.6 *A Banach space X is reflexive if and only if the unit ball of X, $\{u \in X : \|u\| \leqslant 1\}$, is weakly compact.*

Theorem 1.1.7 *The unit ball of a reflexive Banach space X is sequentially compact in the weak topology. That is to say, if $\|u_n\| \leqslant 1$ $(n = 1, 2, \ldots)$, a certain subsequence of $\{u_n\}$ converges weakly to an element u of the unit ball of X.*

An operator $T \in B(X, Y)$ is, by definition, continuous in the strong topologies of X and Y. One can easily show that it is also continuous in the weak topologies of X and Y.

Let K be a subset of a linear space X. K is said to be *convex* if, together with any two points of it, K contains the entire line segment which connects them. In other words, the fact that K is a convex set means that $u, v \in K$ and $0 < t < 1$ imply $(1 - t)u + tv \in K$. As an important consequence of the Hahn–Banach theorem, we have the following *Mazur's theorem.*

Theorem 1.1.8 *Suppose that K is a closed convex set of a normed space X and that $u_0 \notin K$. Then there exists an $f_0 \in X^*$ which satisfies $\sup_{u \in K} \operatorname{Re} f_0(u) < \operatorname{Re} f_0(u_0)$.*

Corollary 1 *If K is a closed convex set of a normed space X, then there exists a subset Φ of $X^* \times \mathbb{R}$ or $X^* \times \mathbb{C}$ such that*

$$K = \{u \in X : \operatorname{Re} f(u) \leqslant c \text{ for all } (f, c) \in \Phi\}.$$

Corollary 2   A strongly closed convex set of a normed space is weakly closed. In particular, a strongly closed subspace is weakly closed.

Corollary 3   If $\{u_n\} \subset X$ converges weakly to u, then a sequence of convex combinations $v_k = \sum_{n=k}^{\infty} \lambda_n^k u_n$ converges strongly to u. Here, $\lambda_n^k \geqslant 0$, $\sum_{n=k}^{\infty} \lambda_n^k = 1$, and only a finite number of λ_n^k are non-vanishing for each k.

Corollary 4   Let T be a closed operator from a normed space X into a normed space Y. If $D(T) \ni u_n \rightarrow u$ weakly and $Tu_n \rightarrow v$ weakly, then $u \in D(T)$ and $Tu = v$.

Theorem 1.1.9   Suppose that K is a convex set of a normed space X and that it contains interior points. If $u_0 \notin K$, then there exists a non-zero $f_0 \in X^*$ which satisfies $\sup_{u \in k} \text{Re } f_0(u) \leqslant \text{Re } f_0(u_0)$.

The symbol Re in the theorems and one of the corollaries is superfluous if X is a real normed space.

The Banach space X is a complete metric space and, consequently, by Baire's theorem, the complement of its subset of the first category is dense. This provides the following two important theorems.

Theorem 1.1.10   (The Range Theorem, The Open Mapping theorem) *Let X, Y be two Banach spaces and T a closed operator from X into Y. Suppose $R(T)$ is a set of the second category, then*

  (i) *$R(T) = Y$,*
  (ii) *there exists a positive number δ such that*

$$\{Tu : u \in D(T), \|u\| \leqslant 1\} \supset \{v \in Y : \|v\| \leqslant \delta\}, \tag{1-.5}$$

  (iii) *the inverse operator T^{-1}, if it exists, is continuous.*

Theorem 1.1.11   (The Uniform Boundedness Theorem, The Banach–Steinhaus Theorem)   *Let X be a Banach space, Y a normed space, and H a subset of $B(X, Y)$. Suppose $\{\|Tu\| : T \in H\}$ is bounded, then so is $\{\|T\| : T \in H\}$.*

From Theorem 1.1.10 there immediately follows

Theorem 1.1.12   (The Closed Graph Theorem)   *Let X, Y be two Banach spaces and T a linear operator from X into Y such that $D(T) = X$. Then, T is a continuous operator if and only if it is closed.*

Corollary   For a closable operator A and a bounded operator B with $R(B) \subset D(A)$ the product AB is bounded.

Let X be a Banach space, Y a normed space, and $\{T_n\}$ a sequence of elements of $B(X, Y)$. If $Tu = \text{s-lim}_{n\to\infty} T_n u$ for every $u \in X$, then, by Theorem 1.1.11, we have $T \in B(X, Y)$ and say that $\{T_n\}$ *converges strongly* to T. Furthermore, in this case the inequality $\|T\| \leqslant \lim \inf \|T_n\|$ holds.

Now let T be a closed operator from X into Y. A question may arise as to whether the equation $v = Tu$ has a solution for every $v \in Y$, in other words, whether $R(T) = Y$. Here is a criterion for this problem.

Theorem 1.1.13 *Let X, Y be Banach spaces and T a closed operator from X into Y with a dense domain $D(T)$. Then $R(T) = Y$ if and only if T^* has a continuous inverse.*

Proof Suppose that $R(T) = Y$. Then, by Theorem 1.1.10, there exists a positive number δ such that (1.5) is valid. Hence, for $f \in D(T^*)$, we have

$$\|T^*f\| = \sup_{u \in D(T), \|u\| \leqslant 1} |(T^*f)(u)|$$

$$= \sup_{u \in D(T), \|u\| \leqslant 1} |f(Tu)| \geqslant \sup_{\|v\| \leqslant \delta} |f(v)| = \delta\|f\|.$$

Conversely, assume that T^* has a continuous inverse. We denote the closure of the set on the left-hand side of (1.5) by K and start with proving the existence of a number $\delta > 0$ satisfying $K \supset \{v \in Y : \|v\| \leqslant \delta\}$. Suppose, on the contrary, that there does not exist such a δ. Then for any positive integer n there is a $v_n \notin K$ satisfying $\|v_n\| < 1/n$. Since K is a closed convex set, Theorem 1.1.8 tells us that there is an $f_n \in Y^*$ satisfying $\sup_{v \in K} \text{Re } f_n(v) < \text{Re } f_n(v_n)$. Let $|f_n(v)| e^{i\theta}$ be the polar representation of $f_n(v)$ with v being an arbitrary element of K. Then, since $e^{-i\theta}v \in K$, we have

$$|f_n(v)| = \text{Re } f_n(e^{-i\theta}v) \leqslant \sup_{v \in K} \text{Re } f_n(v)$$

$$< \text{Re } f_n(v_n) \leqslant |f_n(v_n)|$$

and, hence, $\sup_{v \in K} |f_n(v)| < |f_n(v_n)|$. From this it follows that

$$\|T^*f_n\| = \sup_{u \in D(T), \|u\| \leqslant 1} |(T^*f_n)(u)|$$

$$= \sup_{u \in D(T), \|u\| \leqslant 1} |f_n(Tu)| = \sup_{v \in K} |f_n(v)|$$

$$< |f_n(v_n)| \leqslant n^{-1}\|f_n\|,$$

which contradicts the assumption that T^* has a continuous inverse. Therefore, a δ such as mentioned above exists. It is clear that the relation

$$\text{the closure of } \{Tu : u \in D(T), \|u\| \leqslant 2^{-i}\} \supset \{v \in \dot{Y} : \|v\| \leqslant 2^{-i} \delta\} \qquad (1.6)$$

is valid for any $i = 0, 1, 2, \ldots$. Let v be an arbitrary element of Y which satisfies $\|v\| \leqslant \delta$. The relation (1.6) for $i = 0$ implies that there exists a $u_1 \in D(T)$ satisfying $\|v - Tu_1\| < 2^{-1} \delta$. Next, the same relation for $i = 1$ implies that there exists a $u_2 \in D(T)$ satisfying $\|v - Tu_1 - Tu_2\| < 2^{-2} \delta$ and $\|u_2\| \leqslant 2^{-1}$. In this way, it is seen that for any positive integer n there exists a $u_n \in D(T)$ which satisfies $\|v - \sum_{i=1}^{n} Tu_i\| \leqslant 2^{-n} \delta$ and $\|u_n\| \leqslant 2^{1-n}$. Since $u = \sum_{i=1}^{\infty} u_i$ is strongly convergent and T is a closed operator, it is concluded that $u \in D(T)$ and $v = Tu$. $\quad\square$

A similar proposition remains valid if T is replaced by T^*. Indeed, we have

Theorem 1.1.14 *Let X, Y be Banach spaces and T a closed operator from X into Y with a dense domain $D(T)$. Then $R(T^*) = X^*$ if and only if T has a continuous inverse.*

Proof By Theorem 1.1.10, it follows from $R(T^*) = X^*$ that there is a $\delta > 0$ for which

$$\{T^*g : g \in D(T^*), \|g\| \leqslant 1\} \supset \{f \in X^* : \|f\| \leqslant \delta\} \qquad (1.7)$$

holds. Let u be an arbitrary element of $D(T)$. By Theorem 1.1.2, there exists an $f \in X^*$ satisfying $f(u) = \|u\|$ and $\|f\| = 1$. Since, by virtue of (1.7), there is a $g \in D(T^*)$ which satisfies $\delta f = T^*g$ and $\|g\| \leqslant 1$, we get

$$\delta \|u\| = \delta f(u) = (T^*g)(u) = g(Tu) \leqslant \|Tu\|.$$

Conversely, assuming that T^{-1} exists and is continuous, there is an $\alpha > 0$ such that (1.3) is valid for all $u \in D(T)$. Let f be an arbitrary element of X and put $g(Tu) = f(u)$. Then, since

$$|g(Tu)| \leqslant \|f\| \|u\| \leqslant \alpha^{-1} \|f\| \|Tu\|$$

by (1.3), g is a continuous linear functional defined on $R(T)$. By Theorem 1.1.1 the g can be extended to the entire space Y. If the extension is denoted again by g, we obtain $g \in D(T^*)$ and $f = T^*g$. $\quad\square$

Let X be a Banach space and T a bounded operator in X such that $\|T\| < 1$. The *Neumann series expansion* $(I - T)^{-1} = \sum_{n=0}^{\infty} T^n$ assures us that $(I - T)^{-1}$ exists and belongs to $B(X)$. From this it follows that, if

$T \in B(X)$ has a bounded inverse, then so has any $S \in B(X)$ which satisfies $\|S - T\| < \|T^{-1}\|^{-1}$. In addition, we have

Theorem 1.1.15 *Let T be an element of $B(X)$. If T^{-1} exists and belongs to $B(X)$ and if $\|T_n - T\| \to 0$, then, for sufficiently large n, there exists T_n^{-1} which belongs to $B(X)$ and such that $\|T_n^{-1} - T^{-1}\| \to 0$.*

Suppose T is a closed operator in a real (or complex) Banach space. The set of all points λ of $\mathbb{R}$ (or $\mathbb{C}$) for which $T - \lambda$ is an operator mapping $D(T)$ in a one-to-one way onto the entire X is called the *resolvent set* of T and is denoted by $\rho(T)$. $\mathbb{R} \backslash \rho(T)$ (or $\mathbb{C} \backslash \rho(T)$) is called the *spectrum* of T and is denoted by $\sigma(T)$. The *resolvent* of T at λ signifies $(T - \lambda)^{-1}$ for $\lambda \in \rho(T)$. By Theorem 1.1.12 the resolvent is an element of $B(X)$. $\rho(T)$ is an open set of $\mathbb{R}$ (or $\mathbb{C}$) and, hence, $\sigma(T)$ is a closed set.

$$(T - \mu)^{-1} = \sum_{n=0}^{\infty} (\mu - \lambda)^n (T - \lambda)^{-n-1}$$

is the power series expansion of $(T - \mu)^{-1}$ in a neighbourhood of $\lambda \in \rho(T)$ and, if X is a complex Banach space, $(T - \lambda)^{-1}$ is a regular function of λ with values on $B(X)$. From (1.4), Theorems 1.1.13 and 1.1.14 follows:

Theorem 1.1.16 *Let T be a closed operator in a Banach space X with a dense domain $D(T)$. Then, $\lambda \in \rho(T)$ and $\bar{\lambda} \in \rho(T^*)$ are equivalent statements, $((T - \lambda)^{-1})^* = (T^* - \bar{\lambda})^{-1}$, and, hence, $\|(T - \lambda)^{-1}\| = \|(T^* - \bar{\lambda})^{-1}\|$ is valid for any $\lambda \in \rho(T)$.*

Let T be a linear operator in X and $S \in B(X)$. If $TS \supset ST$, then S is said to be *commutative* with T. Evidently, $(T - \lambda)^{-1}$ for $\lambda \in \rho(T)$ is commutative with T.

A densely-defined linear operator T in a Hilbert space X is said to be *symmetric* if $T \subset T^*$; in particular, it is *self-adjoint* if $T = T^*$. A symmetric operator T in a complex Hilbert space is self-adjoint if and only if all the complex but non-real numbers belong to $\rho(T)$. If T is a symmetric operator, then (Tu, u) is real for every $u \in D(T)$. Let T be self-adjoint and suppose that $m \|u\|^2 \leq (Tu, u)$ for some number m and all $u \in D(T)$, then T is said to be *bounded from below*. In particular, if one can take $m > 0$, T is said to be *positive definite*. T is called a *positive operator* if $(Tu, u) > 0$ for all non-zero u. Similarly, the *boundedness from above*, *negative definite operators* and *negative operators* can be defined. A self-adjoint operator bounded from both above and below is bounded symmetric.

Suppose that to every real number λ there corresponds an orthogonal projection $E(\lambda)$ having the properties

   (i) $E(\lambda)E(\mu) = E(\mu)E(\lambda) = E(\min(\lambda, \mu))$,
   (ii) $E(\lambda + 0) = \text{s-lim}_{\mu \to \lambda + 0} E(\mu) = E(\lambda)$,
   (iii) $E(-\infty) = \text{s-lim}_{\lambda \to -\infty} E(\lambda) = 0$, $E(\infty) = \text{s-lim}_{\lambda \to \infty} E(\lambda) = I$.

$\{E(\lambda)\}$ is called the *resolution of the identity*. For each $u, v \in X$, $(E(\lambda)u, v)$ is a function of bounded variation of λ and $\|E(\lambda)u\|^2$ is an increasing function of λ. Let H be a self-adjoint operator, then the resolution of the identity $\{E(\lambda)\}$ is determined uniquely so that H can be represented as

$$H = \int_{-\infty}^{\infty} \lambda \, dE(\lambda),   \tag{1.8}$$

which means that

$$(Hu, v) = \int_{-\infty}^{\infty} \lambda \, d(E(\lambda)u, v)   \tag{1.19}$$

holds for every $u \in D(H)$ and $v \in X$. The integral on the right-hand side of (1.9) is to be understood in the Riemann–Stieltjes sense. The domain of H is

$$D(H) = \left\{ u \in X : \|Hu\|^2 = \int_{-\infty}^{\infty} \lambda^2 \, d \|E(\lambda)u\|^2 < \infty \right\}.   \tag{1.10}$$

Conversely, once the resolution of the identity $\{E(\lambda)\}$ is given, a self-adjoint operator H is defined by (1.9) and (1.10). Each $E(\lambda)$ is commutative with H. If H is bounded from below so that $m = \inf_{\|u\| \le 1} (Hu, u)$ is determined, then $E(\lambda) = 0$ for $\lambda < m$ and $H = \int_{m-0}^{\infty} \lambda \, dE(\lambda)$. Similarly, if H is bounded from above, by putting $M = \sup_{\|u\| \le 1} (Hu, u)$ we obtain that $E(\lambda) = I$ for $\lambda > M$ and $H = \int_{-\infty}^{M} \lambda \, dE(\lambda)$. Let $\varphi(\lambda)$ be a complex-valued function continuous in $-\infty < \lambda < \infty$, then

$$(\varphi(H)u, v) = \int_{-\infty}^{\infty} \varphi(\lambda) \, d(E(\lambda)u, v),   \tag{1.11}$$

$$D(\varphi(H)) = \left\{ u \in X; \int_{-\infty}^{\infty} |\varphi(\lambda)|^2 \, d \|E(\lambda)u\|^2 < \infty \right\}   \tag{1.12}$$

define an operator $\varphi(H)$. We shall write simply $\varphi(H) = \int_{-\infty}^{\infty} \varphi(\lambda) \, dE(\lambda)$. $\varphi(H)$ is self-adjoint if φ is real-valued and it is bounded if φ is bounded. Let H be positive and $\alpha > 0$. By considering a function $\varphi(\lambda)$ given by $\varphi(\lambda) = \lambda^\alpha$ for $\lambda = 0$ and $\varphi(\lambda) = 0$ for $\lambda < 0$, the power of degree α of H is defined by

$$H^\alpha = \int_0^{\infty} \lambda^\alpha \, dE(\lambda).   \tag{1.13}$$

Again, let H be an arbitrary self-adjoint operator given by (1.8). In this case,

$$(H - \mu)^{-1} = \int_{-\infty}^{\infty} (\lambda - \mu)^{-1} \, dE(\lambda) \tag{1.14}$$

for non-real μ. If $\mu \in \rho(H)$, though it may be real, we know that in some neighbourhood of μ the orthogonal projection $E(\lambda)$ is independent of λ and (1.14) remains valid. Let U be an isometric operator which maps a Hilbert space X on the entire X, i.e. let U be an operator such that $D(U) = R(U) = X$ and $\|Uu\| = \|u\|$ for every $u \in X$. Such an operator U is said to be *unitary*. If $|\varphi(\lambda)| \equiv 1$, the operator $\varphi(H)$ defined by (1.11) is unitary.

1.2 Function spaces

Functions and distributions in this section will be complex-valued, unless otherwise stated explicitly. Let Ω be a region in an n-dimensional Euclidean space $\mathbb{R}^n$ and $1 \le p \le \infty$. As usual, the collection of all functions for which the pth power of the modulus is integrable in Ω is denoted by $L^p(\Omega)$. $L^p(\Omega)$ for $p = \infty$ is the collection of all functions which are essentially bounded and measurable in Ω. The norm of a function u in $L^p(\Omega)$ is denoted by $\|u\|_{p,\Omega}$. Let m be a positive integer. $W_p^m(\Omega)$ consists of all functions which, together with their derivatives to order m in the sense of distribution, belong to $L^p(\Omega)$. The norm of a function $u \in W_p^m(\Omega)$ is defined by

$$\begin{cases} \|u\|_{m,p,\Omega} = \left(\sum_{|\alpha| \le m} \|D^\alpha u\|_{p,\Omega}^p \right)^{1/p}, & 1 \le p < \infty, \\[2ex] \|u\|_{m,\infty,\Omega} = \max_{|\alpha| \le m} \|D^\alpha u\|_{\infty,\Omega}. \end{cases}$$

here α stands for a set of n non-negative integers $(\alpha_1, \ldots, \alpha_n)$ and we have designated symbols as follows: $|\alpha| = \alpha_1 + \cdots + \alpha_n$, $D = (i^{-1} \partial/\partial x_1, \ldots, i^{-1} \partial/\partial x_n)$ and $D^\alpha = (i^{-1} \partial/\partial x_1)^{\alpha_1} \cdots (i^{-1} \partial/\partial x_n)^{\alpha_n}$. $W_2^m(\Omega)$ is usually denoted by $H_m(\Omega)$. It is evident that $W_p^0(\Omega) = L^p(\Omega)$. $W_p^m(\Omega)$ is a Banach space and, especially, $H_m(\Omega)$ is a Hilbert space. We denote by $C^m(\Omega)$ the collection of all functions which are m-times continuously differentiable in Ω. Among them, the collection of functions whose supports are compact in Ω will be denoted by $C_0^m(\Omega)$. Another set of elements of $C^m(\Omega)$, having bounded derivatives to order m, will be denoted by $B^m(\Omega)$. $B^m(\Omega)$ thus defined is a Banach space with the norm given by

$$|u|_{m,\Omega} = \sum_{|\alpha| \le m} \sup_{x \in \Omega} |D^\alpha u(x)|.$$

When no confusion arises, we shall simply write $\|u\|_p$, $\|u\|_{m,p}$, $|u|_m$ instead of $\|u\|_{p,\Omega}$, $\|u\|_{m,p,\Omega}$, $|u|_{m,\Omega}$, respectively. The closure of $C_0^m(\Omega)$ is denoted by $\mathring{W}_p^m(\Omega)$. As above, $\mathring{W}_2^m(\Omega)$ is denoted, in particular, by $\mathring{H}_m(\Omega)$.

Some lemmas on the above-mentioned function spaces will be stated, but the assumptions they make about the smoothness of the boundary Ω, for example, do not necessarily aim at getting the best results. Principally, we follow the work of Browder [36].

Definition 1.2.1 Let Ω be a region of $\mathbb{R}^n$. Ω is said to be of class C^m or more precisely locally regular of class C^m if, for each point x of the boundary $\partial\Omega$ of Ω, there exists a neighbourhood N and a homeomorphism Φ from N to $\{y \in \mathbb{R}^n : |y| = (y_1^2 + \cdots + y_n^2)^{1/2} < 1\}$ such that $\Phi(N \cap \Omega) = \{y \in \mathbb{R}^n : |y| < 1, y_1 > 0\}$ and $\Phi(N \cap \partial\Omega) = \{y \in \mathbb{R}^n : |y| < 1, y_1 = 0\}$, and that each component of both Φ and Φ^{-1} is m-times continuously differentiable.

If we want to deal with boundary-value problems in an unbounded region, we have to take uniformly in some sense the neighbourhood N and the mapping Φ which appeared in the above definition. To this end we put

Definition 1.2.2 Let Ω be a region of $\mathbb{R}^n$. Ω is said to be uniformly regular of class C^m if there exist a family of open sets $\{N_k\}$ of $\mathbb{R}^n$, a family of homeomorphisms Φ_k from N_k to $\{y \in \mathbb{R}^n : |y| < 1\}$ and a positive integer R which satisfy the following conditions:

(i) *Put $N_k' = \Phi_k^{-1}(\{y : |y| < \frac{1}{2}\})$. Then $\bigcup_k N_k'$ contains the R^{-1}-neighbourhood of $\partial\Omega$.*

(ii) *$\Phi_k(N_k \cap \Omega) = \{y : |y| < 1, y_1 > 0\}$ and $\Phi_k(N_k \cap \partial\Omega) = \{y : |y| < 1, y_1 = 0\}$ for all k.*

(iii) *The intersection of $R + 1$ different sets taken from $\{N_k\}$ is void.*

(iv) *Put $\Psi_k = \Phi_k^{-1}$. Then Φ_k and Ψ_k are both mappings of class C^m. Further, let Φ_{jk} and Ψ_{jk} denote the jth component of Φ_k and Ψ_k, respectively. There exists a number M, independent of x, y, z, such that the inequalities*

$$|D^\beta \Phi_{jk}(x)| \le M, \quad |D^\beta \Psi_{jk}(y)| \le M, \quad \Phi_{1,k}(x) \le M \, \text{dist}\,(x, \partial\Omega)$$

hold for every $x \in N_k$, $|y| < 1$ and $|\beta| \le m$.

Evidently, Ω is uniformly regular of class C^m if it is locally regular of class C^m and, in addition, its boundary is bounded. Let $m > 0$, Ω be a region of class C^m and $u \in W_p^m(\Omega)$, then the restriction $u|_{\partial\Omega}$ of u to the boundary $\partial\Omega$ is defined; in particular, for functions belonging to $C^m(\bar{\Omega})$, this restriction coincides with that in the usual sense. With Ω being uniformly regular of class C^m, the collection of all functions on $\partial\Omega$ obtained in this way will be denoted by $W_p^{m-1/p}(\Omega)$. For $g \in W_p^{m-1/p}(\Omega)$, we put

$$[g]_{m-1/p,\partial\Omega} = [g]_{m-1/p} = \inf \|u\|_{m,p,\Omega}, \tag{1.15}$$

where the infimum is to be taken over all $u \in W_p^m(\Omega)$ satisfying $u|_{\partial\Omega} = g$. Equation (1.15) possesses all the properties of the norm and $W_p^{m-1/p}(\Omega)$ becomes a Banach space with this norm. It is clear that $(D^\alpha u)|_{\partial\Omega} \in W_p^{m-|\alpha|-1/p}(\partial\Omega)$ if $u \in W_p^m(\Omega)$ and $|\alpha| < m$. The result known as *Sobolev's imbedding theorem* asserts that inclusion relations hold between $W_p^m(\Omega)$ for various m and p and other function spaces. Among them, only relations which we use later are mentioned.

Lemma 1.2.1 *Let Ω be a region in $\mathbb{R}^n$ which is uniformly regular of class C^m. Let j denote a non-negative integer and p, r real numbers belonging to $(1, \infty)$.*

(i) *If $0 \leqslant j \leqslant m$ and $p^{-1} - (m-j)n^{-1} \leqslant r^{-1} \leqslant p^{-1}$, then $W_p^m(\Omega) \subset W_r^j(\Omega)$ and there exists a constant C such that*

$$\|u\|_{j,r} \leqslant C \|u\|_{m,p}^\lambda \|u\|_p^{1-\lambda}$$

is valid, where $\lambda = nm^{-1}(p^{-1} - r^{-1} + jn^{-1})$.

(ii) *If $p^{-1} < (m-j)n^{-1}$, then $W_p^m(\Omega) \subset B^j(\Omega) \cap C^j(\bar{\Omega})$ and there exists a constant C such that*

$$|u|_j \leqslant C \|u\|_{m,p}^\mu \|u\|_p^{1-\mu}$$

is valid for all $u \in W_p^m(\Omega)$, where $\mu = nm^{-1}p^{-1} + jm^{-1}$.

As a special case of (i), we set $r = p$. Then, for $0 \leqslant j < m$ and every $u \in W_p^m(\Omega)$, the following holds:

$$\|u\|_{j,p} \leqslant C \|u\|_{m,p}^{j/m} \|u\|_p^{(m-j)/m}. \tag{1.16}$$

This is called the *interpolation inequality*.

Remark 1.2.1 If Ω is bounded, the conclusion of Lemma 1.2.1 and, hence, the interpolation inequality are true under a weaker condition called the restricted cone condition (Agmon [1]).

Lemma 1.2.2 (Rellich's Lemma) *Let Ω be a bounded region of class C^m in $\mathbb{R}^n$ and $0 \leqslant j < m$, $1 < p < \infty$. Then the bounded set of $W_p^m(\Omega)$ is relatively compact in $W_p^j(\Omega)$.*

Remark 1.2.2 For any bounded region Ω, the bounded set of $\mathring{H}_m(\Omega)$ is relatively compact in $H_j(\Omega)$ (Agmon [1]).

Lemma 1.2.3 (Hausdorff–Young's Inequality) *Suppose $f \in L^p(R^n)$ with $1 \leqslant p < \infty$ and $g \in L^1(\mathbb{R}^n)$. Then the convolution of f and g*

$$(f * g)(x) = \int_{\mathbb{R}^n} f(x - y) g(y) \, dy$$

is defined for almost all x and belongs to $L^p(\mathbb{R}^n)$. Moreover, the following inequality holds:

$$\|f * g\|_p \leqslant \|f\|_p \|g\|_1.$$

For a proof of this lemma see Mizohata [17], p. 43.

Lemma 1.2.4 *If $u \in L^p(\mathbb{R}^n)$ and $1 \leqslant p < \infty$, then*

$$\lim_{|h| \to 0} \int_{\mathbb{R}^n} |u(x + h) - u(x)|^p \, dx = 0.$$

Proof The lemma is evident if $u \in C_0^\infty(\mathbb{R}^n)$. In the general case, u may be approached by means of a sequence of functions belonging to $C_0^\infty(\mathbb{R}^n)$. $\square$

Here is a tool, called a mollifier, to approximate a function which is not smooth by a sufficiently smooth one. Let φ be a function, belonging to $C_0^\infty(\mathbb{R}^n)$, such that $\varphi(x) \geqslant 0$ for all $x \in \mathbb{R}^n$ and $\int_{\mathbb{R}^n} \varphi(x) \, dx = 1$. Put $\varphi_\delta(x) = \delta^{-n} \varphi(x/\delta)$. For $u \in L_{\text{loc}}^1(\mathbb{R}^n)$, that is, for u which is absolutely integrable on every compact set of $\mathbb{R}^n$, we define

$$(\varphi_\delta * u)(x) = \int_{\mathbb{R}^n} \varphi_\delta(x - y) u(y) \, dy.$$

The operator $\varphi_\delta *$ is called a *mollifier*. If $u \in C(\mathbb{R}^n)$, then as $\delta \to 0$ the function $\varphi_\delta * u$ converges uniformly in the wider sense to u. Let $u \in L^p(\mathbb{R}^n)$ with $1 \leqslant p < \infty$. In this case, $\varphi_\delta * u$ converges strongly to u in $L^p(\mathbb{R}^n)$. Furthermore, the following lemma holds.

Lemma 1.2.5 *For $1 \leqslant p < \infty$, $a \in B^1(\mathbb{R}^n)$ and $u \in L^p(\mathbb{R}^n)$, we formulate*

$$C_\delta u = \varphi_\delta * (a \partial u / \partial x_j) - a(\varphi_\delta * \partial u / \partial x_j) = [\varphi_\delta *, a \partial / \partial x_j] u,$$

where the differentiation $\partial / \partial x_j$ is to be understood in the sense of distribution.

Then

(i) $C_\delta u \in L^p(R^n)$ *and there exists a constant* C, *independent of* δ *and* u, *such that* $\|C_\delta u\|_p \leqslant C \|u\|_p$,

(ii) $C_\delta u$ *converges strongly to* 0 *in* $L^p(\mathbb{R}^n)$ *as* $\delta \to 0$.

A proof of this lemma can be found in Mizohata [17], pp. 313–315.

1.3 Functions with values in a Banach space

Let X be a Banach space and u a function which is defined on the interval $a \leqslant t \leqslant b$ and takes its values on X. u is said to be *strongly continuous* at a point t_0 of $[a, b]$ if $u(t) \to u(t_0)$ strongly as $t \to t_0$, while it is *weakly continuous* at t_0 if $u(t) \to u(t_0)$ weakly as $t \to t_0$. In similar ways are defined the notions that $u(t)$ is *strongly* or *weakly differentiable*. If u is strongly continuous in $[a, b]$, then the Riemann integral $\int_a^b u(t)\,dt$ is definable in just the same way as in the case of a real-valued function.

Lemma 1.3.1 *Let* T *be a closed operator in a Banach space* X *and* u *a function defined on* $[a, b]$ *such that* $u(t) \in D(T)$ *for each* $t \in [a, b]$. *If* u *and* Tu *are both strongly continuous, then* $\int_a^b u(t)\,dt \in D(T)$ *and the following relation holds*:

$$T\int_a^b u(t)\,dt = \int_a^b Tu(t)\,dt.$$

The relation can be proved immediately if we substitute the integrals on both sides by approximate sums and remember that T is closed.

Let X, Y be two Banach spaces and $T(t)$ a function, defined for $a \leqslant t \leqslant b$, with values in $B(X, Y)$. $T(t)$ is said to be *strongly continuous* at $t_0 \in [a, b]$ if s-$\lim_{t \to t_0} T(t)u = T(t_0)u$ for $u \in X$. $T(t)$ is said to be *norm continuous* at t_0 if $\lim_{t \to t_0} \|T(t) - T(t_0)\| = 0$. Similarly, we can define that $T(t)$ is *strongly differentiable* or *norm differentiable*. From Theorem 1.1.11 follows

Theorem 1.3.1 *If* $T(t)$ *is a* $B(X, Y)$-*valued function which is strongly continuous in the interval* $a \leqslant t \leqslant b$, *then* $\|T(t)\|$ *is bounded.*

$C^m((a, b); X)$ (respectively, $C^m([a, b]; X)$) consists of functions which are m-times strongly continuously differentiable in the interval (a, b) (respectively, $[a, b]$) with values in X. If $[a, b]$ is a finite interval, then $C^m([a, b]; X)$ forms a Banach space whose norm is given by $\sum_{k=0}^m \max_{a \leqslant t \leqslant b} \|d^k u(t)/dt^k\|$. We write simply $C((a, b); X)$ and $C([a, b]; X)$ instead of $C^0((a, b); X)$ and $C^0([a, b]; X)$, respectively.

Definition 1.3.1 *Let $u(t)$ be a function defined on $a \leq t \leq b$ with values in X. $u(t)$ is said to be a* step function *if the interval $[a, b]$ is representable as a sum of a finite number of mutually disjoint measurable sets $A_1, \ldots, A_n$ so that $u(t)$ takes on a constant value on each A_i. The function $u(t)$ is said to be* strongly measurable *if it is represented as $u(t) = \text{s-}\lim_{n \to \infty} u_n(t)$ almost everywhere by choosing a sequence of step functions $u_n(t)$. The function $u(t)$ is said to be* weakly measurable *if $f(u(t))$ for each $f \in X^*$ is a real- or complex-valued measurable function.*

If u is strongly measurable, then clearly it is weakly measurable. It is known, conversely, that u is strongly measurable if, in addition to being weakly measurable, it is *separably valued*, that is, there exists some separable subspace Y such that $u(t) \in Y$ for almost all t. It is readily seen from Corollary 2 of Theorem 1.1.8 that u is strongly measurable if it is weakly continuous. Let $u(t)$ be a step function and u_i its value taken on each measurable set A_i. The integral of $u(t)$ is defined by $\int_a^b u(t)\,dt = \sum_{i=1}^n u_i |A_i|$, where $|A_i|$ is the Lebesgue measure of A_i. Obviously, the value of the integral is independent of the choice of $\{A_i\}$.

Definition 1.3.2 *$u(t)$ is said to be* Bochner integrable *if it is strongly measurable and if $\|u(t)\|$ is a real-valued Lebesgue integrable function. In this case it is possible to choose properly a sequence of step functions $\{u_n(t)\}$ so that $u(t) = \text{s-}\lim_{n \to \infty} u_n(t)$ almost everywhere and the limit*

$$\lim_{n \to \infty} \int_a^b u_n(t)\,dt \tag{1.17}$$

exists. The value of (1.17) is independent of the approximating sequence $\{u_n(t)\}$. This is named the Bochner integral *of $u(t)$ and is denoted by $\int_a^b u(t)\,dt$.*

Remark 1.3.1 A measurable function and the Bochner integral can be defined also for functions defined on a general measure space.

We denote by $L^p(a, b; X)$ the collection of strongly measurable functions u which are defined everywhere in $a \leq t \leq b$ with values in X and for which $\|u(\cdot)\| \in L^p(a, b)$. $L^p(a, b; X)$ is a Banach space with the norm given by

$$\begin{cases} \left(\int_a^b \|u(t)\|^p \, dt \right)^{1/p}, & 1 \leq p < \infty, \\ \\ \operatorname*{ess\,sup}_{a < t < b} \|u(t)\|, & p = \infty. \end{cases}$$

Lemma 1.3.2 Let $u \in L^1(a, b; X)$, then

$$\lim_{h \to 0} h^{-1} \int_0^h \|u(t+s) - u(t)\| \, ds = 0 \tag{1.18}$$

for almost all points t of (a, b).

Proof A similar lemma for real-valued functions is well known. The proof here proceeds similarly, but we shall describe it for the reader's convenience. Since u is separably valued, there exists a separable subspace Y such that $u(t) \in Y$ for almost all $t \in (a, b)$. Let $\{u_n\}$ be a countable set which is dense in Y. Since $\|u(t) - u_n\|$ is integrable, we may put

$$F_n(t) = \int_a^t \|u(s) - u_n\| \, ds$$

and $F_n'(t) = \|u(t) - u_n\|$ except for a null set e_n. The union $e = \bigcup_{n=1}^{\infty} e_n$ is also a null set and $F_n'(t) = \|u(t) - u_n\|$ for every $t \in (a, b) \backslash e$ and $n = 1, 2, \ldots$. Let v be an arbitrary element of Y. Since

$$h^{-1} \int_t^{t+h} \|u(s) - u_n\| \, ds - \|u_n - v\| \leqslant h^{-1} \int_t^{t+h} \|u(s) - v\| \, ds$$

$$\leqslant h^{-1} \int_t^{t+h} \|u(s) - u_n\| \, ds + \|u_n - v\|$$

for $h > 0$ and each $t \in (a, b) \backslash e$, as $h \to 0$ we obtain

$$\|u(t) - u_n\| - \|u_n - v\| \leqslant \liminf_{h \to +0} h^{-1} \int_t^{t+h} \|u(s) - v\| \, ds$$

$$\leqslant \limsup_{h \to +0} h^{-1} \int_t^{t+h} \|u(s) - v\| \, ds$$

$$\leqslant \|u(t) - u_n\| + \|u_n - v\|.$$

By picking out a subsequence of $\{u_n\}$ which converges to v, we have

$$\lim_{h \to +0} h^{-1} \int_t^{t+h} \|u(s) - v\| \, ds = \|u(t) - v\|.$$

The case when $h \to -0$ can be treated similarly. Finally, by taking $v = u(t)$, we arrive at (1.18). $\square$

Definition 1.3.3 The point t which permits (1.18) to hold is called the Lebesgue point of u.

Suppose that $\Omega \in \mathbb{R}^n$ and $f \in L^1(\Omega)$, then it can be proved similarly that

$$\lim_{h \to 0} \frac{1}{V_n h^n} \int_{|y-x|<h} |f(y) - f(x)| \, dy = 0 \tag{1.19}$$

is valid for almost all $x \in \Omega$, where V_n is the volume of the n-dimensional unit sphere. The point x for which (1.19) holds is called the Lebesgue point of f.

2

Dissipative operators and fractional powers of operators

2.1. Dissipative operators

2.1.1 Dissipative operators in Hilbert space

Definition 2.1.1 *Let A be a linear operator in a Hilbert space X and its domain is assumed to be dense. The operator A is called a* dissipative *operator* if $\mathrm{Re}\,(Au, u) \leq 0$ *for all* $u \in D(A)$. *If* $\mathrm{Re}\,(Au, u) \geq 0$ *for all* $u \in D(A)$, *that is,* $-A$ *is a dissipative operator, A is said to be an* accretive *operator. A dissipative operator which extends a dissipative operator A is called a* dissipative extension *of A. An operator A is said to be* maximal dissipative *if its only dissipative extension is A itself. Accretive extensions and* maximal accretive operators *are defined similarly.*

Proposition 2.1.1 *An operator A with its domain dense is dissipative if and only of* $\|(A + 1)u\| \leq \|(A - 1)u\|$ *for all* $u \in D(A)$.

The proof is easy, so it is omitted.

Proposition 2.1.2 *Let A be a linear operator with its domain dense. Then the following three statements are equivalent.*

  (i) *A is a dissipative operator.*
 (ii) $\|(A - \lambda)u\| \geq \mathrm{Re}\,\lambda\,\|u\|$ *for all* $u \in D(A)$ *and all* λ *satisfying* $\mathrm{Re}\,\lambda > 0$.
(iii) $\|(A - \lambda)u\| \geq \lambda\,\|u\|$ *for all* $u \in D(A)$ *and all* $\lambda > 0$.

Proof Assume (i) holds. Let $u \in D(A)$ and $\mathrm{Re}\,\lambda > 0$. Then

$$\mathrm{Re}\,((A - \lambda)u, u) = \mathrm{Re}\,(Au, u) - \mathrm{Re}\,\lambda\,\|u\|^2 \leq -\mathrm{Re}\,\lambda\,\|u\|^2.$$

Hence, we have

$$\|(A - \lambda)u\|\,\|u\| \geq -\mathrm{Re}\,((A - \lambda)u, u) \geq \mathrm{Re}\,\lambda\,\|u\|^2.$$

This implies (ii). It is obvious that (ii) implies (iii). Assume (iii) holds. For each $u \in D(A)$ and $\lambda > 0$, we obtain

$$\|Au\|^2 - 2\lambda \operatorname{Re}(Au, u) = \|(A - \lambda)u\|^2 - \lambda^2 \|u\|^2 \geq 0$$

and, hence, $2\lambda \operatorname{Re}(Au, u) \leq \|Au\|^2$. Since $\lambda > 0$ is arbitrary, it follows that $\operatorname{Re}(Au, u) \leq 0$. $\square$

Remark 2.1.1 If a dissipative operator A is closed, it follows from (ii) that $R(A - \lambda)$ is a closed subspace for all λ satisfying $\operatorname{Re} \lambda > 0$.

Theorem 2.1.1 *Any dissipative operator has a closed extension. The minimum closed extension of a dissipative operator is again a dissipative operator. Hence, a maximal dissipative operator is closed.*

Proof Suppose $u_n \to 0$ and $Au_n \to v$ as $n \to \infty$, where $u_n \in D(A)$. For all $u \in D(A)$ and any complex number α, the inequality $\operatorname{Re}(A(u + \alpha u_n), u + \alpha u_n) \leq 0$ is satisfied. By letting $n \to \infty$, we have $\operatorname{Re}(Au, u) + \operatorname{Re} \alpha(v, u) \leq 0$. Since α is arbitrary, $(v, u) = 0$, which implies $v = 0$. The rest of the proof is obvious. $\square$

Proposition 2.1.3 *If A is a dissipative operator and $R(A - \lambda) = X$ for some λ satisfying $\operatorname{Re} \lambda > 0$, then A is maximal dissipative.*

Proof Let $\tilde{A}$ be a dissipative extension of A. Also, let u be an arbitrary element of $D(\tilde{A})$ and put $(\tilde{A} - \lambda)u = v$. From the assumption, there exists a $w \in D(A)$ satisfying $v = (A - \lambda)w$. The equality $(\tilde{A} - \lambda)u = (\tilde{A} - \lambda)w$ and Proposition 2.1.2 lead to $u = w \in D(A)$. Therefore, we have $\tilde{A} = A$. $\square$

Theorem 2.1.2 *Every dissipative operator has a maximal dissipative extension.*

Proof Let A be a dissipative operator. By Theorem 2.1.1, it is enough to show that a maximal dissipative extension of A exists if A is closed. Let $\operatorname{Re} \lambda > 0$ and put $N = R(A - \lambda)^{\perp}$. If $v \in N$, we have $v \in D(A^*)$ and $A^* v = \bar{\lambda} v$, since $((A - \lambda)u, v) = 0$ for each $u \in D(A)$. If $v \in D(A) \cap N$, then $((A - \lambda)v, v) = 0$, from which it follows that $0 \geq \operatorname{Re}(Av, v) = \operatorname{Re} \lambda \|v\|^2$, i.e., $v = 0$. Hence, $D(A) \cap N = \{0\}$. If we put

$$\tilde{D} = \{u + v : u \in D(A), V \in N\}$$

$$\tilde{A}(u + v) = Au - \bar{\lambda} v \quad \text{for each} \quad u \in D(A) \quad \text{and} \quad v \in N,$$

$\tilde{A}$ is defined on $\tilde{D}$ as a linear operator and $\tilde{A} \supset A$. When $u \in D(A)$ and

$v \in N$, we obtain

$$
\begin{aligned}
\operatorname{Re}(\tilde{A}(u+v), u+v) &= \operatorname{Re}(Au - \bar{\lambda}v, u+v) \\
&= \operatorname{Re}(Au, u) + \operatorname{Re}(Au, v) - \operatorname{Re} \bar{\lambda}(v, u) - \operatorname{Re} \lambda \|v\|^2 \\
&= \operatorname{Re}(Au, u) + \operatorname{Re}(u, A^*v - \bar{\lambda}v) - \operatorname{Re} \lambda \|v\|^2 \\
&= \operatorname{Re}(Au, u) - \operatorname{Re} \lambda \|v\|^2 \leq 0.
\end{aligned}
$$

Therefore, $\tilde{A}$ is a dissipative extension of A. To prove that $\tilde{A}$ is a maximal dissipative operator, it is enough to show that $R(\tilde{A} - \lambda) = X$, by Proposition 2.1.3. Since A is a dissipative closed operator, $R(A - \lambda)$ is a closed subspace by Remark 2.1.1. Therefore, for any element w of X, there exists a $u \in D(A)$ and a $v \in N$ such that $w = (A - \lambda)u + v$. Putting $v_1 = -(\lambda + \bar{\lambda})^{-1}v$ we find that $v_1 \in N$ and

$$
w = (A - \lambda)u - (\lambda + \bar{\lambda})v_1 = (\tilde{A} - \lambda)(u + v_1) \in R(\tilde{A} - \lambda). \quad \square
$$

Proposition 2.1.4 *When A is a dissipative operator, the following three conditions are equivalent.*

  (i) *A is a maximal dissipative operator.*
 (ii) *$R(A - \lambda) = X$ for all λ satisfying $\operatorname{Re} \lambda > 0$.*
(iii) *$R(A - \lambda) = X$ for some λ satisfying $\operatorname{Re} \lambda > 0$.*

Proof (i) $\Rightarrow$ (ii). If A is maximal dissipative, it is a closed operator by Theorem 2.1.1. From Remark 2.1.1, if $\operatorname{Re} \lambda > 0$, then $R(A - \lambda)$ is a closed subspace. Suppose $R(A - \lambda) \neq X$. $N = R(A - \lambda)^{\perp}$ contains a non-zero element and the operator $\tilde{A}$ in the proof of the preceding theorem becomes a proper dissipative extension of A, which is a contradiction.

  (ii) $\Rightarrow$ (iii) is obvious. (iii) $\Rightarrow$ (i) is nothing but Proposition 2.1.3. $\quad \square$

Theorem 2.1.3 *Let A be a densely-defined linear operator. A is maximal dissipative if and only if it is closed, its resolvent set $\rho(A)$ contains the half-plane $\{\lambda : \operatorname{Re} \lambda > 0\}$ and $\|(A - \lambda)^{-1}\| \leq (\operatorname{Re} \lambda)^{-1}$ holds there.*

The proof follows immediately from Propositions 2.1.2 and 2.1.4.

Theorem 2.1.4 *Let A be a closed dissipative operator. A is maximal dissipative if and only if A^* is a dissipative operator. In this case, A^* is also maximal dissipative.*

Proof $D(A^*)$ is dense by Theorem 1.1.4. If A is maximal dissipative, so is A^* by Theorems 1.1.16 and 2.1.3. Conversely, if A^* is a dissipative operator, $A^* - 1$ has a continuous inverse by Proposition 2.1.2. Therefore, $R(A - 1) = X$ by Theorem 1.1.13. This, together with Proposition

2.1.3, implies that A is maximal dissipative. The fact that $R(A-1)=X$ can be shown without recourse to Theorem 1.1.13 as follows. If we let $v \in R(A-1)^{\perp}$, then $v \in D(A^*)$ and $A^*v = v$. Hence, $\|v\|^2 = \mathrm{Re}\,(A^*v, v) \leq 0$, which gives $v = 0$. $\quad\square$

2.1.2 Dissipative operators in Banach space

A generalization of dissipative operators in Hilbert space to those in Banach space is explained. Most of the material is due to Lumer and Phillips [123]. Let X be a complex Banach space and let F denote the duality mapping in X (*see* Definition 1.1.1). That is, for any $u \in X$ and $f \in Fu \subset X^*$, we have

$$\|u\|^2 = \|f\|^2 = (u, f). \tag{2.1}$$

Definition 2.1.2 *Let A be a linear operator in X. If for any $u \in D(A)$ there exists an $f \in Fu$ satisfying $\mathrm{Re}\,(Au, f) \leq 0$, A is called a* dissipative *operator. If $-A$ is dissipative, A is called an* accretive *operator.*

In Definition 2.1.1, $D(A)$ was assumed to be dense, but not in the present definition, which, however, has no particular meaning. If X is a Hilbert space, by letting $X = X^*$ by the Riesz theorem, Fu consists of only u, so that Definition 2.1.2 provides a generalization of dissipative operators in Hilbert space to those in Banach space. To derive results similar to Theorems 2.1.3 and 2.1.4 in the preceding subsection, we need the following lemma, since the inner product is not defined in Banach space, unlike in Hilbert space. The lemma was proved by Kato [84] and is fundamental in the theory of non-linear semigroups.

Lemma 2.1.1 *Let $u, v \in X$. $\|u\| \leq \|u + \alpha v\|$ for all $\alpha > 0$ if and only if there exists an $f \in Fu$ satisfying $\mathrm{Re}\,(v, f) \geq 0$.*

Proof Since the lemma is obvious for $u = 0$, let us assume $u \neq 0$. If $f \in Fu$ and $\mathrm{Re}\,(v, f) \geq 0$, then, for all $\alpha > 0$,

$$\|u\|^2 = (u, f) = \mathrm{Re}\,(u, f) \leq \mathrm{Re}\,(u + \alpha v, f)$$
$$\leq \|u + \alpha v\|\,\|f\| = \|u + \alpha v\|\,\|u\|.$$

Thus, we obtain $\|u\| \leq \|u + \alpha v\|$. Conversely, suppose $\|u\| \leq \|u + \alpha v\|$ for all $\alpha > 0$. If we let $f_\alpha \in F(u + \alpha v)$ and put $g_\alpha = f_\alpha / \|f_\alpha\|$, then $\|g\| = 1$ and

$$\|u\| \leq \|u + \alpha v\| = (u + \alpha v, g_\alpha) = \mathrm{Re}\,(u, g_\alpha) + \alpha\,\mathrm{Re}\,(v, g_\alpha)$$
$$\leq \|u\| + \alpha\,\mathrm{Re}\,(v, g_\alpha).$$

Therefore, we have

$$\mathrm{Re}\,(u, g_\alpha) \geq \|u\| - \alpha\,\|v\|, \tag{2.2}$$

$$\mathrm{Re}\,(v, g_\alpha) \geq 0. \tag{2.3}$$

Since the unit ball of X^* is compact in the w^* topology by Theorem 1.1.5, there exists a $g \in X^*$ with $\|g\| \leq 1$ such that, for any neighbourhood V of g in the w^* topology and for any $\alpha > 0$, we find a $\beta \in (0, \alpha)$ satisfying $g_\beta \in V$. Since, in particular, for any $\varepsilon > 0$ and any $\alpha > 0$, there exists a β satisfying $|(u, g_\beta - g)| < \varepsilon$ and $0 < \beta < \alpha$, from (2.2) we obtain

$$\mathrm{Re}\,(u, g) = \mathrm{Re}\,(u, g_\beta) + \mathrm{Re}\,(u, g - g_\beta)$$
$$\geq \|u\| - \beta\,\|v\| - \varepsilon \geq \|u\| - \alpha\,\|v\| - \varepsilon.$$

This leads to $\mathrm{Re}\,(u, g) \geq \|u\|$. Similarly, from (2.3) it follows that $\mathrm{Re}\,(v, g) \geq 0$. On the other hand, $(u, g) = \|u\|$ since $\mathrm{Re}\,(u, g) \leq |(u, g)| \leq \|u\|$. Hence, by letting $f = \|u\|\,g$, we have $f \in Fu$ and $\mathrm{Re}\,(v, f) \geq 0$. $\square$

Proposition 2.1.5 *For any linear operator A, the following three conditions are equivalent.*

(i) *A is a dissipative operator.*
(ii) *$\|(A - \lambda)u\| \geq \mathrm{Re}\,\lambda\,\|u\|$ for all $u \in D(A)$ and all λ satisfying $\mathrm{Re}\,\lambda > 0$.*
(iii) *$\|(A - \lambda)u\| \geq \lambda\,\|u\|$ for all $u \in D(A)$ and all $\lambda > 0$.*

Proof (i) $\Rightarrow$ (ii). Let A be a dissipative operator. Let $u \in D(A)$ and $\mathrm{Re}\,\lambda > 0$. Choose an $f \in Fu$ satisfying $\mathrm{Re}\,(Au, f) \leq 0$, and it will be found that

$$\mathrm{Re}\,((A - \lambda)u, f) = \mathrm{Re}\,(Au, f) - \mathrm{Re}\,\lambda(u, f) \leq -\mathrm{Re}\,\lambda\,\|u\|^2,$$

so that

$$\|(A - \lambda)u\|\,\|u\| = \|(A - \lambda)u\|\,\|f\| \geq -\mathrm{Re}\,((A - \lambda)u, f) \geq \mathrm{Re}\,\lambda\,\|u\|^2,$$

which implies $\|(A - \lambda)u\| \geq \mathrm{Re}\,\lambda\,\|u\|$.

(ii) $\Rightarrow$ (iii) is obvious. Suppose (iii) holds, Then, $\|u\| \leq \|u - \alpha Au\|$ for all $\alpha > 0$; from Lemma 2.1.1, there exists an $f \in Fu$ such that $\mathrm{Re}\,(Au, f) \leq 0$. Therefore, A is a dissipative operator. $\square$

From this proposition, it is seen that for a closed dissipative operator A the range $R(A - \lambda)$ is a closed subspace if $\mathrm{Re}\,\lambda > 0$.

Theorem 2.1.5 *Let A be a closed dissipative operator. If $R(A - \lambda) = X$ for some λ satisfying $\mathrm{Re}\,\lambda > 0$, then the same is true for all λ satisfying*

Re $\lambda > 0$. If, in addition, $D(A)$ is dense, then Re $(Au, f) \leqslant 0$ for all $u \in D(A)$ and all $f \in Fu$.

Proof Let Re $\lambda > 0$ and $R(A - \lambda) = X$. If $|\mu - \lambda| < $ Re λ, then $\|(\mu - \lambda)(A - \lambda)^{-1}\| < 1$ by Proposition 2.1.5, so that the range of

$$A - \mu = \{I - (\mu - \lambda)(A - \lambda)^{-1}\}(A - \lambda)$$

coincides with X. Employing a similar argument, involving the replacement of λ by μ and repeating the process shows that $R(A - \lambda) = X$ for all λ on the right half-plane Re $\lambda > 0$. Assume, in addition, that $D(A)$ is dense. We have $(1 - \alpha A)^{-1} = \alpha^{-1}(\alpha^{-1} - A)^{-1} \in B(X)$ for all $\alpha > 0$. From Proposition 2.1.5, it then follows that $\|(1 - \alpha A)^{-1}\| \leqslant 1$. Hence, if $u \in D(A)$,

$$\|(1 - \alpha A)^{-1} u - u\| = \|\alpha(1 - \alpha A)^{-1} Au\| \leqslant \alpha \|Au\| \to 0$$

as $\alpha \to 0$. Since $D(A)$ is dense, $(1 - \alpha A)^{-1} u \to u$ strongly for all $u \in X$. Let $u \in D(A)$ and $f \in Fu$, then

$$\text{Re} \, ((1 - \alpha A)^{-1} Au, f) = \alpha^{-1} \, \text{Re} \, ((1 - \alpha A)^{-1} u - u, f)$$

$$\leqslant \alpha^{-1} \|(1 - \alpha A)^{-1} u\| \|f\| - \alpha^{-1} \|u\|^2 \leqslant 0.$$

By taking the limit $\alpha \to 0$, we obtain Re $(Au, f) \leqslant 0$. $\square$

Theorem 2.1.6 Let A be a closed operator with its domain $D(A)$ dense. Both A and A^* are dissipative if and only if the half-plane $\{\lambda : \text{Re} \, \lambda > 0\}$ is contained in $\rho(A)$ and $\|(A - \lambda)^{-1}\| \leqslant 1/\text{Re} \, \lambda$ holds in the half-plane.

Proof Let A and A^* be dissipative operators. From Proposition 2.1.5, if Re $\lambda > 0$, we have $\|(A - \lambda)u\| \geqslant \text{Re} \, \lambda \|u\|$ and $\|(A^* - \bar{\lambda})f\| \geqslant \text{Re} \, \lambda \|f\|$ for all $u \in D(A)$ and $f \in D(A^*)$. Therefore, by Theorem 1.1.13, $\lambda \in \rho(A)$ and, moreover, we have $\|(A - \lambda)^{-1}\| \leqslant 1/\text{Re} \, \lambda$. The converse is evident from Proposition 2.1.5 and Theorem 1.1.16.

2.2 Regularly dissipative operators

Let X be a complex Hilbert space; its inner product and norm will be denoted by $(\cdot, \cdot)$ and $|\cdot|$, respectively. Let V be another Hilbert space with inner product and norm denoted by $((\cdot, \cdot))$ and $\|\cdot\|$, respectively. We assume that V is embedded in X as a dense subspace and that V has a stronger topology than X. Therefore, there exists an M_0 such that $|u| \leqslant M_0 \|u\|$ for all $u \in V$. Let $a(u, v)$ be a *quadratic form* defined on $V \times V$. That is, to each $u, v \in V$ there corresponds a complex number

$a(u, v)$ which is linear in u and antilinear in v:

$$a(u_1 + u_2, v) = a(u_1, v) + a(u_2, v),$$
$$a(u, v_1 + v_2) = a(u, v_1) + a(u, v_2)$$
$$a(\lambda u, v) = \lambda a(u, v), \quad a(u, \lambda v) = \bar{\lambda} a(u, v).$$

We assume that $a(u, v)$ is bounded, i.e., there exists a certain number M such that

$$|a(u, v)| \leq M \|u\| \|v\| \tag{2.4}$$

for all $u, v \in V$. We further assume that there exist a positive number δ and a real number k such that the following inequality holds for all $u \in V$:

$$\mathrm{Re}\, a(u, u) \geq \delta \|u\|^2 - k |u|^2. \tag{2.5}$$

This inequality is called *Gårding's inequality*. Using $a(u, v)$, an operator A is defined as follows:

$$\begin{cases} \text{Given } u \in V. \text{ If there exists an element } f \text{ of } X \text{ so that} \\ a(u, v) = (f, v) \text{ for all } v \in V, \text{ then } u \in D(A) \text{ and } Au = f. \end{cases} \tag{2.6}$$

The quadratic form $a(u, v)$, considered as a functional of v, is continuous in V-topology. If, in particular, it is also continuous in the topology of V induced by X, $a(u, v)$ can be extended to X as a continuous functional. Hence, by the Riesz theorem, there exists an element f of X so that $a(u, v) = (f, v)$ for all $v \in V$. In this case, we interpret $u \in D(A)$ and $Au = f$. In studying such an operator A, it is often convenient to extend it in the following way. In this section, the space of all continuous antilinear functionals defined on V and X are denoted by V^* and X^*, respectively. That is, V^* and X^* are the spaces of all continuous functionals l on V and X which satisfy $l(u + v) = l(u) + l(v)$ and $l(\lambda u) = \bar{\lambda} l(u)$ for all $u, v \in V$ and X, and for all complex numbers λ, respectively. For any element of V^* or X^*, its norm is defined similarly to be a continuous linear functional. That is, the norms of l as elements of V^* and X^* are given by

$$\|l\|_* = \sup_{\|v\| \leq 1} |l(v)|, \qquad |l|_* = \sup_{|f| \leq 1} |l(f)|,$$

respectively. Let $l|_V$ denote the restriction of $l \in X^*$ to V, then

$$|(l|_V)(v)| = |l(v)| \leq |l|_* |v| \leq |l|_* M_0 \|v\|. \tag{2.7}$$

Hence, $l|_V \in V^*$. Since V is dense in X, the correspondence $l \to l|_V$ is one-to-one, so that by identifying l with $l|_V$ we may consider $X^* \subset V^*$. Since $\|l|_V\|_* \leq M_0 |l|_*$ by (2.7), X^* has a stronger topology than V^*. By considering $X^* = X$ on the basis of the Riesz theorem applied to X we may conclude that $V \subset X \subset V^*$. Furthermore, the embeddings $V \to X$ and $X \to V^*$ are both continuous. We can show that V is dense in V^* as follows. If $v \in V$ satisfies $(u, v) = 0$ for all u, it follows by taking $u = v$ that $v = 0$. Accordingly, by the reflexivity of V and Theorem 1.1.3, V is dense in V^*, and, hence, X is also dense in V^*. For $l \in V^*$ the value $l(v)$ of l at v is also denoted by (l, v). The use of this notation is convenient, because if, in particular, $l = f \in X$, it is seen from the meaning of $X \subset V^*$ that the notation represents just the inner product of f and v in X. From now on, we denote elements of V^* by f, g and so on, and sometimes (f, v) by (v, f). When $a(u, v)$ with $u \in V$ fixed is considered as a functional of v, it is an element of V^* by (2.4). Therefore, using an element $f \in V^*$, we can express $a(u, v) = (f, v)$. Since f so obtained is determined by u, we write $\tilde{A}u = f$. That is, $\tilde{A}$ is an operator defined by

$$a(u, v) = (\tilde{A}u, v) \quad \text{for all} \quad u, v \in V. \tag{2.8}$$

It is obvious that $\tilde{A}$ is an extension of the operator A defined by (2.6). More precisely,

$$D(A) = \{u \in V : \tilde{A}u \in X\}. \tag{2.9}$$

We prove the following result, known as the *Lax–Milgram theorem*, which has been found to be useful in solving the Dirichlet problem in elliptic equations.

Lemma 2.2.1 *Let H be a Hilbert space, whose inner product and norm will be denoted by $(\cdot, \cdot)$ and $\|\cdot\|$, respectively. Assume that $B[u, v]$ is a quadratic form defined on $H \times H$ and that there exist positive constants C and c such that*

$$|B[u, v]| \leq C \|u\| \|v\|, \tag{2.10}$$

$$|B[u, u]| \geq c \|u\|^2 \tag{2.11}$$

for all $u, v \in H$. Under these conditions, if $F \in H^*$, i.e., if F is a continuous antilinear functional on H, there exists an element u such that $F(v) = B[u, v]$ for all $v \in H$. Furthermore, u is uniquely determined by F.

Proof With u fixed, $B[u, v]$ as a functional of v is continuous and antilinear by (2.10); there exists a $w \in H$ such that $B[u, v] = (w, v)$ for all $v \in H$. Since w is uniquely determined by u, let $w = Su$; S is a linear mapping from H into itself. From (2.10), it follows that

$$|(Su, v)| = |B[u, v]| \leq C \|u\| \|v\|.$$

Hence $\|Su\| \leq C\|u\|$, i.e., S is bounded. By (2.11) it holds that $C\|u\|^2 \leq |B[u, u]| = |(Su, u)| \leq \|Su\| \|u\|$. Since we have $c\|u\| \leq \|Su\|$, S has a continuous inverse. Hence $R(S)$ is a closed subspace of H. Let $v \in R(S)^{\perp}$, then $(Su, v) = 0$ for every $u \in H$. In particular, by taking $u = v$, we have

$B[v, v] = (Sv, v) = 0$. This, together with (2.11), implies $v = 0$. Therefore, $R(S) = H$. If we let $F \in H^*$, from the Riesz theorem there exists a $w \in H$ such that $F(v) = (w, v)$ for all v. Since there is a u satisfying $w = Su$, the equality

$$F(v) = (Su, v) = B[u, v]$$

holds for all $v \in H$. It is obvious that such a u is unique. $\square$

Let us return to our main discussion. Assume that (2.5) holds for $k = 0$, i.e.,

$$\text{Re } a(u, u) \geqslant \delta \|u\|^2. \tag{2.12}$$

If $f \in V^*$, by (2.4) and (2.12) we can apply Lemma 2.2.1 to show that there exists a $u \in V$ such that $(f, v) = a(u, v)$ for all $v \in V$, i.e., $f = \tilde{A}u$ and, hence, $R(\tilde{A}) = V$. Combining this result with (2.9), we also have $R(A) = X$. From (2.4) and (2.12), it is easy to see that

$$\delta \|u\| \leqslant \|\tilde{A}u\|_* \leqslant M \|u\|. \tag{2.13}$$

Accordingly $\tilde{A}$ is an isomorphism from V onto V^*. There also exists an A^{-1} and it is a bounded operator on X. Hence, we have $0 \in \rho(A)$.

A quadratic form $a^*(u, v)$ defined by $a^*(u, v) = \overline{a(v, u)}$ is called an *adjoint quadratic form*. If $a(u, v)$ satisfies (2.4), (2.5) or (2.12), so, correspondingly, does $a^*(u, v)$. Let A' and $\tilde{A}'$ be operators defined by $a^*(u, v)$ in ways similar to (2.6) and (2.8), respectively:

$$\begin{cases} \text{Let } u \in V. \text{ If there exists an } f \in X \text{ such that } a^*(u, v) = (f, v) \text{ for} \\ \text{all } v \in V, \text{ then } u \in D(A') \text{ and } A'u = f. \tag{2.14} \\ a^*(u, v) = (\tilde{A}'u, v) \quad \text{for all} \quad u, v \in V. \tag{2.15} \end{cases}$$

Assume again that (2.12) is satisfied. Then, as in the cases of A and $\tilde{A}$, we have $R(A') = X$ and $R(\tilde{A}') = V^*$ for A' and $\tilde{A}'$.

Lemma 2.2.2 $D(A)$ *is dense in* V. *Therefore, it is also dense in* X.

Proof It is enough to show that $f \in V^*$ and, if $(f, v) = 0$ for all $v \in D(A)$, then $f = 0$. Since $R(\tilde{A}') = V^*$, there exists a $u \in V$ such that $f = \tilde{A}'u$. If $v \in D(A)$, we have

$$(Av, u) = a(v, u) = \overline{a^*(u, v)} = \overline{(\tilde{A}'u, v)} = \overline{(f, v)} = 0.$$

This, together with $R(A) = X$, implies $u = 0$. Hence, $f = 0$. $\square$

Since $\text{Re } (Au, u) = \text{Re } a(u, u) \geqslant \delta \|u\|^2 \geqslant 0$ for any $u \in D(A)$, the operator A is accretive.

Definition 2.2.1 *An operator defined by* (2.6), *using a quadratic form* *satisfying* (2.4) *and* (2.12), *is called a* regularly accretive operator. *If* $-A$ *is regularly accretive,* A *is called a* regularly dissipative operator.

Lemma 2.2.3 *Let* A^* *be an adjoint of* A *when the latter is viewed as an operator in* X. *Then* $A' = A^*$.

Proof Let $u \in D(A)$ and $v \in D(A')$. Then we find

$$(Au, v) = a(u, v) = \overline{a^*(v, u)} = \overline{(A'v, u)} = (u, A'v).$$

This shows $A' \subset A^*$. Let $u \in D(A^*)$ and put $A^*u = f$. Since $R(A') = X$, there exists a $w \in D(A')$ such that $f = A'w$. The relation $A' \subset A^*$ implies $f = A^*w$. Since $0 \in \rho(A)$ we have $0 \in \rho(A^*)$. Therefore, $u = w \in D(A')$ and, hence, $A' = A^*$. $\square$

Like A, the operator A' is also accretive. Hence, from Theorem 2.1.4 and Lemma 2.2.3, we obtain

Theorem 2.2.1 *A regularly accretive operator is maximal accretive.*

Next, coming back to the general case, suppose that (2.4) and (2.5) are satisfied. If $k < 0$, the inequality (2.12) is obviously satisfied. For $k > 0$, if we put

$$a_k(u, v) = a(u, v) + k(u, v),$$

then $a_k(u, v)$ satisfies (2.12). $a_k(u, v)$ also satisfies (2.4) if M is replaced by $M + kM_0^2$. We denote by A_k and $\tilde{A}_k$ the operators defined by $a_k(u, v)$. Evidently, $A_k = A + k$ and $\tilde{A}_k = \tilde{A} + k$. Also, an adjoint quadratic form of $a_k(u, v)$ is given by $a_k^*(u, v) = a^*(u, v) + k(u, v)$. Let A_k' and $\tilde{A}_k'$ be operators defined by $a_k^*(u, v)$, then, by Lemma 2.2.2, $D(A) = D(A_k)$ is dense in V and also in X. From Lemma 2.2.3, it follows that $A_k' = A_k^*$, and, hence, $A' = A^*$. Thus, we have obtained the following theorem.

Theorem 2.2.2 *Let* $a(u, v)$ *be a quadratic form on* $V \times V$ *satisfying* (2.4) *and* (2.5), *and let* A *be the operator defined by* (2.6). *The domain* $D(A)$ *is dense in* V *and also in* X, *and* $0 \in \rho(A + k)$. *Also let* $\tilde{A}$ *be the operator defined by* (2.8). *Then* $\tilde{A} + k$ *is an isomorphism from* V *onto* V^*. *Let* $a^*(u, v)$ *be the adjoint of* $a(u, v)$ *and* A' *the operator defined by* (2.14). *Then the adjoint operator* A^* *of* A *in* X *coincides with* A'. $A + k$ *is a regularly accretive operator.*

From now on, both A and $\tilde{A}$ are denoted simply by A. We also denote $\tilde{A}'$ by A^*. Therefore, for any $u, v \in V$, we have

$$a(u, v) = (Au, v), \qquad a^*(u, v) = (A^*u, v).$$

This notation will not cause any confusion.

When $a^*(u, v) = a(u, v)$ holds for all $u, v \in V$, the quadratic form $a(u, v)$ is said to be symmetric. In this case, by Theorem 2.2.2, an operator A in X is self-adjoint. It is evident that $a(u, u)$ is a real number for each $u \in V$. Since, by (2.5), we have

$$(Au, u) = a(u, u) \geqslant -k |u|^2$$

for all $u \in D(A)$, the operator A is bounded from below. In particular, A is positive definite if (2.12) is satisfied.

Theorem 2.2.3 *If $a(u, v)$ is a symmetric quadratic form satisfying (2.4) and (2.12), then A is positive definite and self-adjoint, $D(A^{1/2}) = V$, and*

$$a(u, v) = (A^{1/2}u, A^{1/2}v), \quad u, v \in V \tag{2.16}$$

Proof According to this assumption, for each $u \in D(A)$ we have

$$\delta \|u\|^2 \leqslant a(u, u) = (Au, u) = |A^{1/2}u|^2. \tag{2.17}$$

Let u be an arbitrary element of $D(A^{1/2})$. For each natural number n we put $u_n = (1 + n^{-1}A)^{-1}u$. Then $u_n \in D(A)$ and we can show by the use of the spectral resolution that $u_n \to u$ and $A^{1/2}u_n = (1 + n^{-1}A)^{-1}A^{1/2}u \to A^{1/2}u$ in X as $n \to \infty$. By applying (2.17) to $u_n - u_m$, it is found that $\{u_n\}$ is a Cauchy sequence in V. Since $u_n \to u$ in X, so it does in V; hence, $D(A^{1/2}) \subset V$. Apply (2.17) to u_n and let $n \to \infty$, and we obtain $\delta \|u\|^2 \leqslant |A^{1/2}u|^2$. On the other hand, if we let $u \in V$, by Lemma 2.2.2 there exists a sequence $\{u_j\}$ of elements of $D(A)$ such that $\|u_j - u\| \to 0$. Since

$$|A^{1/2}(u_j - u_k)|^2 = a(u_j - u_k, u_j - u_k) \leqslant M \|u_j - u_k\|^2,$$

$\{A^{1/2}u_j\}$ is a Cauchy sequence in X. Since $A^{1/2}$ is a closed operator, $u \in D(A^{1/2})$ and thus we have obtained $D(A^{1/2}) = V$. Equation (2.17) can be easily verified. $\square$

Remark 2.2.1 For $k > 0$, replace $a(u, v)$ by $a(u, v) + k(u, v)$ and A by $A + k$; then the conclusion of Theorem 2.2.3 still holds.

Example 1 Let Ω be a region in $\mathbb{R}^n$ and a_{ij} a real-valued function for each $i, j = 1, \ldots, n$. Assume that $a_{ij} = a_{ji} \in B^1(\bar{\Omega})$ and $\{a_{ij}(x)\}$ is positive definite uniformly in Ω, i.e., there exists a positive number δ such that

$$\sum_{i,j=1}^n a_{ij}(x)\xi_i\xi_j \geqslant \delta |\xi|^2 \tag{2.18}$$

for all $x \in \bar{\Omega}$ and all real vectors ξ. Let $b_i \in L^\infty(\Omega)$ and $c \in L^\infty(\Omega)$. Put

$\beta_i = \sum_{j=1}^{n} \partial a_{ij}/\partial x_j + b_i$, then $\beta_i \in L^{\infty}(\Omega)$. For each $u, v \in H_1(\Omega)$, we put

$$a(u, v) = \int_{\Omega} \left\{ \sum_{i,j=1}^{n} a_{ij} \frac{\partial u}{\partial x_i} \frac{\overline{\partial v}}{\partial x_j} + \sum_{i=1}^{n} \beta_i \frac{\partial u}{\partial x_i} \bar{v} + cu\bar{v} \right\} dx. \qquad (2.19)$$

Since $\{a_{ij}\}$ is real symmetric, by (2.18) the inequality

$$\sum_{i,j=1}^{n} a_{ij}(x)\zeta_i\bar{\zeta}_j \geq \delta |\zeta|^2 \qquad (2.20)$$

holds for all complex vectors $\zeta = (\zeta_1, \ldots, \zeta_n)$. On the other hand, by this hypothesis, there exists a certain number K such that $|\beta_i(x)| \leq K$ and $|c(x)| \leq K$ hold almost everywhere. Hence,

$$\begin{aligned} \operatorname{Re} a(u, u) &\geq \int_{\Omega} \delta \sum_{i=1}^{n} \left| \frac{\partial u}{\partial x_i} \right|^2 dx - K \int_{\Omega} \sum_{i=1}^{n} \left| \frac{\partial u}{\partial x_i} \right| |u| \, dx - K \int_{\Omega} |u|^2 \, dx \\ &\geq \delta \int_{\Omega} \sum_{i=1}^{n} \left| \frac{\partial u}{\partial x_i} \right|^2 dx - K \int_{\Omega} \sum_{i=1}^{n} \left(\frac{\varepsilon}{2} \left| \frac{\partial u}{\partial x_i} \right|^2 + \frac{1}{2\varepsilon} |u|^2 \right) dx \\ &\qquad\qquad - K \int_{\Omega} |u|^2 \, dx \\ &= \left(\delta - \frac{\varepsilon}{2} K \right) \sum_{i=1}^{n} \int_{\Omega} \left| \frac{\partial u}{\partial x_i} \right|^2 dx - \left(\frac{nK}{2\varepsilon} + K \right) \int_{\Omega} |u|^2 \, dx. \end{aligned}$$

By choosing $\varepsilon = \delta K^{-1}$, we obtain

$$\begin{aligned} \operatorname{Re} a(u, u) &\geq \frac{\delta}{2} \sum_{i=1}^{n} \int_{\Omega} \left| \frac{\partial u}{\partial x_i} \right|^2 dx - \left(\frac{nK^2}{2\delta} + K \right) \int_{\Omega} |u|^2 \, dx \\ &= \frac{\delta}{2} \|u\|_1^2 - \left(\frac{nK^2}{2\delta} + K + \frac{\delta}{2} \right) \|u\|^2. \end{aligned}$$

Therefore, it follows that, for any closed subspace V of $H_1(\Omega)$ containing $\mathring{H}_1(\Omega)$, the quadratic form $a(u, v)$ satisfies (2.4) and (2.5), where $X = L^2(\Omega)$. If we put

$$\mathscr{A} = -\sum_{i,j=1}^{n} a_{ij}(x) \frac{\partial^2}{\partial x_i \partial x_j} + \sum_{i=1}^{n} b_i(x) \frac{\partial}{\partial x_i} + c(x),$$

$\mathscr{A}$ is elliptic.

   Case (i) $V = \mathring{H}_1(\Omega)$: Let $Au = f$. Assuming that $\partial\Omega$ and u are smooth and applying formally partial integrations to $a(u, v) = (f, v)$, we obtain $(\mathscr{A}u, v) = (f, v)$. Since $v \in \mathring{H}_1(\Omega)$ is arbitrary, we have

$$\mathscr{A}u = f \quad \text{in} \quad \Omega, \qquad (2.21)$$

$$u = 0 \quad \text{on} \quad \partial\Omega. \qquad (2.22)$$

Therefore, u is a solution in the wider sense of the Dirichlet problem, (2.21) and (2.22).

Case (ii) $V = H_1(\Omega)$: Let Ω be locally regular of class C^1 and $Au = f$. A formal calculation leads to

$$a(u, v) = \int_{\partial\Omega} \sum_{i,j=1}^{n} a_{ij}\nu_j \frac{\partial u}{\partial x_i} \bar{v} \, d\sigma + (\mathscr{A}u, v) = (f, v). \tag{2.23}$$

where $\nu = (\nu_1, \ldots, \nu_n)$ is a normal vector of $\partial\Omega$ pointing outward and $d\sigma$ is an area element of $\partial\Omega$. First, letting $v \in C_0^\infty(\Omega)$, we obtain (2.21), since $(\mathscr{A}u, v) = (f, v)$. Then, substituting it into (2.23), we get

$$\int_{\partial\Omega} \sum_{i,j=1}^{n} a_{ij}\nu_j \frac{\partial u}{\partial x_i} \cdot \bar{v} \, d\sigma = 0.$$

Since $v \in H_1(\Omega)$ is arbitrary, we have

$$\sum_{i,j=1}^{n} a_{ij}\nu_j \frac{\partial u}{\partial x_i} = 0 \quad \text{on} \quad \partial\Omega. \tag{2.24}$$

Therefore, u is a solution in the wider sense of the Neumann problem, (2.21) and (2.24).

If Ω is of class C^2 uniformly, it is known that, by the smoothness of solutions of elliptic equations, $D(A) = H_2(\Omega) \cap \mathring{H}_1(\Omega)$ in the case of $V = \mathring{H}_1(\Omega)$ and that $D(A)$ becomes the set of all $u \in H_2(\Omega)$ satisfying (2.24) in the case of $V = H_1(\Omega)$.

In other cases, where $V = \{u \in H_1(\Omega) : u = 0 \text{ on } \Gamma \subset \partial\Omega\}$, a similar calculation shows that u satisfying $Au = f$ gives a solution in the wider sense of the following problem: $\mathscr{A}u = f$ in Ω, $u = 0$ on Γ, and

$$\sum_{i,j=1}^{n} a_{ij}\nu_j \frac{\partial u}{\partial x_i} = 0 \quad \text{on} \quad \partial\Omega - \Gamma.$$

Let h be a continuous bounded function on $\partial\Omega$ and define $a(u, v)$ anew by adding $\int_{\partial\Omega} hu\bar{v} \, d\sigma$ to the right-hand side of (2.19). Then it is also known that

$$\left| \int_{\partial\Omega} h |u|^2 \, d\sigma \right| \le \varepsilon \|u\|_1^2 + C_\varepsilon \|u\|^2$$

holds for any $\varepsilon > 0$ (*see*, e.g., Mizohata [17], p. 194, Theorem 3.16). In this case, too, $a(u, v)$ satisfies (2.4) and (2.5) with $V = H_1(\Omega)$ and the solution u of $Au = f$ is a solution in the wider sense of the problem

$$\mathscr{A}u = f \quad \text{in} \quad \Omega,$$

$$\sum_{i,j=1}^{n} a_{ij}\nu_j \frac{\partial u}{\partial x_i} + hu = 0 \quad \text{on} \quad \partial\Omega.$$

If c is a real-valued function and $\beta_1 \equiv 0$ for each $i = 1, \ldots, n$, then $a(u, v)$ defined by (2.19) is symmetric. If h is real valued, the new quadratic form obtained by adding $\int_{\partial\Omega} hu\bar{v} \, d\sigma$ to $a(u, v)$ is also symmetric. Hence, in these cases, the corresponding operators A in $L^2(\Omega)$ are all self-adjoint and bounded from below.

Example 2   Let Ω be a bounded region in $\mathbb{R}^n$. Assume that $a_{\alpha\beta} \in C(\bar{\Omega})$ for $|\alpha| = |\beta|$ and $\alpha_{\alpha\beta} \in L^\infty(\Omega)$ for $|\alpha| \leq m$, $|\beta| \leq m$ and $|\alpha| + |\beta| < 2m$, and put

$$a(u, v) = \int_\Omega \sum_{|\alpha|, |\beta| \leq m} a_{\alpha\beta} D^\alpha u \overline{D^\beta v} \, dx$$

for each $u, v \in \mathring{H}_m(\Omega)$, where $D = (-i\partial/\partial x_1, \ldots, -i\partial/\partial x_n)$. If the inequality

$$\text{Re} \sum_{|\alpha| = |\beta| = m} a_{\alpha\beta}(x) \xi^\alpha \xi^\beta > 0$$

holds for each $x \in \bar{\Omega}$ and each real vector $\xi \neq 0$, it is well known in the theory of elliptic equations that $a(u, v)$ satisfies Gårding's inequality with $X = L^2(\Omega)$ and $V = \mathring{H}_m(\Omega)$ (Agmon [1], Mizohata [17]). If Ω is of class C^{2m} and $a_{\alpha\beta} \in C^{|\beta|}(\bar{\Omega})$, it is also well known as a consequence of the smoothness of the solution of an elliptic equation that $D(A) = H_{2m}(\Omega) \cap \mathring{H}_m(\Omega)$. If $a_{\alpha\beta}(x) = \overline{a_{\beta\alpha}(x)}$, then $a(u, v)$ is symmetric, so that the operator A is bounded from below and self-adjoint.

2.3   Fractional powers of operators

In this section X denotes a complex Banach space.

Definition 2.3.1   *Let A be a closed operator densely defined in X. The operator A is said to be of* type (ω, M) *if there exist $0 < \omega < \pi$ and $M \geq 1$ such that $\rho(A) \supset \{\lambda : |\arg \lambda| > \omega\}$ and $\|\lambda(A - \lambda)^{-1}\| \leq M$ for $\lambda < 0$, and if there exists a number M_ε such that $\|\lambda(A - \lambda)^{-1}\| \leq M_\varepsilon$ holds in $|\arg \lambda| > \omega + \varepsilon$ for all $\varepsilon > 0$.*

By Theorem 2.1.3, if A is a maximal accretive operator in a Hilbert space, then it is of type $(\pi/2, 1)$. That the converse is also true can be observed as follows. Let A be of type $(\pi/2, 1)$ and $\text{Re}\,\lambda > 0$. For a sufficiently large μ satisfying $|\lambda - \mu| < \mu$, we have

$$\|(A + \lambda)^{-1}\| = \left\| \sum_{n=0}^\infty (\mu - \lambda)^n (A + \mu)^{-n-1} \right\|$$

$$\leq \sum_{n=0}^\infty |\mu - \lambda|^n \mu^{-n-1} = (\mu - |\mu - \lambda|)^{-1}.$$

By taking the limit $\mu \to \infty$, we obtain $\|(A+\lambda)^{-1}\| \le (\mathrm{Re}\,\lambda)^{-1}$. Similarly, let A be a closed operator in a Banach space with a dense domain. Then, both A and A^* are accretive if and only if A is of type $(\pi/2, 1)$.

When A is of type (ω, M), we will attempt to define A^α for each $\alpha > 0$. The discussion is mainly due to Kato [78, 80].

Lemma 2.3.1 *Let A be of type (ω, M). Then, $(1+\lambda^{-1}A)^{-1} \in B(X)$ for $\lambda > 0$ and $(1+\lambda^{-1}A)^{-1} \to I$ strongly as $\lambda \to \infty$.*

Proof The first statement is obvious from the assumption:

$$\|(1+\lambda^{-1}A)^{-1}\| = \|(-\lambda)(A-(-\lambda))^{-1}\| \le M. \tag{2.25}$$

Hence, if $u \in D(A)$,

$$\|(1+\lambda^{-1}A)^{-1}u - u\| = \|(1+\lambda^{-1}A)^{-1}\{u - (1+\lambda^{-1}A)u\}\|$$
$$= \|(1+\lambda^{-1}A)^{-1}\lambda^{-1}Au\| \le \lambda^{-1}M\|Au\| \to 0$$

as $\lambda \to \infty$. Since $D(A)$ is dense and (2.25) holds, $(1+\lambda^{-1}A)^{-1}u \to u$ strongly for all $u \in X$.

Corollary *If A is of type (ω, M), the domain $D(A^n)$ for any natural number n is dense.*

Proof Let u be an arbitrary element of X. $u_\lambda = (1+\lambda^{-1}A)^{-n}u \in D(A^n)$ and, by the lemma, $u_\lambda \to u$ strongly as $\lambda \to \infty$. $\quad\square$

2.3.1 The case in which A has a bounded inverse

In this case, there exists a neighbourhood U of 0 such that

$$\rho(A) \supset S \equiv \{\lambda: |\arg \lambda| > \omega\} \cup U.$$

Hence, we can choose a positive number a and an angle φ in $\omega < \varphi < \pi$ so that the contour Γ consisting of two half-lines $\arg(\lambda - a) = \pm\varphi$ is contained in S. For each $\alpha > 0$, we put

$$A^{-\alpha} = \frac{1}{2\pi i} \int_\Gamma \lambda^{-\alpha}(A-\lambda)^{-1}\,d\lambda, \tag{2.26}$$

where the integration is taken from $\infty e^{-i\varphi}$ to $\infty e^{i\varphi}$ along the contour Γ and $\lambda^{-\alpha}$ is chosen to be positive for $\lambda > 0$. By assumption, $A^{-\alpha}$ is a bounded operator. Since the integrand is regular on a region obtained from S by eliminating the non-positive part of the real axis, the value of the integral is independent of the choice of a and φ. If $\alpha = n$ is a natural number, λ^{-n}

is regular except at the origin and we can transform the integral in (2.26) to one along a small smooth closed curve around the origin. Since $(A - \lambda)^{-1}$ is regular in the neighbourhood of $\lambda = 0$, by the residue theorem the integral turns out equal to A^{-n} in the usual sense. Thus, the definition of $A^{-\alpha}$ by (2.27) does not cause inconvenience. Let α and β be two positive numbers. Define

$$\Gamma = \{\lambda : \arg (\lambda - a) = \pm \varphi\} \subset S,$$

as above and

$$\Gamma' = \{\lambda : \arg (\lambda - a') = \pm \varphi'\} \subset S,$$

where $0 < a' < a$ and $\omega < \varphi < \varphi' < \pi$, then $\Gamma \cap \Gamma'$ is empty. If we use the representations

$$A^{-\alpha} = \frac{1}{2\pi i} \int_\Gamma \lambda^{-\alpha} (A - \lambda)^{-1} \, d\lambda, \qquad A^{-\beta} = \frac{1}{2\pi i} \int_{\Gamma'} \mu^{-\beta} (A - \mu)^{-1} \, d\mu,$$

then

$$A^{-\alpha} A^{-\beta} = \left(\frac{1}{2\pi i}\right)^2 \int_\Gamma \int_{\Gamma'} \lambda^{-\alpha} \mu^{-\beta} (A - \lambda)^{-1} (A - \mu)^{-1} \, d\mu \, d\lambda$$

$$= \left(\frac{1}{2\pi i}\right)^2 \int_\Gamma \int_{\Gamma'} \lambda^{-\alpha} \mu^{-\beta} (\lambda - \mu)^{-1}$$

$$\times \{(A - \lambda)^{-1} - (A - \mu)^{-1}\} \, d\mu \, d\lambda \qquad (2.27)$$

$$= \frac{1}{2\pi i} \int_\Gamma \lambda^{-\alpha} (A - \lambda)^{-1} \left\{ \frac{1}{2\pi i} \int_{\Gamma'} \frac{\mu^{-\beta}}{\lambda - \mu} \, d\mu \right\} d\lambda$$

$$- \frac{1}{2\pi i} \int_{\Gamma'} \mu^{-\beta} (A - \mu)^{-1} \left\{ \frac{1}{2\pi i} \int_\Gamma \frac{\lambda^{-\alpha}}{\lambda - \mu} \, d\lambda \right\} d\mu.$$

We put

$$\Gamma'_R = \{\mu \in \Gamma' : |\mu - a'| \leq R\} \cup \{\mu : |\mu - a'| = R, |\arg (\mu - a')| < \varphi'\}$$

for $R > 0$. Γ'_R is a closed curve whose direction we take as being clockwise. For each $\lambda \in \Gamma$, and for R sufficiently large, we have

$$\frac{1}{2\pi i} \int_{\Gamma'_R} \frac{\mu^{-\beta}}{\lambda - \mu} \, d\mu = \lambda^{-\beta},$$

and, hence,

$$\frac{1}{2\pi i} \int_{\Gamma'} \frac{\mu^{-\beta}}{\lambda - \mu} \, d\mu = \lim_{R \to \infty} \frac{1}{2\pi i} \int_{\Gamma'_R} \frac{\mu^{-\beta}}{\lambda - \mu} \, d\mu = \lambda^{-\beta}.$$

Similarly, if we put

$$\Gamma_R = \{\lambda \in \Gamma : |\lambda - a| \leqslant R\} \cup \{\lambda : |\lambda - a| = R, |\arg(\lambda - a)| < \varphi\},$$

we have

$$\frac{1}{2\pi i} \int_\Gamma \frac{\lambda^{-\alpha}}{\lambda - \mu} \, d\lambda = \lim_{R \to \infty} \frac{1}{2\pi i} \int_{\Gamma_R} \frac{\lambda^{-\alpha}}{\lambda - \mu} \, d\lambda = 0$$

since, for each $\mu \in \Gamma'$, the function $\lambda^{-\alpha}/(\lambda - \mu)$ is regular on a closed set surrounded by Γ_R. By substituting these results into (2.27), we obtain

$$A^{-\alpha} A^{-\beta} = A^{-\alpha-\beta}. \tag{2.28}$$

If $\alpha = n$ is a natural number, there exists an inverse A^n of A^{-n}. If α is not a natural number, we can make the following argument for the existence of an inverse of $A^{-\alpha}$. Suppose $A^{-\alpha} u = 0$, then, for any natural number n greater than α, by (2.28), we have

$$A^{-n} u = A^{-(n-\alpha)} A^{-\alpha} u = 0,$$

which implies that $u = 0$.

Definition 2.3.2 *Let A be of type (ω, M) and suppose that $0 \in \rho(A)$. For each $\alpha \geqslant 0$, we define an operator A^α as follows:*

$$A^\alpha = \begin{cases} (A^{-\alpha})^{-1} & \text{for} \quad \alpha > 0, \\ 1 & \text{for} \quad \alpha = 0. \end{cases}$$

Proposition 2.3.1
  (i) A^α *is a closed operator with its domain dense.*
  (ii) *If $0 < \alpha < \beta$, then $D(A^\alpha) \supset D(A^\beta)$.*
  (iii) $A^{\alpha+\beta} = A^\alpha A^\beta = A^\beta A^\alpha$ *for all $\alpha > 0$ and $\beta > 0$.*

Proof Since A^α is the inverse of a bounded operator, it is a closed operator. If $0 < \alpha < \beta$ and $u \in D(A^\beta)$, from (2.28) we obtain

$$u = A^{-\beta} A^\beta u = A^{-\alpha-(\beta-\alpha)} A^\beta u = A^{-\alpha} A^{-(\beta-\alpha)} A^\beta u \in D(A^\alpha).$$

Let n be a natural number greater than α. Then, by the corollary of Lemma 2.3.1, $D(A^n)$ is dense and, hence, $D(A^\alpha)$ is also dense. Next, let $\alpha > 0$, $\beta > 0$ and $u \in D(A^\alpha A^\beta)$, i.e., $u \in D(A^\beta)$ and $A^\beta u \in D(A^\alpha)$. If we put $v = A^\alpha A^\beta u$, then $A^\beta u = A^{-\alpha} v$ and $u = A^{-\beta} A^{-\alpha} v = A^{-\alpha-\beta} v$. Hence, we have $u \in D(A^{\alpha+\beta})$ and $A^{\alpha+\beta} u = v$. On the other hand, let us assume $u \in D(A^{\alpha+\beta})$. If we put $A^{\alpha+\beta} u = v$, then $u \in D(A^\alpha A^\beta)$, because of $u = A^{-\alpha-\beta} v = A^{-\beta} A^{-\alpha} v$. Therefore, $A^{\alpha+\beta} = A^\alpha A^\beta$. $\square$

Lemma 2.3.2 *For* $0 < \alpha < 1$, *each* $\lambda \leq 0$ *belongs to* $\rho(A^\alpha)$ *and*

$$(A^\alpha - \lambda)^{-1} = \frac{1}{2\pi i} \int_\Gamma \frac{1}{\mu^\alpha - \lambda} (A - \mu)^{-1} \, d\mu, \tag{2.29}$$

where Γ *is the same contour as in* (2.26).

Proof Let B be an operator defined by the right-hand side of (2.29). Since $\mu^\alpha - \lambda \neq 0$ for $\mu \in \Gamma$, the operator B is bounded. As in the proof of (2.28), we find

$$A^{\alpha-1} B = BA^{\alpha-1} = \frac{1}{2\pi i} \int_\Gamma \frac{\mu^{\alpha-1}}{\mu^\alpha - \lambda} (A - \mu)^{-1} \, d\mu. \tag{2.30}$$

By the use of the identity

$$\frac{\mu^{\alpha-1}}{\mu^\alpha - \lambda} = \frac{1}{\mu} + \frac{\lambda}{\mu(\mu^\alpha - \lambda)},$$

(2.30) becomes

$$A^{\alpha-1} B = BA^{\alpha-1}$$

$$= \frac{1}{2\pi i} \int_\Gamma \mu^{-1} (A - \mu)^{-1} \, d\mu + \frac{1}{2\pi i} \int_\Gamma \frac{\lambda}{\mu(\mu^\alpha - \lambda)} (A - \mu)^{-1} \, d\mu.$$

The first term on the right-hand side of this equation is equal to A^{-1} and the second term to $\lambda A^{-1} B = \lambda BA^{-1}$. Thus, we obtain

$$A^{\alpha-1} B = BA^{\alpha-1} = A^{-1}(1 + \lambda B) = (1 + \lambda B) A^{-1}.$$

From this it follows that if $u \in D(A)$ we have $BA^\alpha u = BA^{\alpha-1} Au = (1 + \lambda B)u$, i.e.,

$$B(A^\alpha - \lambda)u = u. \tag{2.31}$$

Next, we put $u_\mu = (1 + \mu^{-1} A)^{-1} u$ for each $\mu > 0$, where $u \in D(A^\alpha)$, then $u_\mu \in D(A)$. Since $A^{-\alpha}$ and $(1 + \mu^{-1} A)^{-1}$ are obviously commutative, $A^\alpha u_\mu = (1 + \mu^{-1} A)^{-1} A^\alpha u$. By Lemma 2.3.1, we have that $u_\mu \to u$ and $A^\alpha u_\mu \to A^\alpha u$ as $\mu \to \infty$. Hence, by applying (2.31) to u_μ and taking the limit $\mu \to \infty$, we can observe that (2.31) holds for every $u \in D(A^\alpha)$. On the other hand, B and $A^{-\alpha}$ are commutative, so that, from (2.31), for any $u \in D(A^\alpha)$, we obtain

$$Bu = BA^{-\alpha} A^\alpha u = A^{-\alpha} BA^\alpha u = A^{-\alpha} (\lambda Bu + u),$$

which implies that $B = A^{-\alpha}(\lambda B + 1)$ since B and $A^{-\alpha}$ are bounded. Hence, $R(B) \subset D(A^\alpha)$ and $A^\alpha B = \lambda B + 1$, i.e., $(A^\alpha - \lambda)B = I$. This, together with (2.31), gives $B = (A^\alpha - \lambda)^{-1}$. $\square$

If we let $a \to 0$ and $\varphi \to \pi$ in (2.29), we can observe that

$$(A^\alpha - \lambda)^{-1} = \frac{\sin \pi \alpha}{\pi} \int_0^\infty \frac{\mu^\alpha (A + \mu)^{-1}}{\mu^{2\alpha} - 2\lambda \mu^\alpha \cos \pi \alpha + \lambda^2} \, d\mu \qquad (2.32)$$

holds for each $\lambda \leq 0$. In particular, for $\lambda = 0$, we have the formula

$$A^{-\alpha} = \frac{\sin \pi \alpha}{\pi} \int_0^\infty \mu^{-\alpha} (A + \mu)^{-1} \, d\mu. \qquad (2.33)$$

Proposition 2.3.2 *If $0 < \alpha < 1$, then A^α is of type $(\alpha \omega, M)$.*

Proof First, we will see that the right-hand side of (2.32) has an analytic continuation to the region $\pi \alpha < |\arg \lambda| \leq \pi$. For $\lambda < 0$, we may write

$$(A^\alpha - \lambda)^{-1} = \frac{\sin \pi \alpha}{\pi} \int_0^\infty \frac{\mu^\alpha (A + \mu)^{-1}}{(\mu^\alpha e^{-\pi \alpha i} - \lambda)(\mu^\alpha e^{\pi \alpha i} - \lambda)} \, d\mu. \qquad (2.34)$$

When $\lambda = |\lambda| e^{i\theta}$ and $\pi \alpha < \theta \leq \pi$, the norm of the right-hand side of (2.34) is not greater than

$$\frac{\sin \pi \alpha}{\pi} \int_0^\infty \frac{M \mu^{\alpha - 1} \, d\mu}{|\mu^\alpha e^{-\pi \alpha i} - |\lambda| e^{i\theta}| \, |\mu^\alpha e^{\pi \alpha i} - |\lambda| e^{i\theta}|}$$

$$= \frac{\sin \pi \alpha}{\pi} \frac{M}{|\lambda|} \int_0^\infty \frac{s^{\alpha - 1} \, ds}{|s^\alpha e^{-\pi \alpha i} - e^{i\theta}| \, |s^\alpha e^{\pi \alpha i} - e^{i\theta}|}. \qquad (2.35)$$

It is clear that the integral on the right-hand side of (2.35) converges. Moreover, if $\lambda < 0$, from (2.32) and

$$\frac{\sin \pi \alpha}{\pi} \int_0^\infty \frac{\mu^{\alpha - 1}}{\mu^{2\alpha} - 2\lambda \mu^\alpha \cos \pi \alpha + \lambda^2} \, d\mu = -\frac{1}{\lambda} \qquad (2.36)$$

the result $\|(A^\alpha - \lambda)^{-1}\| \leq M/|\lambda|$ follows; hence, we may conclude that A^α is of type $(\pi \alpha, M)$. Next, let ε be an arbitrary positive number smaller than $\min ((\pi - \omega)\alpha, 2\pi(1 - \alpha))$ and let

$$\kappa = \min (\pi - \omega - \varepsilon/\alpha, 2\pi(1 - \alpha)/\alpha - \varepsilon/\alpha),$$

then we have $0 < \kappa < \pi - \omega$, $(\pi - \kappa)\alpha \geq \alpha\omega + \varepsilon$, and $(\pi + \kappa)\alpha \leq 2\pi - (\pi \alpha + \varepsilon)$. Since $\omega + \varepsilon/\alpha < \pi - \kappa < \pi$, we obtain $-re^{-i\kappa} = re^{i(\pi - \kappa)} \in \rho(A)$ and $\|(A + re^{-i\kappa})^{-1}\| \leq M_{\varepsilon/\alpha}/r$, where $r > 0$. If we put λ in the sector $\pi \alpha < \arg \lambda < \pi \alpha + \varepsilon$, we can transform the integration path on the right-hand side of (2.34) to $re^{-i\kappa}$, $0 < r < \infty$, obtaining

$$(A^\alpha - \lambda)^{-1} = \frac{\sin \pi \alpha}{\pi e^{i(\alpha + 1)\kappa}} \int_0^\infty \frac{r^\alpha (A + re^{-i\kappa})^{-1} \, dr}{(r^\alpha e^{-i(\pi + \kappa)\alpha} - \lambda)(r^\alpha e^{i(\pi - \kappa)\alpha} - \lambda)}. \qquad (2.37)$$

Thus, the integral on the right-hand side of (2.37) can be analytically continued to $(\pi - \kappa)\alpha < \arg \lambda < \pi\alpha + \varepsilon$. Furthermore, let $\lambda = |\lambda| e^{i\theta}$ and $(\pi - \kappa)\alpha < \theta < \pi\alpha + \varepsilon$, then, by a calculation similar to (2.35), the norm of the right-hand side of (2.37) is bounded by

$$\frac{\sin \pi\alpha}{\pi} \frac{M_{\varepsilon/\alpha}}{|\lambda|} \int_0^\infty \frac{s^{\alpha-1}\, ds}{|s^\alpha e^{-i(\pi+\kappa)\alpha} - e^{i\theta}|\,|s^\alpha e^{i(\pi-\kappa)\alpha} - e^{i\theta}|} = \frac{M'_\varepsilon}{|\lambda|}.$$

If $\pi - \omega \leq 2\pi(1-\alpha)/\alpha$, then $(\pi - \kappa)\alpha = \alpha\omega + \varepsilon$. If $\pi - \omega > 2\pi(1-\alpha)/\alpha$, by repeating a similar process, we can observe that $(A^\alpha - \lambda)^{-1}$ has an analytic continuation to $\alpha\omega < \arg \lambda \leq \pi$; similarly, it can be analytically continued to $-\alpha\omega > \arg \lambda \geq -\pi$. Also, by a calculation similar to the one above, it can be shown that A^α is of type $(\alpha\omega, M)$. $\square$

Lemma 2.3.3 *For each α satisfying $0 \leq \alpha \leq 1$, there exists a constant C_α, depending only on M and α, such that, for all $\mu > 0$,*

$$\|A^\alpha(A+\mu)^{-1}\| \leq C_\alpha \mu^{\alpha-1}.$$

Proof For $\alpha = 0$ or 1, the lemma follows immediately from the assumption. For $0 < \alpha < 1$, it can be verified as follows:

$$\|A^\alpha(A+\mu)^{-1}\| = \|A^{\alpha-1}A(A+\mu)^{-1}\|$$

$$= \frac{\sin \pi\alpha}{\pi} \left\| \int_0^\infty \lambda^{\alpha-1}(A+\lambda)^{-1}A(A+\mu)^{-1}\, d\lambda \right\|$$

$$\leq \frac{\sin \pi\alpha}{\pi} \left\{ \int_0^\mu \lambda^{\alpha-1}\|A(A+\lambda)^{-1}\|\,\|(A+\mu)^{-1}\|\, d\lambda \right.$$

$$\left. + \int_\mu^\infty \lambda^{\alpha-1}\|(A+\lambda)^{-1}\|\,\|A(A+\mu)^{-1}\|\, d\lambda \right\}$$

$$\leq \frac{\sin \pi\alpha}{\pi} M(M+1) \left(\mu^{-1} \int_0^\mu \lambda^{\alpha-1}\, d\lambda + \int_\mu^\infty \lambda^{\alpha-2}\, d\lambda \right)$$

$$= \frac{\sin \pi\alpha}{\pi} M(M+1) \frac{\mu^{\alpha-1}}{\alpha(1-\alpha)}. \quad \square$$

Proposition 2.3.3 *For $0 \leq \alpha < \beta \leq 1$, there exists a constant $c_{\alpha,\beta}$, depending only on M, α and β, such that, for all $u \in D(A^\beta)$,*

$$\|A^\alpha u\| \leq c_{\alpha,\beta} \|A^\beta u\|^{\alpha/\beta} \|u\|^{1-\alpha/\beta}. \tag{2.38}$$

Proof Observe that $u \in D(A^\beta)$ implies that $u \in D(A^\alpha)$ by Proposition 2.3.1, (ii). Since the proposition is trivial for $\alpha = 0$, we may assume $\alpha > 0$.

With $u \in D(A)$, we have

$$\|A^\alpha u\| = \|A^{\alpha-1} Au\| = \left\| \frac{\sin \pi\alpha}{\pi} \int_0^\infty \mu^{\alpha-1} (A+\mu)^{-1} Au \, d\mu \right\|$$

$$\leq \frac{\sin \pi\alpha}{\pi} \left\{ \int_0^\delta \mu^{\alpha-1} \|A(A+\mu)^{-1} u\| \, d\mu \right.$$

$$\left. + \int_\delta^\infty \mu^{\alpha-1} \|A^{1-\beta}(A+\mu)^{-1} A^\beta u\| \, d\mu \right\}.$$

By means of Lemma 2.3.3, this can be estimated as

$$\leq \frac{\sin \pi\alpha}{\pi} \left\{ (M+1) \|u\| \int_0^\delta \mu^{\alpha-1} \, d\mu + C_{1-\beta} \|A^\beta u\| \int_\delta^\infty \mu^{\alpha-\beta-1} \, d\mu \right\}$$

$$= \frac{\sin \pi\alpha}{\pi} \left\{ (M+1) \|u\| \frac{\delta^\alpha}{\alpha} + C_{1-\beta} \|A^\beta u\| \frac{\delta^{\alpha-\beta}}{\beta-\alpha} \right\}.$$

Thus, by letting $\delta = \{C_{1-\beta} \|A^\beta u\| / ((M+1) \|u\|)\}^{1/\beta}$, the inequality (2.38) has been shown to hold for $u \in D(A)$. For $u \in D(A^\beta)$, we may approach u by a sequence of elements of $D(A)$, as in the proof of Lemma 2.3.2. $\square$

Remark 2.3.1 The inequality (2.38) is called a *moment inequality*. A more general form of the moment inequality can be proved. Indeed, for any $\alpha < \beta < \gamma$ and for any $u \in D(A^\gamma)$, the inequality

$$\|A^\beta u\| \leq c(\alpha, \beta, \gamma) \|A^\gamma u\|^{(\beta-\alpha)/(\gamma-\alpha)} \|A^\alpha u\|^{(\gamma-\beta)/(\gamma-\alpha)}$$

holds. For more details the reader may refer to Kreĭn [9], p. 115.

Corollary If A is of type (ω, M) and $0 < \alpha < 1$, then $\omega + \varepsilon \leq |\arg \lambda| \leq \pi$ for all $\varepsilon > 0$ and

$$\|A^\alpha (A-\lambda)^{-1}\| \leq c_{\alpha,1} (M_\varepsilon + 1)^\alpha M_\varepsilon^{1-\alpha} |\lambda|^{\alpha-1}, \tag{2.39}$$

where $c_{\alpha,1}$ is the value of $c_{\alpha,\beta}$ in the proposition at $\beta = 1$.

Proof For all $u \in X$, we have

$$\|A^\alpha (A-\lambda)^{-1} u\| \leq c_{\alpha,1} \|A(A-\lambda)^{-1} u\|^\alpha \|(A-\lambda)^{-1} u\|^{1-\alpha}.$$

(2.39) follows from this immediately. $\square$

2.3.2 General case

In the preceding subsection, we assumed that $0 \in \rho(A)$ for an operator A of type (ω, M). We will attempt to define fractional powers of A without such an assumption. In the present subsection, we always assume $0 < \alpha < 1$. With the expectation that, once A^α is defined, its resolvent can be expressed by (2.32), for each $\lambda < 0$ we consider the operator

$$I(\lambda) = \frac{\sin \pi\alpha}{\pi} \int_0^\infty \frac{\mu^\alpha (A + \mu)^{-1}}{\mu^{2\alpha} - 2\lambda\mu^\alpha \cos \pi\alpha + \lambda^2} \, d\mu. \tag{2.40}$$

It is easily seen that $I(\lambda)$ is a bounded operator. Let δ be an arbitrary positive number. Then it is easy to show that $A_\delta = A + \delta$ is also of type (ω, M) and that the number M_ε in Definition 2.3.1 can be chosen independently of δ. Since $0 \in \rho(A_\delta)$, the operator A_δ^α is well defined and, using (2.32), its resolvent is given by

$$(A_\delta^\alpha - \lambda)^{-1} = \frac{\sin \pi\alpha}{\pi} \int_0^\infty \frac{\mu^\alpha (A + \delta + \mu)^{-1}}{\mu^{2\alpha} - 2\lambda\mu^\alpha \cos \pi\alpha + \lambda^2} \, d\mu.$$

For all $\lambda < 0$, we have

$$\|I(\lambda) - (A_\delta^\alpha - \lambda)^{-1}\| = \left\| \frac{\sin \pi\alpha}{\pi} \int_0^\infty \frac{\mu^\alpha \{(A + \mu)^{-1} - (A + \delta + \mu)^{-1}\}}{\mu^{2\alpha} - 2\lambda\mu^\alpha \cos \pi\alpha + \lambda^2} \, d\mu \right\|$$

$$\leqslant \frac{\sin \pi\alpha}{\pi} \int_0^\eta \frac{\mu^\alpha \{\|(A + \mu)^{-1}\| + \|(A + \delta + \mu)^{-1}\|\}}{\mu^{2\alpha} - 2\lambda\mu^\alpha \cos \pi\alpha + \lambda^2} \, d\mu$$

$$+ \frac{\sin \pi\alpha}{\pi} \delta \int_\eta^\infty \frac{\mu^\alpha \|(A + \mu)^{-1}(A + \delta + \mu)^{-1}\|}{\mu^{2\alpha} - 2\lambda\mu^\alpha \cos \pi\alpha + \lambda^2} \, d\mu$$

$$\leqslant \frac{\sin \pi\alpha}{\pi} 2M \int_0^\eta \frac{\mu^{\alpha-1}}{\mu^{2\alpha} - 2\lambda\mu^\alpha \cos \pi\alpha + \lambda^2} \, d\mu$$

$$+ \frac{\sin \pi\alpha}{\pi} M^2 \delta \int_\eta^\infty \frac{\mu^{\alpha-2}}{\mu^{2\alpha} - 2\lambda\mu^\alpha \cos \pi\alpha + \lambda^2} \, d\mu,$$

which implies that

$$\lim_{\delta \to 0} (A_\delta^\alpha - \lambda)^{-1} = I(\lambda). \tag{2.41}$$

Since $(A_\delta^\alpha - \lambda)^{-1}$ satisfies the resolvent equation, so does $I(\lambda)$:

$$I(\lambda) - I(\mu) = (\lambda - \mu) I(\lambda) I(\mu). \tag{2.42}$$

Definition 2.3.3 *If a bounded operator-valued function $I(\lambda)$, defined on a set $\Omega \subset \mathbb{C}$, satisfies the resolvent equation (2.42) on Ω, then it is called a pseudo-resolvent on Ω.*

Lemma 2.3.4 *Let $I(\lambda)$ be a pseudo-resolvent on Ω. The range $R(I(\lambda))$ and the null set $N(I(\lambda))$ are independent of λ. When $N(I(\lambda))$ consists of only 0, there exists a closed operator T such that $\rho(T) \supset \Omega$ and $I(\lambda) = (T - \lambda)^{-1}$ for all $\lambda \in \Omega$.*

Proof Since the first part of the lemma is easy to prove, it is omitted. When $N(I(\lambda))$ consists of only 0, it is readily seen that $T = I(\lambda)^{-1} + \lambda$ is a closed operator and, by (2.42), it is independent of λ. $\square$

$I(\lambda)$ defined by (2.40) is a pseudo-resolvent on $(-\infty, 0)$. Now we will show that $N(I(\lambda))$ consists of only 0. By (2.34), we have

$$\|\lambda I(\lambda)u + u\| = \left\| \frac{\lambda \sin \pi\alpha}{\pi} \int_0^\infty \frac{\mu^{\alpha-1}\{\mu(A+\mu)^{-1}u - u\}}{\mu^{2\alpha} - 2\mu^\alpha\lambda \cos \pi\alpha + \lambda^2} d\mu \right\|$$

$$\leq |\lambda| \frac{\sin \pi\alpha}{\pi} \int_0^N \frac{\mu^{\alpha-1}}{\mu^{2\alpha} - 2\mu^\alpha\lambda \cos \pi\alpha + \lambda^2} d\mu (M+1)\|u\|$$

$$+ |\lambda| \frac{\sin \pi\alpha}{\pi} \int_N^\infty \frac{\mu^{\alpha-1}}{\mu^{2\alpha} - 2\mu^\alpha\lambda \cos \pi\alpha + \lambda^2}$$

$$\times \|\mu(A+\mu)^{-1}u - u\| d\mu.$$

Let ε be an arbitrary positive number. By Lemma 2.3.1 we have $\|\mu(A+\mu)^{-1}u - u\| < \varepsilon$ for $\mu > N$ with N sufficiently large, so that the second term on the right-hand side of the above inequality does not exceed ε. Fix N as above and let $\lambda \to -\infty$, then the first term converges to 0. Therefore, we find

$$\lim_{\lambda \to -\infty} \|\lambda I(\lambda)u + u\| = 0. \tag{2.43}$$

If $I(\lambda)u = 0$ for some λ, then, by (2.42), $I(\lambda)u = 0$ for all λ; we conclude $u = 0$ from (2.43). Hence, by Lemma 2.3.4, $I(\lambda)$ is the resolvent of a closed operator. We define A^α by

$$I(\lambda) = (A^\alpha - \lambda)^{-1}, \qquad \lambda < 0. \tag{2.44}$$

It is clear that $-\lambda I(\lambda)u \in D(A^\alpha)$ for all u. This, together with (2.43), implies that $D(A^\alpha)$ is dense. Equation (2.32) can be seen from the definition; it can also be shown by repeating the proof of Proposition 2.3.2 that A^α is of type $(\alpha\omega, M)$. Next, we study the relation between A_δ^α and A^α.

Lemma 2.3.5 $D((A+\delta)^\alpha) = D(A^\alpha)$ *for all $\delta > 0$ and*

$$\|(A+\delta)^\alpha u - A^\alpha u\| \leq c\delta^\alpha \|u\| \tag{2.45}$$

for every $u \in D(A^\alpha)$, where c is a constant determined only by M and α.

Proof When $u \in D(A)$,

$$A_\delta^\alpha u = A_\delta^{\alpha-1} A_\delta u = \frac{\sin \pi\alpha}{\pi} \int_0^\infty \mu^{\alpha-1}(A_\delta + \mu)^{-1} A_\delta u \, d\mu.$$

For $0 < \eta < \delta$, by expressing $A_\eta^\alpha u$ similarly, we have

$$A_\delta^\alpha u - A_\eta^\alpha u = \frac{\sin \pi\alpha}{\pi}\bigg[\int_0^\varepsilon \mu^{\alpha-1}\{(A_\delta + \mu)^{-1}A_\delta u - (A_\eta + \mu)^{-1}A_\eta u\}\, d\mu$$

$$+ (\delta - \eta) \int_\varepsilon^\infty \mu^\alpha (A_\delta + \mu)^{-1}(A_\eta + \mu)^{-1}u\, d\mu \bigg],$$

$$\|(A_\delta + \mu)^{-1}A_\delta u\| \leq (M+1)\|u\|, \qquad \|(A_\varepsilon + \mu)^{-1}\| \leq M/\mu,$$

so that

$$\|A_\delta^\alpha u - A_\eta^\alpha u\| \leq \frac{\sin \pi\alpha}{\pi}\bigg\{ 2(M+1)\frac{\varepsilon^\alpha}{\alpha} + M^2(\delta - \eta)\frac{\varepsilon^{\alpha-1}}{1-\alpha}\bigg\}\|u\|.$$

On choosing $\varepsilon = (\delta - \eta)M^2/(2(M+1))$, this gives

$$\|A_\delta^\alpha u - A_\eta^\alpha u\| \leq c(\delta - \eta)^\alpha \|u\|,$$

$$c = \frac{\sin \pi\alpha}{\pi}\, \frac{1}{\alpha(1-\alpha)}\, \frac{M^{2\alpha}}{(2(M+1))^{\alpha-1}}.$$

Hence, the limit $\lim_{\delta \to 0} A_\delta^\alpha u = Bu$ exists and

$$\|A_\delta^\alpha u - Bu\| \leq c\delta^\alpha \|u\|$$

holds for all $u \in D(A)$. From this, it follows that the smallest closed extension $\bar{B}$ of B exists, $D(\bar{B}) = D(A_\delta^\alpha)$, and that

$$\|A_\delta^\alpha u - \bar{B}u\| \leq c\delta^\alpha \|u\| \tag{2.46}$$

holds for all $u \in D(\bar{B})$. For $\lambda < 0$, from (2.46), we obtain,

$$\|(\bar{B} - A_\delta^\alpha)(A_\delta^\alpha - \lambda)^{-1}\| \leq c\delta^\alpha M/|\lambda|.$$

Hence, if δ is sufficiently small, a bounded inverse of

$$\bar{B} - \lambda = \{1 + (\bar{B} - A_\delta^\alpha)(A_\delta^\alpha - \lambda)^{-1}\}(A_\delta^\alpha - \lambda)$$

exists and it satisfies

$$(\bar{B} - \lambda)^{-1} = \lim_{\delta \to 0}(A_\delta^\alpha - \lambda)^{-1}. \tag{2.47}$$

Equations (2.41), (2.47) and the definition of A^α lead to $\bar{B} = A^\alpha$. $\square$

Theorem 2.3.1 *Let A be an operator of type (ω, M). By means of (2.40) and (2.44), an operator A^α of type $(\alpha\omega, M)$ is defined for each $0 < \alpha < 1$. If $0 < \alpha < \beta \leq 1$, then $D(A^\alpha) \supset D(A^\beta)$; if $\alpha > 0$, $\beta > 0$ and $\alpha + \beta \leq 1$, then $u \in D(A^\alpha A^\beta)$ and $A^\alpha A^\beta u = A^{\alpha+\beta}u$ for all $u \in D(A^{\alpha+\beta})$. For each α, β*

satisfying $0 \le \alpha < \beta \le 1$, *there exists a positive constant* $c_{\alpha,\beta}$ *such that*

$$\|A^\alpha u\| \le c_{\alpha,\beta} \|A^\beta u\|^{\alpha/\beta} \|u\|^{1-\alpha/\beta} \tag{2.48}$$

for all $u \in D(A^\beta)$.

Proof Let $0 < \alpha < \beta \le 1$. From Proposition 2.3.1, (ii) and Lemma 2.3.5, it follows that $D(A^\alpha) = D((A+\delta)^\alpha) \supset D((A+\delta)^\beta) = D(A^\beta)$ for any $\delta > 0$. Next, let $\alpha > 0$, $\beta > 0$ and $\alpha + \beta \le 1$. If $u \in D(A^{\alpha+\beta})$, from Proposition 2.3.1, (iii) and Lemma 2.3.5, we find that $u \in D(A_\delta^{\alpha+\beta}) = D(A_\delta^\alpha A_\delta^\beta)$ for $\delta > 0$, so that $A_\delta^\beta u \in D(A_\delta^\alpha) = D(A^\alpha)$. From (2.45), as $\delta \to 0$, we have $A_\delta^\beta u \to A^\beta u$, $A_\delta^\alpha A_\delta^\beta u = A_\delta^{\alpha+\beta} u \to A^{\alpha+\beta} u$ and

$$\|A^\alpha A_\delta^\beta u - A_\delta^\alpha A_\delta^\beta u\| \le c\delta^\alpha \|A_\delta^\beta u\| \to 0.$$

Therefore, $A^\alpha A_\delta^\beta u \to A^{\alpha+\beta} u$. Since A^α is a closed operator, $A^\beta u \in D(A^\alpha)$ and $A^\alpha A^\beta u = A^{\alpha+\beta} u$. Finally, let $0 \le \alpha < \beta \le 1$. From Proposition 2.3.3, with A_δ replacing A, Eq. (2.48) holds for all $u \in D(A^\beta) = D(A_\delta^\beta)$. By letting $\delta \to 0$, the proof is completed. $\square$

2.3.3 Heinz–Kato's theorem

In this subsection, X is assumed to be a Hilbert space. Let A be a positive operator in X, and its spectral resolution be represented by

$$A = \int_0^\infty \lambda \, dE(\lambda). \tag{2.49}$$

Since $A + \delta$ is positive definite for $\delta > 0$, its fractional powers are calculated by means of (2.26) and (1.14) as follows:

$$(A+\delta)^{-\alpha} = \frac{1}{2\pi i} \int_\Gamma \lambda^{-\alpha} (A+\delta-\lambda)^{-1} \, d\lambda$$

$$= \frac{1}{2\pi i} \int_\Gamma \lambda^{-\alpha} \int_0^\infty \frac{1}{\mu+\delta-\lambda} \, dE(\mu) \, d\lambda$$

$$= \int_0^\infty \frac{1}{2\pi i} \int_\Gamma \frac{\lambda^{-\alpha}}{\mu+\delta-\lambda} \, d\lambda \, dE(\mu) = \int_0^\infty (\mu+\delta)^{-\alpha} \, dE(\mu).$$

Thus, we have

$$(A+\delta)^\alpha = \int_0^\infty (\mu+\delta)^\alpha \, dE(\mu), \tag{2.50}$$

which shows that the fractional powers of $A + \delta$ coincide with those defined by the usual method of spectral resolution. The same applies to the fractional powers of A itself, as is easily seen from Lemma 2.3.5 by letting $\delta \to 0$ in (2.50). Therefore, we have the following theorem.

Theorem 2.3.2 *If A is a positive self-adjoint operator, the fractional power A^α coincides with that defined by (1.13) using spectral resolution.*

Also, in this case, by the use of Hölder's inequality, we obtain

$$\|A^\alpha u\|^2 = \int_0^\infty \lambda^{2\alpha} \, d\,\|E(\lambda)u\|^2$$

$$\leq \left\{ \int_0^\infty \lambda^{2\beta} \, d\,\|E(\lambda)u\|^2 \right\}^{\alpha/\beta} \left\{ \int_0^\infty d\,\|E(\lambda)u\|^2 \right\}^{1-\alpha/\beta}$$

$$= \|A^\beta u\|^{2\alpha/\beta} \|u\|^{2(1-\alpha/\beta)}.$$

That is, the inequality (2.48) holds for $c_{\alpha,\beta} = 1$.

Theorem 2.3.3 (Heinz [71]) *Let X_1 and X_2 be two Hilbert spaces and let A and B be self-adjoint positive operators in X_1 and X_2, respectively. Also, let T be a bounded operator from X_1 into X_2, which maps $D(A)$ into $D(B)$. Assume that there exists a number M such that, for all $u \in D(A)$,*

$$\|BTu\| \leq M\,\|Au\|. \tag{2.51}$$

Then, for each α satisfying $0 < \alpha < 1$, the image of $D(A^\alpha)$ under T is included in $D(B^\alpha)$ and

$$\|B^\alpha Tu\| \leq M^\alpha \|T\|^{1-\alpha} \|A^\alpha u\| \tag{2.52}$$

holds for all $u \in D(A^\alpha)$.

Proof First, we assume A is positive definite. Let $u \in D(A^\alpha)$ and $v \in D(B)$. $A^z u$ is regular in $\operatorname{Re} z < \alpha$ and continuous in $\operatorname{Re} z \leq \alpha$; $B^{\alpha-z}v$ is regular in $\alpha - 1 < \operatorname{Re} z < \alpha$ and continuous in $\alpha - 1 \leq \operatorname{Re} z \leq \alpha$. Hence, $f(z) = (TA^z u, B^{\alpha-\bar{z}}v)$ is regular in $\alpha - 1 < \operatorname{Re} z < \alpha$ and continuous in $\alpha - 1 \leq \operatorname{Re} z \leq \alpha$. Next, we will estimate $|f(z)|$ on $\operatorname{Re} z = \alpha - 1$ and $\operatorname{Re} z = \alpha$. Since $A^{\alpha-1+iy}u = A^{-1}A^{\alpha+iy}u \in D(A)$ for y real, we have $TA^{\alpha-1+iy} \in D(B)$ and

$$|f(\alpha - 1 + iy)| = |(TA^{\alpha-1+iy}u, B^{1+iy}v)|$$

$$= |(BTA^{\alpha-1+iy}u, B^{iy}v)| \leq \|BTA^{\alpha-1+iy}u\| \|B^{iy}v\|.$$

Equation (2.51) and the fact that A^{iy} and B^{iy} are unitary imply that

$$\leq M\,\|A^{\alpha+iy}u\| \|B^{iy}v\| = M\,\|A^\alpha u\| \|v\|.$$

Similarly,

$$|f(\alpha + iy)| = |(TA^{\alpha+iy}u, B^{iy}v)|$$
$$\leq \|TA^{\alpha+iy}u\| \|B^{iy}v\| \leq \|T\| \|A^{\alpha}u\| \|v\|.$$

Hence, by the three-line theorem in the theory of functions, we get

$$|(Tu, B^{\alpha}v)| = |f(0)|$$

$$\leq \left\{\sup_{y} |f(\alpha - 1 + iy)|\right\}^{\alpha} \left\{\sup_{y} |f(\alpha + iy)|\right\}^{1-\alpha}$$

$$\leq M^{\alpha} \|T\|^{1-\alpha} \|A^{\alpha}u\| \|v\|. \tag{2.53}$$

Thus, we have found that $Tu \in D(B^{\alpha})$ and (2.52) holds. When A is merely positive, the assumption of the theorem is still satisfied if we replace A by $A + \varepsilon$ with $\varepsilon > 0$ since $\|Au\| \leq \|(A + \varepsilon)u\|$, so that T maps $D(A^{\alpha}) = D((A + \varepsilon)^{\alpha})$ into $D(B^{\alpha})$ and

$$\|B^{\alpha}Tu\| \leq M^{\alpha} \|T\|^{1-\alpha} \|(A + \varepsilon)^{\alpha}u\|$$

holds for every $u \in D(A^{\alpha})$. By letting $\varepsilon \to 0$, we have (2.52). $\square$

Corollary Let A and B be positive self-adjoint operators in a Hilbert space X, and suppose $D(A) \subset D(B)$. For all α satisfying $0 < \alpha < 1$, we have $D(A^{\alpha}) \subset D(B^{\alpha})$. If, in addition, there is a certain number M such that $\|Bu\| \leq M \|Au\|$ for all $u \in D(A)$, then $\|B^{\alpha}u\| \leq M^{\alpha} \|A^{\alpha}u\|$ for all $u \in D(A^{\alpha})$.

Proof From the assumption and the corollary of Theorem 1.1.12, it follows that $B(A + 1)^{-1} \in B(X)$. Let $X_1 = X_2 = X$, $T = I$ and replace A by $A + 1$; the assumption of the theorem is still satisfied. Hence, we have $D(A^{\alpha}) = D((A + 1)^{\alpha}) \subset D(B^{\alpha})$. For the proof of the latter part of the corollary, we only need to let $X_1 = X_2 = X$ and $T = I$ and to apply the theorem. $\square$

Theorem 2.3.3 is important in applications, and it is hoped that it can be generalized to operators defined in Banach spaces. This is an open question yet to be solved. It is known, however, that if A and B are both of type (ω, M) and $D(A) \subset D(B)$, then $D(A^{\beta}) \subset D(B^{\alpha})$ for $0 < \alpha < \beta \leq 1$ (Kreĭn [9], Lemma 7.3, Chapter 1).

When A and B are maximal accretive operators in a Hilbert space, both are of type $(\pi/2, 1)$, as explained previously in this section. In this case, a generalization of Theorem 2.3.3 by Kato [81, 82] is available.

Theorem 2.3.4 Let X_1 and X_2 be Hilbert spaces, A and B maximal accretive operators in X_1 and X_2, respectively, and T a bounded operator *from X_1 into X_2. Assume that T maps $D(A)$ into $D(B)$ and that there is a certain number M such that, for all $u \in D(A)$,*

$$\|BTu\| \leq M \|Au\|. \tag{2.54}$$

Then T maps $D(A^\alpha)$ into $D(B^\alpha)$ for each α satisfying $0 \leq \alpha \leq 1$, and

$$\|B^\alpha Tu\| \leq e^{\pi\sqrt{\alpha(1-\alpha)}} M^\alpha \|T\|^{1-\alpha} \|A^\alpha u\| \tag{2.55}$$

holds for all $u \in D(A^\alpha)$.

Though we will give a proof of this theorem along the lines of [81], it is shown in [82] that (2.55) still holds when $\pi\sqrt{\alpha(1-\alpha)}$ is replaced by a smaller number $\pi^2\alpha(1-\alpha)/2$.

Lemma 2.3.6 *Let A be a maximal accretive operator. Suppose that there exists a $\delta > 0$ such that $\operatorname{Re}(Au, u) \geq \delta \|u\|^2$ for every $u \in D(A)$. Then, for $0 < \alpha < 1$ and for all $u \in D(A^\alpha)$,*

$$\operatorname{Re}(A^\alpha u, u) \geq \delta^\alpha \|u\|^2. \tag{2.56}$$

Proof From the assumption, it follows that $\|(A + \mu)^{-1}\| \leq (\mu + \delta)^{-1}$ for $\mu > 0$, so that

$$\operatorname{Re}(A(A + \mu)^{-1}u, u) = \|u\|^2 - \mu \operatorname{Re}((\mu + A)^{-1}u, u)$$

$$\geq \|u\|^2 - \mu(\mu + \delta)^{-1} \|u\|^2 = \delta(\mu + \delta)^{-1} \|u\|^2.$$

Therefore, for $u \in D(A)$, we have

$$\operatorname{Re}(A^\alpha u, u) = \operatorname{Re}(A^{\alpha-1}Au, u)$$

$$= \frac{\sin \pi\alpha}{\pi} \int_0^\infty \mu^{\alpha-1} \operatorname{Re}((A + \mu)^{-1}Au, u) \, d\mu$$

$$\geq \frac{\sin \pi\alpha}{\pi} \int_0^\infty \mu^{\alpha-1}\delta(\mu + \delta)^{-1} \|u\|^2 \, d\mu = \delta^\alpha \|u\|^2.$$

For general $u \in D(A^\alpha)$, we may use an approximating sequence in $D(A)$, as in the proof of Lemma 2.3.2. $\square$

Lemma 2.3.7 *Let A be an accretive operator. If an inverse A^{-1} exists, it is also accretive. In particular, if there exists a $\delta > 0$ such that $\operatorname{Re}(Au, u) \geq \delta \|u\|^2$ for every u, then $\operatorname{Re}(A^{-1}v, u) \geq \delta \|A\|^{-2} \|u\|^2$.*

Proof Since

$$\operatorname{Re}(A^{-1}u, u) = \operatorname{Re}(A^{-1}u, AA^{-1}u) \geq 0$$

for every $u \in D(A)$, the operator A^{-1} is accretive. The rest of the proof is evident from

$$\mathrm{Re}\,(A^{-1}u, AA^{-1}u) \geqslant \delta\,\|A^{-1}u\|^2 \geqslant \delta\,\|A\|^{-2}\,\|u\|^2. \quad \square$$

Suppose A is a bounded accretive operator. If there exists a $\delta > 0$ such that $\mathrm{Re}\,(Au, u) \geqslant \delta \|u\|^2$ for every u, then, from Theorem 1.1.13, $\sigma(A)$ is contained in $\{\lambda : \mathrm{Re}\,\lambda \geqslant \delta, |\lambda| \leqslant \|A\|\}$. Choose a smooth closed curve Γ which encircles $\sigma(A)$ and does not touch the non-positive part of the real axis. For all complex numbers α, we put

$$A^\alpha = \frac{-1}{2\pi i} \int_\Gamma \lambda^\alpha (A - \lambda)^{-1}\, d\lambda,$$

where the integration is performed counterclockwise. For all α, A^α is bounded, $A^{\alpha+\beta} = A^\alpha A^\beta$ and, moreover, A^α is an entire function of α. If $\alpha > 0$, it coincides with that given by Definition 2.3.2. By taking Γ symmetric with respect to the real axis, we can also observe that $A^{*\alpha} = A^{\bar\alpha *}$ holds.

Lemma 2.3.8 *Let A be a bounded accretive operator. If there exists a $\delta > 0$ such that $\mathrm{Re}\,(Au, u) \geqslant \delta \|u\|^2$ for every u, then*

$$\|A^{*\alpha}A^{-\alpha}\| \leqslant \left(1 + \left|\tan\frac{\pi\alpha}{2}\right|\right)\left(1 - \left|\tan\frac{\pi\alpha}{2}\right|\right)^{-1} \tag{2.57}$$

holds in $|\mathrm{Re}\,\alpha| \leqslant \frac{1}{2}$. Also, let $i\eta$ be a pure imaginary number, then the following inequality holds:

$$\|A^{i\eta}\| \leqslant e^{\pi|\eta|/2}. \tag{2.58}$$

Proof Put $H_\alpha = (A^\alpha + A^{*\alpha})/2$ and $K_\alpha = (A^\alpha - A^{*\alpha})/2i$. It is easy to see that

$$\|H_\alpha u\|^2 - \|K_\alpha u\|^2 = \mathrm{Re}\,(A^\alpha u, A^{*\alpha}u) = \mathrm{Re}\,(A^{\alpha+\bar\alpha}u, u). \tag{2.59}$$

for each u. Write $\mathrm{Re}\,\alpha = \xi$. $A^{2\xi}$ is accretive for $0 \leqslant \xi \leqslant \frac{1}{2}$, since it is of type $(\pi\xi, 1)$, so that $\|K_\alpha u\| \leqslant \|H_\alpha u\|$ by (2.59). Since A^{-1} is also accretive by Lemma 2.3.7, $A^{2\xi}$ is an accretive operator for $-\frac{1}{2} \leqslant \xi \leqslant 0$ as well, so that $\|K_\alpha u\| \leqslant \|H_\alpha u\|$ by (2.59). Therefore, in $|\mathrm{Re}\,\alpha| \leqslant \frac{1}{2}$, it holds that

$$\|K_\alpha u\| \leqslant \|H_\alpha u\|. \tag{2.60}$$

From Lemma 2.3.6 and (2.59), it follows that

$$\|H_\alpha u\|^2 \geqslant \mathrm{Re}\,(A^{2\xi}u, u) \geqslant \delta^{2\xi}\,\|u\|^2$$

for $0 \leqslant \xi \leqslant \frac{1}{2}$. On the other hand, for $-\frac{1}{2} \leqslant \xi \leqslant 0$, from (2.59) and Lemmas

2.3.6 and 2.3.7 we obtain

$$\|H_\alpha u\|^2 \leqslant \operatorname{Re}(A^{2\xi}u, u) \geqslant (\delta \|A\|^{-2})^{-2\xi} \|u\|^2.$$

Therefore, H_α has a continuous inverse for $|\operatorname{Re}\alpha| \leqslant \frac{1}{2}$. In particular, since H_α is symmetric for α real, H_α^{-1} is a bounded operator. Since H_α is continuous with respect to α in the norm topology, H_α^{-1} is bounded for all α satisfying $|\operatorname{Re}\alpha| \leqslant \frac{1}{2}$; since H_α is a regular function of α, so is H_α^{-1}. Therefore, by (2.60),

$$K_\alpha H_\alpha^{-1} \text{ is regular in } |\operatorname{Re}\alpha| \leqslant \frac{1}{2} \quad \text{and} \quad \|K_\alpha H_\alpha^{-1}\| \leqslant 1. \tag{2.61}$$

Put $T(\alpha) = K_\alpha H_\alpha^{-1}/\tan(\pi\alpha/2)$. $\tan(\pi\alpha/2)$ is regular in $|\operatorname{Re}\alpha| \leqslant \frac{1}{2}$ and has its only zero at $\alpha = 0$, $\tan(\pi\alpha/2) = 1$ for $|\operatorname{Re}\alpha| = \frac{1}{2}$, $\tan(\pi\alpha/2) \to \pm i$ as $\operatorname{Im}\alpha \to \pm\infty$; from (2.61) we obtain that $T(\alpha)$ is regular in $|\operatorname{Re}\alpha| \leqslant \frac{1}{2}$, $\|T(\alpha)\| \leqslant 1$ for $|\operatorname{Re}\alpha| = \frac{1}{2}$, and $\limsup \|T(\alpha)\| \leqslant 1$ as $\operatorname{Im}\alpha \to \pm\infty$. Hence, by the maximum principle, we have $\|T(\alpha)\| \leqslant 1$ in $|\operatorname{Re}\alpha| \leqslant \frac{1}{2}$, and, therefore, $\|K_\alpha u\| \leqslant |\tan(\pi\alpha/2)| \|H_\alpha u\|$. From this it follows immediately that

$$(1 - |\tan(\pi\alpha/2)|) \|A^{*\alpha}u\| \leqslant (1 + |\tan(\pi\alpha/2)|) \|A^\alpha u\|,$$

which is equivalent to (2.57). When $\alpha = -i\eta$ is purely imaginary, from (2.57) we obtain $\|A^{i\eta}\|^2 = \|A^{*-i\eta}A^{i\eta}\| \leqslant e^{\pi|\eta|}$, which proves (2.58). $\quad\square$

Lemma 2.3.9 *Let X_1 and X_2 be Hilbert spaces, A and B bounded accretive operators in X_1 and X_2, respectively, and T a bounded operator from X_1 into X_2. Then, for $0 \leqslant \xi \leqslant 1$, the result*

$$\|B^\xi T A^\xi\| \leqslant e^{\pi\sqrt{\xi(1-\xi)}} \|T\|^{1-\xi} \|BTA\|^\xi \tag{2.62}$$

holds.

Proof By virtue of Lemma 1.3.5, it is enough to prove the lemma for the case in which A and B have a bounded inverse. Let k be a positive number and put $F(\alpha) = e^{k\alpha(\alpha-1)}B^\alpha T A^\alpha$, then F is an entire function of α. Write $\alpha = \xi + i\eta$. By Lemma 2.3.8, for $0 \leqslant \xi = \operatorname{Re}\alpha \leqslant 1$, we have

$$\|F(\alpha)\| \leqslant e^{k\xi(\xi-1)-k\eta^2} \|B^\xi B^{i\eta} T A^\xi A^{i\eta}\|$$

$$\leqslant e^{k\xi(\xi-1)-k\eta^2+\pi|\eta|} \|B^\xi T A^\xi\|.$$

Hence, $F(\alpha)$ is uniformly bounded in $0 \leqslant \operatorname{Re}\alpha \leqslant 1$. In particular, when $\alpha = i\eta$ and $\alpha = 1 + i\eta$, similar calculations lead to

$$\|F(i\eta)\| \leqslant e^{\pi^2/4k} \|T\|$$

and

$$\|F(1 + i\eta)\| \leqslant e^{\pi^2/4k} \|BTA\|,$$

respectively. Hence, by the three-line theorem, for $0 \leqslant \xi \leqslant 1$, we have

$$\|F(\xi)\| \leqslant e^{\pi^2/4k} \|T\|^{1-\xi} \|BTA\|^{\xi}$$

and, therefore,

$$\|B^{\xi}TA^{\xi}\| \leqslant e^{\pi^2/4k+k\xi(1-\xi)} \|T\|^{1-\xi} \|BTA\|^{\xi},$$

which proves (2.62) by letting $k = \pi/2\sqrt{\xi(1-\xi)}$.

Proof of Theorem 2.3.4 Case (i): A and B are bounded, and $\mathrm{Re}\,(Au, u) \geqslant \delta \|u\|^2$ with $\delta > 0$. Since A^{-1} is also accretive by Lemma 2.3.7, we may apply Lemma 2.3.9 in replacing A by A^{-1} to get

$$\|B^{\alpha}TA^{-\alpha}\| \leqslant e^{\pi\sqrt{\alpha(1-\alpha)}} \|T\|^{1-\alpha} \|BTA^{-1}\|^{\alpha} \leqslant e^{\pi\sqrt{\alpha(1-\alpha)}} M^{\alpha} \|T\|^{1-\alpha}$$

for $0 \leqslant \alpha \leqslant 1$. From this follows (2.55).

Case (ii): A and B are bounded. Let $\varepsilon > 0$. By observing that

$$\|(A+\varepsilon)u\|^2 = \|Au\|^2 + 2\varepsilon\,\mathrm{Re}\,(Au, u) + \varepsilon^2 \|u\|^2 \geqslant \|Au\|^2,$$

we have $\|BTu\| \leqslant M \|Au\| \leqslant M \|(A+\varepsilon)u\|$. Hence, from (i),

$$\|B^{\alpha}Tu\| \leqslant e^{\pi\sqrt{\alpha(1-\alpha)}} \|T\|^{1-\alpha} M^{\alpha} \|(A+\varepsilon)^{\alpha}u\|.$$

By letting $\varepsilon \to 0$, by virtue of Lemma 2.3.5, we obtain (2.55).

Case (iii): A and B are not necessarily bounded, but there exists a $\delta > 0$ such that $\mathrm{Re}\,(Au, u) \geqslant \delta \|u\|^2$ and $\mathrm{Re}\,(Bu, u) \geqslant \delta \|u\|^2$ for every $u \in D(A)$ and $u \in D(B)$, respectively. The operators A^{-1} and B^{-1} are not only accretive by Lemma 2.3.7 but also maximal accretive, since they are bounded; the same applies to A^{*-1} and B^{*-1}. Furthermore, it is not hard to prove that, under the assumptions stated above, we have $(A^{-1})^{\alpha} = A^{-\alpha}$ and $(B^{-1})^{\alpha} = B^{-\alpha}$. From

$$|(A^{*-1}T^*v, u)| = |(v, TA^{-1}u)| = |(B^{*-1}v, BTA^{-1}u)| \leqslant M \|B^{*-1}v\| \|u\|,$$

it follows that $\|A^{*-1}T^*v\| \leqslant M \|B^{*-1}v\|$. Hence, by replacing A, B and T by B^{*-1}, A^{*-1} and T^*, respectively, in (ii), we obtain

$$\|A^{*-\alpha}T^*v\| \leqslant e^{\pi\sqrt{\alpha(1-\alpha)}} \|T\|^{1-\alpha} M^{\alpha} \|B^{*-\alpha}v\|.$$

Therefore, if $u \in D(A^{\alpha})$, for every $v \in D(B^{*\alpha})$ it holds that

$$|(Tu, B^{*\alpha}v)| = |(A^{\alpha}u, A^{*-\alpha}T^*B^{*\alpha}v)|$$
$$\leqslant \|A^{\alpha}u\| \|A^{*-\alpha}T^*B^{*\alpha}v\| \leqslant \|A^{\alpha}u\| e^{\pi\sqrt{\alpha(1-\alpha)}} \|T\|^{1-\alpha} M^{\alpha} \|v\|,$$

which shows that $Tu \in D(B^{\alpha})$ and (2.55) holds.

General case (iv): Let $\varepsilon > 0$. For every $u \in D(A)$, we have

$$\|(B+\varepsilon^2)Tu\| \leqslant M \|Au\| + \varepsilon^2 \|T\| \|u\|$$
$$\leqslant (M^2 + \varepsilon^2 \|T\|^2)^{1/2} (\|Au\|^2 + \varepsilon^2 \|u\|^2)^{1/2}.$$

Hence, if in (iii) we replace A, B and M by $A + \varepsilon$, $B + \varepsilon^2$ and $(M^2 + \varepsilon^2 \|T\|^2)^{1/2}$, respectively, then $T \cdot D(A^\varepsilon) \subset D(B^\varepsilon)$ holds and we obtain a replaced equation corresponding to (2.55). By letting $\varepsilon \to 0$, we complete the proof. $\square$

The main references to this chapter are Phillips [148] and chapter 1, §4 of Kreĭn's book [9].

3

Semigroup of linear operators

3.1. Semigroups

In this section, we will describe fundamental facts of the C_0-semigroup. The reader will find many books dealing with semigroups, so that the description here will be as concise as possible. Throughout this chapter we only consider complex Banach spaces.

Definition 3.1.1 Let $T(t)$ for each $t \in [0, \infty)$ be a bounded linear operator in a Banach space X. $\{T(t)\}$ is called a C_0-semigroup, or simply a semigroup, of bounded operators if the following conditions are satisfied:

(I) $T(t+s) = T(t)T(s) = T(s)T(t)$ for any $t \geqslant 0$ and any $s \geqslant 0$, and $T(0) = I$;

(II) $T(t)$ is strongly continuous in t in the interval $0 \leqslant t < \infty$, namely, s-$\lim_{t \to s} T(t)u = T(s)u$ for every $u \in X$ and $s \in [0, \infty)$.

In particular, if $T(t)$ is defined on $-\infty < t < \infty$ and satisfies there the conditions (I) and (II), then $\{T(t)\}$ is said to be a C_0-group, or simply a group, of bounded operators. If $\{T(t)\}$ is a group, we have $T(-t) = T(t)^{-1}$ by (I).

Remark 3.1.1 By Theorem 1.3.1, if $\{T(t)\}$ is a C_0-semigroup, then $\|T(t)\|$ is bounded as $t \to 0$. In general, however, the semigroup does not satisfy such a condition. For the details about related problems see Hille and Phillips [7], Miyadera, Oharu and Okazawa [138], Okazawa [145], Lions [114], Ushijima [173, 174] and the other references cited there. In this book, we will deal only with C_0-semigroups and we will call them simply semigroups.

Example 1 Suppose $X = B^0([0, \infty))$ or $X = L^p(0, \infty)$ $(1 \leqslant p < \infty)$ and define $(T(t)u)(x) = u(x+t)$ for each $u \in X$. Then $\{T(t)\}$ is a semigroup by Lemma 1.2.4.

Example 2 Suppose $X = B^0((-\infty, \infty))$ or $X = L^p(-\infty, \infty)$ $(1 \leq p < \infty)$ and define $(T(t)u)(x) = u(x + t)$ for each $u \in X$. Then $\{T(t)\}$ is a group.

Example 3 Suppose $X = L^p(\mathbb{R}^n)$, $1 \leq p < \infty$, put

$$G(t, x) = (2\sqrt{\pi t})^{-n} \exp(-|x|^2/4t)$$

for each $t > 0$ and each $x \in \mathbb{R}^n$, and define

$$(T(t)u)(x) = \int_{\mathbb{R}^n} G(t, x - y)u(y)\, dy,$$

then $\|T(t)\| \leq 1$ by Lemma 1.2.3. $G(t, x)$ is the fundamental solution of the heat equation $\partial/\partial t - \Delta$ and satisfies

$$G(t + s, x) = \int_{\mathbb{R}^n} G(t, x - y)G(s, y)\, dy,$$

from which it follows immediately that $T(t + s) = T(t)T(s)$. By the use of the definite integral $\int_{-\infty}^{\infty} e^{-x^2}\, dx = \sqrt{\pi}$ and by a simple substitution of variables, we have

$$|(T(t)u)(x) - u(x)| = \pi^{-n/2} \left| \int \exp(-|y|^2)(u(x - 2\sqrt{t}\, y) - u(x))\, dy \right|$$

$$\leq \pi^{-n/2} \int \exp\left(-\frac{|y|^2}{p'}\right) \exp\left(-\frac{|y|^2}{p}\right)$$

$$\times |u(x - 2\sqrt{t}\, y) - u(x)|^p\, dy,$$

where $p' = p(p - 1)^{-1} \leq \infty$. By virtue of Hölder's inequality this turns into

$$\leq \pi^{-(2p)^{-1}n} \left\{ \int \exp(-|y|^2)\, |u(x - 2\sqrt{t}\, y) - u(x)|^p\, dy \right\}^{1/p}.$$

Hence, by Fubini's theorem, we obtain

$$\int_{\mathbb{R}^n} |(T(t)u)(x) - u(x)|^p\, dx$$

$$\leq \pi^{-n/2} \int \exp(-|y|^2) \int |u(x - 2\sqrt{t}\, y) - u(x)|^p\, dx\, dy.$$

This estimate, together with Lemma 1.2.4, implies that

$$\lim_{t \to 0} \|T(t)u - u\| = 0.$$

It is also easy to see that $\lim_{t \to s} \|T(t)u - T(s)u\| = 0$ for each $s \in (0, \infty)$.

Example 4 Let X be a Hilbert space and $H = \sum_{-\infty}^{a} \lambda\, dE(\lambda)$ a self-adjoint operator bounded from above. Then $T(t) = \int_{-\infty}^{a} e^{\lambda t}\, dE(\lambda)$ is a semigroup. The properties (I) and (II) in this case follow from the basic facts about the spectral resolution which we mentioned in Section 1.2.

Example 5 Let X be a Hilbert space and $H = \int_{-\infty}^{\infty} \lambda\, dE(\lambda)$ a self-adjoint operator, then the family of unitary operators $T(t) = \int_{-\infty}^{\infty} e^{i\lambda t} \, dE(\lambda)$ forms a group.

Theorem 3.1.1 *For a semigroup $\{T(t)\}$ there exist real numbers $M \geq 1$ and β such that*

$$\|T(t)\| \leq M \, e^{\beta t} \tag{3.1}$$

for all $t \geq 0$.

Proof By Remark 3.1.1, $\|T(t)\|$ is bounded in the interval $0 \leq t \leq 1$. With Gauss' symbol $[t]$ used to denote the integral part of $t > 0$, we get

$$\|T(t)\| = \|T([t] + t - [t])\| = \|T([t])T(t - [t])\|$$
$$= \|T(1)^{[t]}T(t - [t])\| \leq \|T(1)\|^{[t]}\|T(t - [t])\|. \tag{3.2}$$

Let us fix β by $\|T(1)\| = e^{\beta}$. If $\beta \geq 0$, putting $M = \sup_{0 \leq t \leq 1}\|T(t)\|$, we obtain $\|T(t)\| \leq Me^{\beta[t]} \leq Me^{\beta t}$ by (3.2). On the other hand, if $\beta < 0$, then, by putting $M = e^{-\beta}\sup_{0 \leq t \leq 1}\|T(t)\|$ and using (3.2), we obtain $\|T(t)\| \leq Me^{\beta(1+[t])} \leq Me^{\beta t}$. $\square$

Remark 3.1.2 If we only required the conclusion of the theorem, we could finish the proof faster by choosing β in such a way that $\|T(t)\| \leq e^{\beta}$ and $\beta \geq 0$. But, since (3.1) can be true for some $\beta < 0$, we prefer the proof given above. Further, it is not possible to take $M < 1$, as may be seen by putting $t = 0$ in (3.1).

Theorem 3.1.2 *Let $\{T(t)\}$ be a semigroup. Denote by D the set of those elements u of X for which the limit $\text{w-}\lim_{h \to +0} h^{-1}(T(h) - I)u$ exists. This weak limit for each $u \in D$ will be written as Au. Then D is a dense subspace of X, and for each $u \in D$, we have*

$$Au = \text{s-}\lim_{h \to +0} h^{-1}(T(h)u - u). \tag{3.3}$$

If $u \in D$, then $T(t)u \in D$ for each $t \geq 0$ and $T(t)u$ is strongly differentiable in $t \geq 0$. Moreover, the following relation holds:

$$(d/dt)T(t)u = AT(t)u = T(t)Au. \tag{3.4}$$

Proof Let u be an arbitary element of X. Apply $T(h)$ to

$$u_{\delta} = \delta^{-1}\int_{0}^{\delta} T(s)u \, ds$$

with $\delta > 0$, obtaining

$$T(h)u_\delta = \delta^{-1} \int_0^\delta T(h)T(s)u \, ds$$

$$= \delta^{-1} \int_0^\delta T(h+s)u \, ds = \delta^{-1} \int_h^{h+\delta} T(s)u \, ds.$$

By letting $h \to +0$, we have

$$h^{-1}(T(h) - I)u_\delta = \delta^{-1} \left\{ h^{-1} \int_\delta^{h+\delta} T(s)u \, ds - h^{-1} \int_0^h T(s)u \, ds \right\}$$

$$\to \delta^{-1} \{ T(\delta)u - u \}$$

strongly, and, hence, $u_\delta \in D$. The denseness of D follows from the fact that $u_\delta \to u$ as $\delta \to 0$. Since the bounded operator is continuous even in the weak topology, for $u \in D$ we get

$$\text{w-lim}_{.h \to +0} h^{-1}(T(h) - I)T(t)u = \text{w-lim}_{h \to +0} T(t)h^{-1}(T(h) - I)u = T(t)Au,$$

and, accordingly, $T(t)u \in D$ and $AT(t)u = T(t)Au$. Let D^+ denote the right differentiation, then

$$D^+ T(t)u = \text{w-lim}_{h \to +0} h^{-1}(T(t+h) - T(t))u$$

$$= T(t) \, \text{w-lim}_{h \to +0} h^{-1}(T(h) - I)u = T(t)Au.$$

From this it follows that, for any $f \in X^*$,

$$D^+ f(T(t)u) = f(D^+ T(t)u) = f(T(t)Au)$$

exists and is continuous. Therefore, we have

$$f(T(t)u - u) = f(T(t)u) - f(u) = \int_0^t D^+ f(T(s)u) \, ds$$

$$= \int_0^t f(T(s)Au) \, ds = f\left(\int_0^t T(s)Au \, ds \right)$$

and, hence,

$$T(t)u - u = \int_0^t T(s)Au \, ds. \tag{3.5}$$

Since $T(s)Au$ is strongly continuous, (3.5) implies that $T(t)u$ is strongly differentiable and that (3.3) and (3.4) hold. $\quad \square$

Definition 3.1.2 *The operator A in the preceding theorem is called the* generator *of the semigroup* $\{T(t)\}$. *It is also said that A* generates *the semigroup* $\{T(t)\}$.

By writing $T(t) = \exp(tA)$, we may specify that A is the generator of a semigroup $\{T(t)\}$. This is due to the result (3.4) and the fact that $T(0) = I$.

Theorem 3.1.3 *The generator A of a semigroup* $\{T(t)\}$ *is a closed operator. Let M and β be numbers such that (3.1) is satisfied, then the half-plane* $\{\lambda : \operatorname{Re} \lambda > \beta\}$ *is contained in* $\rho(A)$ *and, therein,*

$$(A - \lambda)^{-1} = -\int_0^\infty e^{-\lambda t} T(t) \, dt \tag{3.6}$$

and

$$\|(A - \lambda)^{-n}\| \leq M(\operatorname{Re} \lambda - \beta)^{-n}, \quad n = 1, 2, \ldots \tag{3.7}$$

hold.

Remark 3.1.3 In particular, if $M = 1$, then the inequality (3.7) for $n = 2$, $3, \ldots$ can be derived from that for $n = 1$.

Proof of Theorem 3.1.3 (i) The domain of A is the D given in the preceding theorem. If $u_n \in D(A)$, $u_n \to u$ and $Au_n \to v$, then (3.5) applied to u_n, namely,

$$T(t)u_n - u_n = \int_0^t T(t)Au_n \, ds$$

gives

$$T(t)u - u = \int_0^t T(s)v \, ds,$$

as $n \to \infty$, from which it follows immediately that $u \in D(A)$ and $Au = v$. Thus, A is a closed operator.

(ii) We will show that $R(A - \lambda)$ is dense if $\operatorname{Re} \lambda > \beta$. For this purpose it is enough, by Theorem 1.1.3, to verify that any element f of X^* orthogonal to $R(A - \lambda)$ vanishes. Let u be an arbitrary element of $D(A)$, then, because of $f((A - \lambda)u) = 0$, we have $f(Au) = \lambda f(u)$. Theorem 3.1.2 affords a differential equation

$$(d/dt)f(T(t)u) = f(AT(t)u) = \lambda f(T(t)u). \tag{3.8}$$

By solving this equation, we obtain $f(T(t)u) = f(u)e^{\lambda t}$. Thus, with the aid of (3.1), we have

$$|f(u)| = |e^{-\lambda t} f(T(t)u)| \leq M \|f\| \|u\| e^{(\beta - \mathrm{Re}\,\lambda)t}.$$

Letting $t \to \infty$ it can be concluded that $f(u) = 0$ and, hence, $f = 0$.

(iii) We will prove that, for $\mathrm{Re}\,\lambda > \beta$ and for any $u \in D(A)$, the result

$$\|(A - \lambda)u\| \geq M^{-1}(\mathrm{Re}\,\lambda - \beta) \|u\| \tag{3.9}$$

holds. By Theorem 1.1.2, there is an $f \in X^*$ which satisfies $f(u) = \|u\|$ and $\|f\| = 1$. The foregoing argument, which led to (3.8), similarly gives

$$\begin{aligned}
(d/dt)f(T(t)u) &= f(T(t)Au) \\
&= f(T(t)(A - \lambda)u) + \lambda f(T(t)u).
\end{aligned}$$

We solve this equation, regarding it as a differential equation for $f(T(t)u)$, and obtain

$$f(T(t)u) = e^{\lambda t} f(u) + \int_0^t e^{\lambda(t-s)} f(T(s)(A - \lambda)u)\, ds.$$

Thus, we have

$$\begin{aligned}
\|u\| &= f(u) \\
&= \left| e^{-\lambda t} f(T(t)u) - \int_0^t e^{-\lambda s} f(T(s)(A - \lambda)u)\, ds \right| \\
&\leq M e^{(\beta - \mathrm{Re}\,\lambda)t} \|u\| + M(\mathrm{Re}\,\lambda - \beta)^{-1} \|(A - \lambda)u\|
\end{aligned}$$

and, hence, (3.9) by letting $t \to \infty$.

We observe that, since A is a closed operator, by (iii) the range $R(A - \lambda)$ is a closed subspace if $\mathrm{Re}\,\lambda > \beta$. Therefore, combining (ii) and (iii), we find that $\{\lambda : \mathrm{Re}\,\lambda > \beta\} \subset \rho(A)$.

(iv) Suppose $u \in D(A)$ and $\mathrm{Re}\,\lambda > \beta$. By Lemma 1.3.1, for each $a > 0$, we have

$$\begin{aligned}
A \int_0^a e^{-\lambda t} T(t)u\, dt &= \int_0^a e^{-\lambda t} A T(t)u\, dt \\
&= \int_0^a e^{-\lambda t} (d/dt) T(t)u\, dt \\
&= e^{-\lambda a} T(a)u - u + \lambda \int_0^a e^{-\lambda t} T(t)u\, dt.
\end{aligned}$$

Since $\mathrm{Re}\,\lambda > \beta$, the right-hand side converges strongly to $-u + \lambda \int_0^\infty e^{-\lambda t} T(t)u\, dt$ as $a \to \infty$. A being closed, it follows that

$$\int_0^\infty e^{-\lambda t} T(t)u\, dt \in D(A)$$

and

$$A \int_0^\infty e^{-\lambda t} T(t) u \, dt = -u + \lambda \int_0^\infty e^{-\lambda t} T(t) u \, dt.$$

Hence, both members of (3.6) coincide on $D(A)$. Since $D(A)$ is dense, they also coincide on all X. Finally, by repeatedly differentiating both sides of (3.6) $(n-1)$-times, we have

$$(n-1)! \, (A-\lambda)^{-n} = (-1)^n \int_0^\infty t^{n-1} e^{-\lambda t} T(t) \, dt.$$

From this and (3.1) it is easy to deduce (3.7). $\square$

Hitherto we have studied the generator when a semigroup was given. We now consider its inverse, i.e., the problem of how to construct the semigroup from a given generator.

Theorem 3.1.4 *Let A be a densely-defined closed operator. Assume that the half-plane $\{\lambda \in \mathbb{C} : \operatorname{Re} \lambda > \beta\}$ is contained in $\rho(A)$ and that*

$$\|(A-\lambda)^{-n}\| \leq M(\operatorname{Re} \lambda - \beta)^{-n} \tag{3.10}$$

is satisfied for any $\operatorname{Re} \lambda > \beta$ and $n = 1, 2, \ldots$. Then one and only one semigroup $\{T(t)\}$ exists and it satisfies $\|T(t)\| \leq M e^{\beta t}$ for $0 \leq t < \infty$.

Proof Two methods of proof are known for this theorem. They are called Hille's method and Yosida's method, and are important for the study of linear and non-linear equations of evolution, so we will explain both of them.

1 Hille's method First, let us assume $\beta = 0$, so that

$$\|(A-\lambda)^{-n}\| \leq M(\operatorname{Re} \lambda)^{-n} \tag{3.11}$$

is satisfied for any $\operatorname{Re} \lambda > 0$ and $n = 1, 2, \ldots$. In the same way, as in the proof of Lemma 2.3.1, it is found that

$$\text{s-}\lim_{\lambda \to \infty} (1 - \lambda^{-1} A)^{-1} = I. \tag{3.12}$$

(i) Put $u_n(t) = (1 - t n^{-1} A)^{-n} u$ for each $u \in X$. Furthermore, put

$$y_\varepsilon = \int_\varepsilon^{t-\varepsilon} \frac{d}{ds} \left\{ \left(1 - \frac{t-s}{m} A\right)^{-m} \left(1 - \frac{s}{n} A\right)^{-n} u \right\} ds$$

for $0 < \varepsilon < t$. Then, since

$$y_\varepsilon = \left(1 - \frac{\varepsilon}{m} A\right)^{-m} \left(1 - \frac{t-\varepsilon}{n} A\right)^{-n} u - \left(1 - \frac{t-\varepsilon}{m} A\right)^{-m} \left(1 - \frac{\varepsilon}{n} A\right)^{-n} u,$$

by (3.12) we obtain

$$y_\varepsilon \to u_n(t) - u_m(t) \tag{3.13}$$

strongly as $\varepsilon \to 0$. By the formula

$$(d/dt)(1 - tn^{-1}A)^{-n} = A(1 - tn^{-1}A)^{-n-1}, \tag{3.14}$$

which can be verified by an elementary calculation, the integral above becomes

$$y_\varepsilon = \int_\varepsilon^{t-\varepsilon} \left\{ -A\left(1 - \frac{t-s}{m} A\right)^{-m-1} \left(1 - \frac{s}{n} A\right)^{-n} u \right.$$
$$\left. + \left(1 - \frac{t-s}{m} A\right)^{-m} A\left(1 - \frac{s}{n} A\right)^{-n-1} u \right\} ds.$$

The right-hand side can be rearranged by the use of the identity

$$-\left(1 - \frac{t-s}{m} A\right)^{-1} + \left(1 - \frac{s}{n} A\right)^{-1}$$
$$= \left(\frac{s}{n} - \frac{t-s}{m}\right) A \left(1 - \frac{t-s}{m} A\right)^{-1} \left(1 - \frac{s}{n} A\right)^{-1},$$

thus yielding, for $u \in D(A^2)$,

$$y_\varepsilon = \int_\varepsilon^{t-\varepsilon} \left(\frac{s}{n} - \frac{t-s}{m}\right) \left(1 - \frac{t-s}{m} A\right)^{-m-1} \left(1 - \frac{s}{n} A\right)^{-n-1} A^2 u \, ds.$$

This, together with (3.11), implies that

$$\|y_\varepsilon\| \leqslant \int_\varepsilon^{t-\varepsilon} \left| \frac{s}{n} - \frac{t-s}{m} \right| M^2 \|A^2 u\| \, ds$$
$$\leqslant \int_0^t \left(\frac{s}{n} + \frac{t-s}{m}\right) M^2 \|A^2 u\| \, ds = \frac{t^2}{2}\left(\frac{1}{n} + \frac{1}{m}\right) M^2 \|A^2 u\|.$$

Therefore, from (3.13), it follows that

$$\|u_n(t) - u_m(t)\| \leqslant \frac{t^2}{2}\left(\frac{1}{n} + \frac{1}{m}\right) M^2 \|A^2 u\|,$$

which shows that, if $u \in D(A^2)$, the limit s-$\lim_{n \to \infty} u_n(t)$ exists uniformly in the wider sense in $0 \leqslant t < \infty$. When u is an arbitrary element of X, we argue as follows. Put $u_\lambda = (1 - \lambda^{-1}A)^{-2} u$ for $\lambda > 0$, then $u_\lambda \in D(A^2)$ and, since $u_\lambda \to u$ as $\lambda \to \infty$ by (3.12), $D(A^2)$ is dense. Moreover, (3.11) implies that

$$\|(1 - tn^{-1}A)^{-n}\| \le M. \tag{3.15}$$

Therefore, even if u is an arbitrary element of X, the limit

$$T(t)u = \text{s-}\lim_{n\to\infty} u_n(t) = \text{s-}\lim_{n\to\infty} (1 - tn^{-1}A)^{-n}u \tag{3.16}$$

exists uniformly in the wider sense in $0 \le t < \infty$. $T(t)$ is linear and $\|T(t)\| \le M$. Since, by virtue of (3.12), $u_n(t)$ is strongly continuous in $0 \le t < \infty$ and s-$\lim_{t\to 0} u_n(t) = u$, $T(t)u$ is also strongly continuous in $0 \le t < \infty$ and $T(t)u \to u$ strongly as $t \to 0$.

(ii) Suppose $u \in D(A)$. Then, since $Au_n(t) = (1 - tn^{-1}A)^{-n}Au \to T(t)Au$ strongly, we have $T(t)u \in D(A)$ and $AT(t)u = T(t)Au$; A and $T(t)$ are commutative. Also, since $du_n(t)/dt = (1 - tn^{-1}A)^{-n-1}Au$ by (3.14), we get

$$u_n(t) - u = \int_0^t (1 - sn^{-1}A)^{-1}(1 - sn^{-1}A)^{-n}Au \, ds,$$

By letting n tend to infinity and using (3.12), (3.15), (3.16), we have

$$T(t)u - u = \int_0^t T(s)Au \, ds, \tag{3.17}$$

which shows that $T(t)u$ is strongly differentiable and (3.4) holds. Accordingly, it is easy to see that $(d/ds)\{T(t-s)T(s)u\} = 0$ in $0 < s < t$, so that $T(t-s)T(s)u$ is independent of s belonging to $(0, t)$. Since $T(t-s)T(s)u \to T(t)u$ as $s \to 0$, $T(t-s)T(s)u = T(t)u$ holds in $0 < s < t$ for each $u \in D(A)$ and, hence, for each $u \in X$. From this, the relation $T(t)T(s) = T(t+s)$ follows for any $t, s \in [0, \infty)$. Thus it has been proved that $\{T(t)\}$ is a semigroup.

(iii) Let $\tilde{A}$ denote the generator of $\{T(t)\}$. Given any $u \in D(A)$, since, by virtue of (3.17),

$$h^{-1}(T(h) - I)u = h^{-1}\int_0^h T(s)Au \, ds \to Au$$

strongly as $h \to +0$, we obtain $u \in D(\tilde{A})$ and $\tilde{A}u = Au$, i.e., $A \subset \tilde{A}$. Next let u be an arbitrary element of $D(\tilde{A})$ and put $v = (\tilde{A} - 1)u$. As $1 \in \rho(A)$, there exists a $w \in D(A)$ satisfying $v = (A - 1)w$, so that $(\tilde{A} - 1)u = (A - 1)w = (\tilde{A} - 1)w$. Since $\|T(t)\| \le M$, as was shown in (i), Theorem 3.1.3 ensures that the half-plane $\{\lambda : \text{Re } \lambda > 0\}$ is contained in $\rho(\tilde{A})$ and so, in particular, $1 \in \rho(\tilde{A})$. This implies $u = w \in D(A)$ and, thus, it has been proved that $A = \tilde{A}$.

In this way it has turned out that for $\beta = 0$ there exists a semigroup having A as its generator. For a non-vanishing real β, by considering $A_1 = A - \beta$ instead of A, the assumptions of the theorem for $\beta = 0$ are satisfied. Let $\{S(t)\}$ denote the semigroup which A_1 generates, then it is easy to see that $T(t) = e^{\beta t}S(t)$ is the semigroup having A as its generator and that $\|T(t)\| \leqslant Me^{\beta t}$. Lastly, to see the uniqueness, suppose that A generates another semigroup $\{U(t)\}$. For any $u \in D(A)$ and $t > 0$, we have

$$(d/ds)(T(t-s)U(s)u) = -AT(t-s)U(s)u + T(t-s)AU(s)u = 0$$

in $0 < s < t$. Therefore, $T(t-s)U(s)u$ is independent of s in $0 < s < t$. Since $T(t-s)U(s)u \to T(t)u$ or $U(t)u$ according as $s \to 0$ or $s \to t$, the equality $T(t)u = U(t)u$ holds for every $u \in D(A)$ and, hence, for every $u \in X$.

2 Yosida's method Assume $\beta = 0$. Put $I_n = (1 - n^{-1}A)^{-1}$ and $A_n = AI_n$ for each natural number n, then A_n is bounded and I_n commutes with A. Since, by (3.12), $I_n \to I$ strongly as $n \to \infty$, for $u \in D(A)$ we have

$$\text{s-}\lim_{n \to \infty} A_n u = Au. \tag{3.18}$$

Put $T_n(t) = \exp(tA_n) = \sum_{m=0}^{\infty}(m!)^{-1}(tA_n)^m$. A_n being bounded, $T_n(t)$ is an entire function of t with values in $B(X)$, and satisfies $(d/dt)T_n(t) = A_nT_n(t)$ and $T_n(0) = I$. Since A_n can be expressed as $A_n = n(I_n - I)$, $T_n(t)$ becomes

$$T_n(t) = \exp(tn(I_n - I)) = e^{-nt}\exp(tnI_n).$$

By (3.11), we have $\|(nI_n)^m\| = n^{2m}\|(n - A)^{-m}\| \leqslant Mn^m$ and, hence, for $t \geqslant 0$,

$$\|\exp(tnI_n)\| \leqslant \sum_{m=0}^{\infty}(m!)^{-1}t^m\|(nI_n)^m\|$$

$$\leqslant M\sum_{m=0}^{\infty}(m!)^{-1}(nt)^m = Me^{nt}.$$

Therefore, the estimate

$$\|T_n(t)\| \leqslant M \tag{3.19}$$

holds for all $t \geqslant 0$ and $n = 1, 2, \ldots$. Since A_n and $T_m(t)$ are commutative for all natural numbers n, m, for $u \in D(A)$ we obtain

$$\|T_n(t)u - T_m(t)u\| = \left\| -\int_0^t (\partial/\partial s)(T_n(t-s)T_m(s)u)\,ds \right\|$$

$$= \left\| \int_0^t T_n(t-s)(A_n - A_m)T_m(s)u\,ds \right\|$$

$$= \left\| \int_0^t T_n(t-s)T_m(s)(A_n - A_m)u\,ds \right\|$$

$$\leqslant M^2 \|(A_n - A_m)u\|\,t,$$

which implies that, as $n \to \infty$, $T_n(t)u$ converges uniformly in the wider sense in $0 \leqslant t < \infty$. By (3.19), this is also true for any element u of X and there exists the limit $T(t)u = \text{s-lim}_{n \to \infty} T_n(t)u$ uniformly in the wider sense in $0 < t < \infty$. $T_n(t+s) = T_n(t)T_n(s)$ is evident and thus, combining it with $T_n(0) = I$, it turns out that $\{T(t)\}$ is a semigroup. The remainder of the proof is the same as Hille's method.

Remark 3.1.4 In the proof given above, we have used (3.10) only in the case of real λ, but a direct verification is possible in order to show that (3.10) is also true for any complex λ with $\text{Re }\lambda > \beta$ if it is true for real $\lambda > \beta$. More precisely, we can prove the following. If there exists a certain $\gamma \geqslant \beta$ such that the half-line $\{\lambda \in R : \lambda > \gamma\}$ is contained in $\rho(A)$ and that

$$\|(A - \lambda)^{-n}\| \leqslant M(\lambda - \beta)^{-n} \tag{3.20}$$

is satisfied for every $\lambda > \gamma$ and $n = 1, 2, \ldots$, then $\{\lambda \in \mathbb{C} : \text{Re }\lambda > \beta\} \subset \rho(A)$ and (3.10) holds. Indeed, we first note that, for $\mu > \gamma$, the power series expansion

$$(A - \lambda)^{-1} = \sum_{m=0}^{\infty} (\lambda - \mu)^m (A - \mu)^{-m-1} \tag{3.21}$$

at μ is, by (3.20), convergent inside the circle $|\lambda - \mu| < \mu - \beta$. Whatever the complex number λ with $\text{Re }\lambda > \beta$ may be, $|\lambda - \mu| < \mu - \beta$ for sufficiently large μ and, accordingly, $\{\lambda : \text{Re }\lambda > \beta\}$ is contained in $\rho(A)$, and (3.21) is satisfied therein. By differentiating (3.21) $(n-1)$-times repeatedly with respect to λ, we obtain

$$(n-1)!\,(A - \lambda)^{-n} =$$

$$\sum_{m=n-1}^{\infty} m!\,((m-n+1)!)^{-1}(\lambda - \mu)^{m-n+1}(A - \mu)^{-m-1}.$$

Therefore, by (3.20), we have

$$\|(A-\lambda)^{-n}\| \leqslant \frac{M}{(\mu-\beta)^n} \sum_{m=n-1}^{\infty} \frac{m!}{(n-1)!\,(m-n+1)!} \left(\frac{|\lambda-\mu|}{\mu-\beta}\right)^{m-n+1}$$

$$= M(\mu-\beta-|\lambda-\mu|)^{-n},$$

which gives (3.10) by letting $\mu \to \infty$. From this, it follows that all that has been stated above is equally true for a real Banach space.

The collection of all generators of those semigroups in X which satisfy (3.1) will be denoted by $G(X, M, \beta)$. Put $G(X, \beta) = \bigcup_{M \geqslant 1} G(X, M, \beta)$ and $G(X) = \bigcup_{-\infty < \beta < \infty} G(X, \beta)$, then $G(X)$ is the collection of all generators of semigroups in X.

Definition 3.1.3 *A semigroup $\{T(t)\}$ satisfying $\|T(t)\| \leqslant 1$ for each t is called a contraction semigroup.*

In terms of the notation introduced above, the collection of all generators of contraction semigroups in X is expressed as $G(X, 1, 0)$. By Remark 3.1.3, $A \in G(X, 1, 0)$ if and only if $\rho(A)$ includes the half-plane $\{\lambda \in \mathbb{C}: \operatorname{Re} \lambda > 0\}$, in which the result $\|(A-\lambda)^{-1}\| \leqslant (\operatorname{Re} \lambda)^{-1}$ holds. This, combined with Theorems 2.1.3 and 2.1.6, provides the following:

Theorem 3.1.5 (i) *Let X be a Hilbert space. An operator A is the generator of a contraction semigroup in X if and only if it is maximal dissipative.*

(ii) *Let A be a closed operator defined densely in a Banach space. A is the generator of a contraction semigroup if and only if A and A^* are both dissipative.*

Remark 3.1.5 The proof that the generator A of a contraction semigroup $\{T(t)\}$ in a Hilbert space X is dissipative may proceed as follows. Since $\|T(t)u\| = \|T(t-s)T(s)u\| \leqslant \|T(s)u\|$ for $0 \leqslant s < t$, $\|T(t)u\|$ is a non-increasing function of t, so that $2 \operatorname{Re}(Au, u) = D^+ \|T(t)u\|^2|_{t=0} \leqslant 0$ if $u \in D(A)$. Similarly, if $\{T(t)\}$ is a contraction semigroup in a Banach space and A is its generator,

$$\operatorname{Re}(h^{-1}(T(h)-I)u, f) = h^{-1}\{\operatorname{Re}(T(h)u, f) - \|u\|^2\}$$

$$\leqslant h^{-1}(\|T(h)u\| - \|u\|)\|u\| \leqslant 0$$

for every $u \in D(A)$ and $f \in Fu$, so that we have $\operatorname{Re}(Au, f) \leqslant 0$ by letting $h \to 0$.

Theorem 3.1.6 *Let X be a reflexive Banach space. If an operator A generates a semigroup $\{T(t)\}$, then $A^* \in G(X^*)$ and the semigroup generated by A^* consists of $T^*(t) = (T(t))^*$.*

Proof By Theorems 1.1.4, 1.1.16, 3.1.3 and 3.1.4, the fact that $A^* \in G(X^*)$ and $(1 - tn^{-1}A)^{-n} \to T(t)$ strongly implies that $(1 - tn^{-1}A^*)^{-n}f = ((1 - tn^{-1}A)^{-n})^*f \to T^*(t)f$ weakly for every $f \in X^*$. From this the conclusion of the theorem follows easily. $\square$

Examples of generators We will illustrate the generators of the semigroups mentioned earlier.

Example 1 Suppose $(T(t)u)(x) = u(x + t)$ for $u \in X = B^0([0, \infty))$. It is readily seen that $D(A) = B^1([0, \infty))$ and $Au = du/dx$. If we assume $X = L^p(0, \infty)$, $1 \le p < \infty$, then, first, by putting $Au = v$ with $u \in D(A)$, it can be shown by an elementary calculation that the derivative u' of u in the sense of distribution coincides with v. Thus we have $D(A) \subset W_p^1(0, \infty)$ and $Au = u'$. Conversely, suppose $u \in W_p^1(0, \infty)$. Since $u' \in L^p(0, \infty)$, u is absolutely continuous and, accordingly, it can be expressed as

$$u(x) = u(0) + \int_0^x u'(y)\, dy.$$

By the use of Hölder's inequality we get

$$|h^{-1}(u(x + h) - u(x)) - u'(x)| = \left| h^{-1} \int_0^h (u'(x + y) - u'(x))\, dy \right|$$

$$\le h^{-1/p} \left(\int_0^h |u'(x + y) - u'(x)|^p\, dy \right)^{1/p}.$$

Therefore, by Lemma 1.2.4, letting $h \to +0$,

$$\int_0^\infty |h^{-1}(u(x + h) - u(x)) - u'(x)|^p\, dx$$

$$\le h^{-1} \int_0^h \int_0^\infty |u'(x + y) - u'(x)|^p\, dx\, dy$$

converges to 0. Thus it can be concluded that $D(A) = W_p^1(0, \infty)$ and $Au = u'$.

Example 2 Similarly to the above, the semigroup $(T(t)u)(x) = u(x + t)$ for $u \in X = B^0((-\infty, \infty))$ or $X = L^p(-\infty, \infty)$, $1 \le p < \infty$, has the generator A given by $Au = u'$ with the domain $D(A) = B^1((-\infty, \infty))$ or $D(A) = W_p^1(-\infty, \infty)$.

Example 3 The semigroup has the generator given by $Au = \Delta u$ in the sense of a distribution, with the domain $D(A) = W_p^2(\mathbb{R}^n)$, $1 < p < \infty$. The proof is omitted, since it is difficult to describe it in an elementary way.

Examples 4 and 5   In these cases it is easy to see that the generators are given by H and iH, respectively.

3.2   Temporally homogenous equations of evolution

Let X be a Banach space and $A \in G(X)$. Consider the initial-value problem

$$du(t)/dt = Au(t) + f(t), \qquad 0 \leqslant t \leqslant T, \tag{3.22}$$

$$u(0) = u_0, \tag{3.23}$$

where u_0 and f are given elements of $D(A)$ and $C([0, T]; X)$, respectively. A function u is called a *solution* of this initial-value problem if, besides satisfying (3.22) and (3.23), it has the properties $u \in C^1([0, T]; X)$, $u(t) \in D(A)$ for each $t \in [0, T]$ and $Au \in C([0, T]; X)$. Let $\{T(t)\}$ denote the semigroup generated by A. Then, by Theorem 3.1.2, $u(t) = T(t)u_0$ is a solution of the homogeneous equation

$$du(t)/dt = Au(t)$$

with the initial condition (3.23). Thus it is expected that

$$u(t) = T(t)u_0 + \int_0^t T(t-s)f(s)\,ds \tag{3.24}$$

is a solution of (3.22), (3.23).

Theorem 3.2.1   If a solution of the initial-value problem (3.22), (3.23) exists, it is represented by (3.24). Therefore, the solution is unique.

Proof   In the interval $0 < s < t$, we have

$$(d/ds)(T(t-s)u(s)) = T(t-s)u'(s) - T(t-s)Au(s)$$
$$= T(t-s)f(s). \tag{3.25}$$

By integrating both sides of this equality from 0 to t, we obtain (3.24). $\square$

Whether or not the function u given by (3.24) is the solution of (3.22), (3.23) will not be obvious only from the fact that $f \in C([0, T]; X)$, even if $u_0 \in D(A)$. A sufficient condition for this to be affirmative is given by the following:

Theorem 3.2.2   If $u_0 \in D(A)$ and $f \in C^1([0, T]; X)$, *then the function* u *defined by (3.24) is the solution of (3.22), (3.23).*

Proof Suppose $A \in G(X, M, \beta)$. Changing the unknown $u(t)$ by writing $v(t) = e^{-kt}u(t)$, with $k > \beta$, we have

$$dv(t)/dt = (A - k)v(t) + e^{-kt}f(t),$$

$$v(0) = u_0.$$

Since $A - k \in G(X, M, \beta - k)$, $\beta - k < 0$, we obtain $0 \in \rho(A - k)$ and hence, without loss of generality, we may assume that $0 \in \rho(A)$. We need only examine the second term on the right-hand side of (3.24), which will be denoted by $w(t)$. Since $(\partial/\partial s)T(t - s)A^{-1} = -T(t - s)$, $w(t)$ can be calculated as

$$w(t) = -\int_0^t (\partial/\partial s)T(t - s)A^{-1}f(s)\,ds$$

$$= -A^{-1}f(t) + T(t)A^{-1}f(0) + \int_0^t T(t - s)A^{-1}f'(s)\,ds,$$

from which it is easy to see that

$$dw(t)/dt = T(t)f(0) + \int_0^t T(t - s)f'(s)\,ds$$

$$= Aw(t) + f(t).$$

This completes the proof of the theorem. $\square$

Theorem 3.2.3 *Assume that $u_0 \in D(A)$, $f \in C([0, T]; X)$, $f(t) \in D(A)$ for each $t \in [0, T]$ and $Af \in C([0, T]; X)$. Then the function u defined by (3.24) is the solution of (3.22), (3.23).*

The proof of this theorem is easy if one takes note of Lemma 1.3.1.

3.3 Analytic semigroups

Suppose there exist two real numbers β, M and an angle $\omega \in [0, \pi/2]$ such that $-A + \beta$ is of the type (ω, M) (*see* Definition 2.3.1). We will show that A is then the generator of a semigroup which can be continued holomorphically into the sector $|\arg t| < \pi/2$ containing $(0, \infty)$. We may assume that $-A$ is of the type (ω, M) by considering $A + \beta$ instead of A. Further, if necessary, we replace A by $A - \varepsilon$ $(\varepsilon > 0)$ to ensure that $0 \in \rho(A)$. Thus we may suppose that

$$\rho(A) \supset \{\lambda \in \mathbb{C} : |\arg \lambda| < \pi - \omega\} \cup \{0\}.$$

Let θ be an arbitrary angle satisfying $0 < \theta < \pi/2 - \omega$, so that $\pi/2 <$

$\pi/2 + \theta < \pi - \omega$, then the closed sector

$$\Sigma_\theta = \{\lambda \in \mathbb{C}: |\arg \lambda| \leq \pi/2 + \theta\} \cup \{0\}$$

is contained in $\rho(A)$. Let Γ be a smooth curve in Σ_θ running from $\infty e^{-i(\pi/2+\theta)}$ to $\infty e^{i(\pi/2+\theta)}$, and, for each $t > 0$, put

$$T(t) = \frac{-1}{2\pi i} \int_\Gamma e^{\lambda t}(A - \lambda)^{-1} d\lambda. \tag{3.26}$$

Further, put $C_\theta = M_{\pi/2-\theta}$, where $M_{\pi/2-\theta}$ is the number which appeared in the definition of the operator of the type (ω, M). Since $\|(A - \lambda)^{-1}\| \leq C_\theta/|\lambda|$ for $\lambda \in \Sigma_\theta$, the integral on the right-hand side of (3.26) converges in the norm of $B(X)$ and, the integrand being a regular function of λ, its values are independent of the choice of Γ, θ. For $t > 0$ fixed, we define $\Gamma = \Gamma_1 \cup \Gamma_2 \cup \Gamma_3$, where

$$\Gamma_1 = \{re^{-i(\theta+\pi/2)}: t^{-1} \leq r < \infty\},$$

$$\Gamma_2 = \{t^{-1}e^{i\varphi}: -(\theta + \pi/2) \leq \varphi \leq \theta + \pi/2\},$$

$$\Gamma_3 = \{re^{i(\theta+\pi/2)}: t^{-1} \leq r < \infty\}.$$

Then we have the estimate for the integral on Γ_3:

$$\left\| \frac{-1}{2\pi i} \int_{\Gamma_3} e^{\lambda t}(A - \lambda)^{-1} d\lambda \right\| \leq \frac{1}{2\pi} \int_{t^{-1}}^\infty e^{-rt\sin\theta} C_\theta r^{-1} dr$$

$$= \frac{C_\theta}{2\pi} \int_{\sin\theta}^\infty e^{-s} \frac{ds}{s}.$$

The integral on Γ_1 can be estimated similarly. An estimate for the integral on Γ_2 is given by

$$\left\| \frac{-1}{2\pi i} \int_{\Gamma_2} e^{\lambda t}(A - \lambda)^{-1} d\lambda \right\| \leq \frac{C_\theta}{2\pi} \int_{-\theta-\pi/2}^{\theta+\pi/2} e^{\cos\varphi} d\varphi.$$

Therefore there exists a constant C such that the inequality

$$\|T(t)\| \leq C \tag{3.27}$$

holds for $0 < t < \infty$. Next, with Γ not passing through 0, we calculate

$$T(t)A^{-1} = \frac{-1}{2\pi i} \int_\Gamma e^{\lambda t}(A - \lambda)^{-1}A^{-1} d\lambda$$

$$= \frac{-1}{2\pi i} \int_\Gamma \lambda^{-1}e^{\lambda t}\{(A - \lambda)^{-1} - A^{-1}\} d\lambda$$

$$= \frac{-1}{2\pi i} \int_\Gamma \lambda^{-1}e^{\lambda t}(A - \lambda)^{-1} d\lambda + \frac{1}{2\pi i} \int_\Gamma \lambda^{-1}e^{\lambda t} d\lambda A^{-1}$$

and, on account of the fact that the integral in the second term on the right-hand side is equal to unity, obtain

$$T(t)A^{-1} = \frac{-1}{2\pi i} \int_{\Gamma} \lambda^{-1} e^{\lambda t}(A-\lambda)^{-1} \, d\lambda + A^{-1}.$$

Letting $t \to 0$ and using the residue theorem, we have, in the norm of $B(X)$,

$$T(t)A^{-1} \to \frac{-1}{2\pi i} \int_{\Gamma} \lambda^{-1}(A-\lambda)^{-1} \, d\lambda + A^{-1} = A^{-1}.$$

Thus it has been shown that $T(t)u \to u$ if $u \in D(A)$. This, combined with (3.27), implies that $T(t) \to I$ strongly as $t \to 0$. Just as in the proof of (2.28) it can be seen that

$$T(t+s) = T(t)T(s) \tag{3.28}$$

is satisfied for each $t>0$ and each $s>0$. $T(t)$ is not only strongly continuous but also norm continuous in $t>0$. Therefore, by putting $T(0) = I$, $\{T(t)\}$ forms a semigroup. It is also obvious that $T(t)$ commutes with A. For $u \in D(A)$, we have

$$\frac{d}{dt} T(t)u = \frac{-1}{2\pi i} \int_{\Gamma} \lambda e^{\lambda t}(A-\lambda)^{-1} u \, d\lambda$$

$$= \frac{-1}{2\pi i} \int_{\Gamma} e^{\lambda t}(A-\lambda)^{-1} Au \, d\lambda + \frac{1}{2\pi i} \int_{\Gamma} e^{\lambda t} u \, d\lambda = T(t)Au.$$

From this it follows that A is the generator of $\{T(t)\}$, which can be proved in the same way as part (iii) of the proof of Theorem 3.1.4 by Hille's method. Also, in a manner analogous to the proof of (3.27), it is found that $T(t)$ can be continued holomorphically to the sector $|\arg t| < \pi/2 - \omega$, since the integral on the right-hand side of (3.26) converges uniformly in the wider sense in $|\arg t| < \theta$, and $\theta \in (0, \pi/2 - \omega)$ can be taken arbitrarily. Therefore, by analytic continuation from the positive real axis, it turns out that (3.28) is valid also for t, s satisfying $|\arg t| < \pi/2 - \omega$ and $|\arg s| < \pi/2 - \omega$. Moreover, in the same way as the proof for s-$\lim_{t \to +0} T(t) = I$, it can be shown that $T(t) \to I$ strongly as t approaches 0 in the sector $\{t : |\arg t| \le \theta\}$. For any natural number n we have

$$\left(\frac{d}{dt}\right)^n T(t) = \frac{-1}{2\pi i} \int_{\Gamma} \lambda^n e^{\lambda t}(A-\lambda)^{-1} \, d\lambda.$$

If the boundary of Σ_θ is taken as Γ, the integral can be estimated for $t>0$ as

$$\left\| \left(\frac{d}{dt}\right)^n T(t) \right\| \le \frac{C_\theta}{\pi} \int_0^\infty r^{n-1} e^{-rt\sin\theta} \, dr = \frac{C_\theta(n-1)!}{\pi(t \sin \theta)^n}.$$

Therefore, the power series expansion at each $t > 0$,

$$\sum_{n=0}^{\infty} \frac{(z-t)^n}{n!} \left(\frac{\mathrm{d}}{\mathrm{d}t}\right)^n T(t),$$

converges uniformly in the wider sense inside the circle $|z - t| < t \sin \theta$. This also indicates that $T(t)$ is regular in $\{t : |\arg t| < \pi/2 - \omega\}$. When $-A_1 = -A + \beta$ is of the type (ω, M), the semigroup generated by A is given by $T(t) = \mathrm{e}^{\beta t} S(t)$, where $\{S(t)\}$ is the semigroup having A_1 as its generator. Summarizing, we have

Theorem 3.3.1 *Suppose there exist real numbers β, M and an angle $\omega \in [0, \pi/2]$ such that $-A + \beta$ is of the type (ω, M), then A is the generator of a semigroup $\{T(t)\}$. $T(t)$ can be continued holomorphically with respect to t into the sector $\{t : |\arg t| \leq \pi/2 - \omega\}$, where (3.28) holds. Let θ be an arbitrary angle satisfying $0 < \theta < \pi/2 - \omega$, then $\mathrm{e}^{-\beta t} T(t)$ is uniformly bounded in the closed subsector $\{t \, |\arg t| \leq \theta\}$ and, when t approaches 0 inside this subsector, $T(t)$ converges strongly to I, For any natural number n we have*

$$\limsup_{t \to +0} t^n \left\| \left(\frac{\mathrm{d}}{\mathrm{d}t}\right)^n T(t) \right\| = \limsup_{t \to +0} t^n \|A^n T(t)\| < \infty. \tag{3.29}$$

Definition 3.3.1 *The semigroup described in Theorem 3.3.1 is called an* analytic semigroup *or a* parabolic semigroup.

The term 'parabolic semigroup' is due to the fact that such a semigroup appears in the initial-value problem or the mixed problem for parabolic equations. As a converse of the above theorem, we have

Theorem 3.3.2 *Assume that a semigroup $\{T(t)\}$ can be continued holomorphically into the sector $\{t : |\arg t| < \pi/2 - \omega\}$ and that there exists a real number β such that $\mathrm{e}^{-\beta t} T(t)$ is uniformly bounded in the subsector $|\arg t| \leq \theta$ for each θ satisfying $0 < \theta < \pi/2 - \omega$, and that $T(t)$ converges strongly to I as t approaches 0 inside this subsector. Then, for the generator A of the semigroup $\{T(t)\}$, there exists a real number M such that $-A + \beta$ is of the type (ω, M).*

Proof Assume $0 < \theta < \pi/2 - \omega$. According as $\pm \mathrm{Im}\, \lambda > (\beta - \mathrm{Re}\, \lambda) \cot \theta$, we change the contour of the integral of (3.6) into $\{r\mathrm{e}^{\mp i\theta} : 0 < r < \infty\}$. Then we find that $\{|\mathrm{Im}\, \lambda| \sin \theta + (\mathrm{Re}\, \lambda - \beta) \cos \theta\}(A - \lambda)^{-1}$ is uniformly bounded in $\mathrm{Re}\, \lambda < \beta$ and $|\mathrm{Im}\, \lambda| > (\beta - \mathrm{Re}\, \lambda) \cot \theta$. $\square$

Theorem 3.3.3 *Let $\{T(t)\}$ be a semigroup with a generator A such that $-A$ is of the type (ω, M). Then, for an arbitrary real $\alpha > 0$, we have*

$$\limsup_{t \to +0} t^\alpha \, \|(-A)^\alpha T(t)\| < \infty. \tag{3.30}$$

Proof We use the same notations as in the proof of Theorem 3.3.1. Taking the boundary of Σ_θ as Γ and noting that

$$(-A)^\alpha T(t) = \frac{-1}{2\pi i} \int_\Gamma (-\lambda)^\alpha \, \mathrm{e}t(A - \lambda)^{-1} \, \mathrm{d}\lambda,$$

we immediately obtain (3.30). $\square$

Remark 3.3.1 The semigroup of Example 4, $T(t) = \int_{-\infty}^a \mathrm{e}^{\lambda t} \, \mathrm{d}E(\lambda)$, is an analytic semigroup.

Remark 3.3.2 Let $-A + \beta$ be of the type (ω, M), where $0 \leq \omega < \pi/2$, then, for each $\gamma > \beta$, the half-plane $\{\lambda : \operatorname{Re} \lambda > \gamma\}$ is contained in $\rho(A)$ and, in this half-plane,

$$\|(A - \lambda)^{-1}\| \leq C(1 + |\lambda - \gamma|)^{-1}$$

is valid for some constant C. Conversely, if A is a densely-defined closed operator satisfying this condition, then the power-series expansion of $(A - \lambda)^{-1}$ at each point on the straight line $\operatorname{Re} \lambda = \gamma$ indicates that there exist some $\beta < \gamma$, M and $\omega \in [0, \pi/2]$ such that $-A + \beta$ is of the type (ω, M). This statement is often used to express a necessary and sufficient condition for A to generate an analytic semigroup.

Let $\{T(t)\}$ be a parabolic semigroup and A its generator. Then, by Theorem 3.3.1, even if u_0 is an arbitrary element of X, $T(t)u_0$ is differentiable in $t > 0$, belongs to $D(A)$ and satisfies $(\mathrm{d}/\mathrm{d}t)T(t)u_0 = AT(t)u_0$. Therefore, when A is the generator of a parabolic semigroup $\{T(t)\}$, by the *solution* of

$$\mathrm{d}u(t)/\mathrm{d}t = Au(t) + f(t), \qquad 0 < t \leq T, \tag{3.31}$$

$$u(0) = u_0 \tag{3.32}$$

we mean a function u, satisfying these equations, such that $u \in C([0, T]; X) \cap C^1((0, T]; X)$ and $u(t) \in D(A)$ for each $t \in (0, T]$ and $Au \in C((0, T]; X)$. In (3.31), the function f is an element of $C([0, T]; X)$. The uniqueness of the solution of (3.31), (3.32) can be proved along the line of the proof of Theorem 3.2.1, first by integrating (3.25) from some $\varepsilon > 0$

to t and then by letting $\varepsilon \to 0$. As for the existence of the solution, we have

Theorem 3.3.4 Let u_0 be an arbitrary element of X. Assume that f is Hölder continuous, namely, that there exists a positive number K and some $\gamma \leq 1$ such that the inequality

$$\|f(t) - f(s)\| \leq K |t - s|^{\gamma} \tag{3.33}$$

is satisfied for $0 \leq t \leq T$ and $0 \leq s \leq T$. Then

$$u(t) = T(t)u_0 + \int_0^t T(t-s)f(s)\,ds$$

is the unique solution of (3.31), (3.32).

Proof For $0 < \varepsilon \leq t \leq T$, we put

$$v_\varepsilon(t) = \int_0^{t-\varepsilon} T(t-s)f(s)\,ds,$$

obtaining

$$
\begin{aligned}
Av_\varepsilon(t) &= \int_0^{t-\varepsilon} AT(t-s)f(s)\,ds \\
&= \int_0^{t-\varepsilon} AT(t-s)(f(s)-f(t))\,ds - \int_0^{t-\varepsilon} (\partial/\partial s)T(t-s)f(t)\,ds \\
&= \int_0^{t-\varepsilon} AT(t-s)(f(s)-f(t))\,ds - T(\varepsilon)f(t) + T(t)f(t),
\end{aligned}
$$

which converges strongly as $\varepsilon \to 0$ by virtue of (3.29) with $n = 1$ and (3.33). Since

$$v_\varepsilon(t) \to v(t) \equiv \int_0^t T(t-s)f(s)\,ds$$

and since A is a closed operator, we find that $v(t) \in D(A)$, that

$$Av(t) = \int_0^t AT(t-s)(f(s)-f(t))\,ds - f(t) + T(t)f(t)$$

and, moreover, that the right-hand side of this equation is strongly continuous in $[0, T]$. Thus, we have

$$(d/dt)v_\varepsilon(t) = T(\varepsilon)f(t-\varepsilon) + Av_\varepsilon(t) \to f(t) + Av(t),$$

which shows that $v(t)$ satisfies (3.31). $\square$

Questions may naturally arise about the smoothness of u when f is several times continuously differentiable or about the regularity of u when f is a regular function, but these problems will be postponed until Chapter 5, where we will be concerned with more general cases in which A depends on t.

3.4 Perturbation of semigroups

Let A generate a semigroup. We consider the problem whether or not the sum $A + B$ of A and some linear operator B generates a semigroup. Many results concerning this problem have been obtained by Hille and Phillips [7], Trotter [171, 172], Kato [8], Yosida [179], Okazawa [143, 144] and others. Here we will mention only of the case of bounded B.

Theorem 3.4.1 If $A \in G(X)$ and $B \in B(X)$, then the operator $A + B$ has $D(A)$ as its domain and is an element of $G(X)$. More precisely, $A \in G(X, M, \beta)$ implies $A + B \in G(X, M, \beta + M\|B\|)$.

Proof Evidently,

$$A + B - \lambda = \{1 + B(A - \lambda)^{-1}\}(A - \lambda)$$

for $\lambda > \beta$, but, under a more restrictive condition, $\lambda > \beta + M\|B\|$, we have $\|B(A - \lambda)^{-1}\| \leq \|B\| M(\lambda - \beta)^{-1} < 1$. Hence, such λ belongs to $\rho(A + B)$ and

$$(A + B - \lambda)^{-1} = (A - \lambda)^{-1} \sum_{j=0}^{\infty} (B(A - \lambda)^{-1})^j.$$

For each natural number n,

$$(A + B - \lambda)^{-n} = \left\{ \sum_{j=0}^{\infty} (A - \lambda)^{-1}(B(A - \lambda)^{-1})^j \right\}^n$$

has an expansion whose term

$$(A - \lambda)^{-n_1} B(A - \lambda)^{-n_2} B \cdots (A - \lambda)^{-n_k} B(A - \lambda)^{-n_{k+1}},$$

contains only k factors of B, where $\sum_{i=1}^{k+1} n_i = n + k$ and $n_i > 0$. The norm of this term does not exceed $M^{k+1} \|B\|^k (\lambda - \beta)^{-n-k}$. Also, the number of such terms is equal to the coefficient of x^k in the expansion $(1 - x)^{-n} = \sum_{k=0}^{\infty} c_{n,k} x^k$. Therefore, we have

$$\|(A + B - \lambda)^{-n}\| \leq \sum_{k=0}^{\infty} c_{n,k} M^{k+1} \|B\|^k (\lambda - \beta)^{-n-k}$$

$$= M(\lambda - \beta)^{-n}(1 - M\|B\|(\lambda - \beta)^{-1})^{-n}$$

$$= M(\lambda - \beta - M\|B\|)^{-n}. \quad \square$$

Theorem 3.4.2   Under the assumptions of the preceding theorem, the semigroup $\{S(t)\}$ generated by $A + B$ is represensed as follows:

$$S(t) = \sum_{n=0}^{\infty} T_n(t), \tag{3.34}$$

where

$$T_0(t) = T(t) = \exp(tA) \tag{3.35}$$

and

$$T_n(t) = \int_0^t T(t-s)BT_{n-1}(s)\, ds$$

$$= \int_0^t T_{n-1}(s)BT(s)\, ds, \quad n = 1, 2, \ldots. \tag{3.36}$$

Proof   If $u \in D(A)$, then

$$S(t)u = T(t)u + \int_0^t (\partial/\partial s)\{T(t-s)S(s)u\}\, ds$$

$$= T(t)u + \int_0^t T(t-s)BS(s)u\, ds.$$

Therefore,

$$S(t) = T(t) + \int_0^t T(t-s)BS(s)\, ds$$

is an integral equation for $S(t)$ to satisfy. (3.34) and (3.36) represent none other than a solution of this equation by successive iteration. $\quad\square$

Remark 3.4.1   It can be shown by induction that, for each n,

$$\|T_n(t)\| \leq M^{n+1}\|B\|^n\, e^{\beta t}t^n/n!,$$

and, hence, $S(t)$ has the estimate

$$\|S(t)\| \leq Me^{(\beta + M\|B\|)t}.$$

Therefore, we can prove Theorem 3.4.1 alternatively, by establishing that the $S(t)$ found as a solution of the integral equation above is indeed the semigroup having $A + B$ as its generator.

3.5 Application 1. Initial-value problem of the symmetric hyperbolic system

Consider the following initial-value problem for the system of partial differential equations:

$$\partial u/\partial t = \sum_{j=1}^{n} a_j(x)\partial u/\partial x_j + b(x)u + f(x, t), \quad x \in \mathbb{R}^n, \quad 0 \leqslant t \leqslant T, \quad (3.37)$$

$$u(x, 0) = u_0(x), \quad (3.38)$$

where $u = {}^t(u_1, \ldots, u_N)$ (the superscript t denoting a transposed matrix) is the set of unknowns, and $a_j(x)$, $b(x)$ for each x are square matrices of order N. We will assume the following:

$$\begin{cases} \text{The matrices } a_j(x) \text{ for } j = 1, \ldots, n \text{ and } x \in \mathbb{R}^n \text{ are Hermi-} \\ \text{tian.} \end{cases} \quad (3.39)$$

$$\begin{cases} \text{Each component of } a_j, j = 1, \ldots, n, \text{ and } b \text{ belong to} \\ B^1(\mathbb{R}^n) \text{ and } B^0(\mathbb{R}^n), \text{ respectively,} \end{cases} \quad (3.40)$$

For each $u \in L^2(\mathbb{R}^n)^N$, we put

$$\mathscr{A}u = \sum_{j=1}^{n} a_j(x)\partial u/\partial x_j + b(x)u. \quad (3.41)$$

By the assumption (3.40) the right-hand side of (3.41) makes sense as a distribution. Let us write $X = L^2(\mathbb{R}^n)^N$ and stipulate that $\|\cdot\|$ always means the norm of $L^2(\mathbb{R}^n)^N$. Define an operator A as follows:

$$D(A) = \{u \in X: \mathscr{A}u \in X\}, \quad Au = \mathscr{A}u \quad \text{for} \quad u \in D(A). \quad (3.42)$$

In this way we have rewritten (3.37), (3.38) in an abstract form of (3.22), (3.23). Put $Y = H_1(\mathbb{R}^n)^N$, then, evidently, $D(A) \subset Y$. For two N-dimensional vectors $u = {}^t(u_1, \ldots, u_N)$ and $v = {}^t(v_1, \ldots, v_N)$, we denote their inner product by $u \cdot v = \sum_{j=1}^{N} u_j \bar{v}_j$. Let $u \in Y$ and λ be real, then

$$\|(A - \lambda)u\|^2 = \|Au\|^2 - 2\lambda \operatorname{Re}(Au, u) + \lambda^2 \|u\|^2. \quad (3.43)$$

Integration by parts gives

$$(Au, u) = \int_{\mathbb{R}^n} \sum_{j=1}^{n} a_j \frac{\partial u}{\partial x_j} \cdot u \, dx + (bu, u)$$

$$= -\int_{\mathbb{R}^n} u \cdot \sum_{j=1}^{n} \frac{\partial}{\partial x_j} (a_j u) \, dx + (bu, u)$$

$$= -\int_{\mathbb{R}^n} u \cdot \sum_{j=1}^{n} a_j \frac{\partial u}{\partial x_j} dx - \int_{\mathbb{R}^n} u \cdot \sum_{j=1}^{n} \frac{\partial a_j}{\partial x_j} u \, dx + (bu, u)$$

$$= -(u, Au) - \left(u, \sum_{j=1}^{n} \frac{\partial a_j}{\partial x_j} \cdot u\right) + (u, bu) + (bu, u)$$

and, hence,

$$2 \operatorname{Re}(Au, u) = -\left(u, \sum_{j=1}^{n} \frac{\partial a_j}{\partial x_j} \cdot u\right) + 2 \operatorname{Re}(bu, u).$$

From this and (3.43), it follows that

$$\|(A - \lambda)u\|^2 \geqslant (\lambda^2 - C|\lambda|)\|u\|^2$$

is satisfied for some positive constant C. Therefore, there exists a positive number β such that, if $|\lambda|$ is sufficiently large, we have the estimate

$$\|(A - \lambda)u\| \geqslant (|\lambda| - \beta)\|u\| \tag{3.44}$$

for every $u \in Y$. Let φ_δ be a mollifier introduced in Section 1.2. Take $u \in D(A)$ and consider the identity

$$(A - \lambda)(\varphi_\delta * u) = \varphi_\delta * (A - \lambda)u + \{A(\varphi_\delta * u) - \varphi_\delta * Au\}.$$

The expression in the curly brackets converges to 0 as $\delta \to 0$ by Lemma 1.2.5. By applying (3.44) to $\varphi_\delta * u \in Y$ and letting $\delta \to 0$, it is found that (3.44) holds also for $u \in D(A)$. Let us denote the formal adjoint of $\mathcal{A}$ by $\mathcal{A}'$, namely,

$$\mathcal{A}'u = -\sum_{j=1}^{n} (\partial/\partial x_j)(a_j(x)u) + b^*(x)u.$$

If an operator A' is defined by

$$D(A') = \{u \in X : \mathcal{A}'u \in X\}, \qquad A'u = \mathcal{A}'u \quad \text{for} \quad u \in D(A'),$$

then, as in the case of (3.44), there exists some $\beta' > 0$ such that

$$\|(A' - \lambda)u\| \geqslant (|\lambda| - \beta')\|u\| \tag{3.45}$$

is satisfied for real λ with sufficiently large modulus and for every $u \in D(A')$.

Lemma 3.5.1 $\lambda \in \rho(A)$ if λ is real and $|\lambda|$ is sufficiently large.

Proof First, we will show that A is a closed operator. Suppose $u_j \in D(A)$, $u_j \to u$ in X and $Au_j \to v$. Then $Au = v \in X$, because $\mathcal{A}u_j \to \mathcal{A}u$ and $\mathcal{A}u_j \to v$ in the sense of a distribution. Hence, $u \in D(A)$ and $Au = v$. If $|\lambda|$ is sufficiently large, then, because of (3.44), the operator $A - \lambda$ has a continuous inverse, so that $R(A - \lambda)$ is a closed subspace of X. Next, let v belong to $R(A - \lambda)^\perp$, then $(\mathcal{A}' - \lambda)v = 0$ since $((A - \lambda)u, v)$ for every $u \in D(A)$ and, hence, for every $u \in C_0^\infty(\mathbb{R}^n)$. Thus, $\mathcal{A}'v = \lambda v \in X$. Therefore, we have $v \in D(A')$ and $A'v = \lambda v$. From this and (3.45), it follows that $v = 0$. $\square$

Combining (3.44), Lemma 3.5.1 and Remark 3.1.4, we have

Theorem 3.5.1 *Let A be the operator defined by (3.42). Then both A and $-A$ belong to $G(X, 1, \beta)$. Therefore, Theorems 3.2.1–3.2.3 are applicable not only to (3.37), (3.38) but also to the problem of solving (3.37) in the backwards time direction.*

3.6 Application 2. Regular dissipative operators

We follow the assumptions and the notations in Section 2.2. For simplicity, it is assumed that (2.12) is satisfied, namely,

$$\text{Re } a(u, u) \geq \delta \|u\|^2, \qquad u \in V. \tag{3.46}$$

Lemma 3.6.1 *For Re $\lambda \leq 0$ there exists a bounded inverse of $A - \lambda$ which has various bounds for every f of X or V^*:*

$$|(A - \lambda)^{-1} f| \leq M_1 |\lambda|^{-1} |f|, \tag{3.47}$$

$$|(A - \lambda)^{-1} f| \leq M_2 |\lambda|^{-1/2} \|f\|_*, \tag{3.48}$$

$$\|(A - \lambda)^{-1} f\| \leq M_2 |\lambda|^{-1/2} |f|, \tag{3.49}$$

$$\|(A - \lambda)^{-1} f\| \leq \delta^{-1} \|f\|_*, \tag{3.50}$$

$$\|(A - \lambda)^{-1} f\|_* \leq M_1 |\lambda|^{-1} \|f\|_*, \tag{3.51}$$

where $M_1 = 1 + M/\delta$ and $M_2 = \{(1 + M\delta)/\delta\}^{1/2}$.

Proof It is seen from Lemma 2.2.1 and (3.46) that the inverse $(A - \lambda)^{-1}$ exists and is bounded in V^*. Put $f = (A - \lambda)u$ with Re $\lambda \geq 0$, then, by definition,

$$(f, v) = a(u, v) - \lambda(u, v) \tag{3.52}$$

for every $v \in V$. If we take $v = u$, (3.52) becomes

$$(f, u) = a(u, u) - \lambda(u, u), \tag{3.53}$$

from which it follows that

$$\|f\|_* \|u\| \geq \text{Re } (f, u) \geq \delta \|u\|^2 - \text{Re } \lambda |u|^2 \geq \delta \|u\|^2 \tag{3.54}$$

and, thus, (3.50) has been obtained. Inequalities (3.50), (3.53) and (3.24) imply that

$$|\lambda| |u|^2 \leq \|f\|_* \|u\| + M \|u\|^2 \leq M_2^2 \|f\|_*^2, \tag{3.55}$$

from which (3.48) follows immediately. In a manner similar to that leading to (3.54), it is found that

$$\delta \|u\|^2 \leq |f| \, |u| \tag{3.56}$$

and, hence, from (3.53), there follows the result that

$$|\lambda| \, |u|^2 \leq |f| \, |u| + M \, \|u\|^2 \leq M_1 \, |f| \, |u|,$$

which is merely (3.47). Inequalities (3.47) and (3.56) imply that $\delta \|u\|^2 \leq M_1 |\lambda|^{-1} |f|^2$, i.e., (3.49). Finally, from (3.52) and (3.50), we obtain

$$|\lambda| \, |(u, v)| \leq \|f\|_* \, \|v\| + M \, \|u\| \, \|v\| \leq M_1 \, \|f\|_* \, \|v\|,$$

which gives (3.51), since v is arbitrary. $\square$

Theorem 3.6.1 $\quad -A$ *generates an analytic semigroup in both X and V^*.*

Proof This theorem is obvious from Remark 3.3.2 and the fact that (3.47) and (3.51) hold for Re $\lambda \leq 0$. $\square$

Remark 3.6.1 With other appropriate coefficients, the estimates (3.48), (3.49), (3.50), as well as (3.47), (3.51), are valid for some sector $\Sigma = \{\lambda : |\arg \lambda| \geq \theta\}$ $(0 < \theta < \pi/2)$.

We will again denote the norms of $B(X)$ and $B(V^*)$ by $|\cdot|$ and $\|\cdot\|_*$, respectively. If $\{T(t)\}$ represents the semigroup which $-A$ generates, then there exists a constant C such that the following inequalities are satisfied for all $t > 0$:

$$|T(t)| \leq 1, \tag{3.57}$$

$$\|T(t)\|_* \leq C. \tag{3.58}$$

Lemma 3.6.2 *There exists a constant C such that the following inequalities hold for all $t > 0$ and every $f \in X$ or V^*:*

$$|T(t)f| \leq C t^{-1/2} \|f\|_*, \tag{3.59}$$

$$\|T(t)f\| \leq C t^{-1/2} |f|, \tag{3.60}$$

$$\|T(t)f\| \leq C t^{-1} \|f\|_*, \tag{3.61}$$

$$|AT(t)f| \leq C t^{-3/2} \|f\|_*, \tag{3.62}$$

$$\|AT(t)f\| \leq C t^{-3/2} |f|, \tag{3.63}$$

$$\|AT(t)f\|_* \leq C t^{-1/2} |f|. \tag{3.64}$$

The proof of this lemma is easily obtained by using the representation (3.26) of $T(t)$, Lemma 3.6.1 and (2.13).

3.7 Elliptic boundary-value problems

In the preceding section, we have applied the results of Section 3.3 to the parabolic equations. Aiming at another form of their application, we will sketch the theory of elliptic boundary-value problems due to Agmon, Douglis and Nirenberg [22] and Schechter [153, 154, 155]. All the proofs will be omitted and for the details the reader should refer to these papers and also to Lions and Magenes [15].

Let Ω be a bounded region of class C^m in $\mathbb{R}^n$. Furthermore, let $A(x, D) = \sum_{|\alpha| \leq m} a_\alpha(x) D^\alpha$ be a linear differential operator whose coefficients $a_\alpha(x)$ are defined in Ω and belong to $L^\infty(\Omega)$; in particular, the coefficients of mth order, $a_\alpha(x)$ with $|\alpha| = m$, are assumed to be continuous in $\bar{\Omega}$. Here $D = (i^{-1}\partial/\partial x_1, \ldots, i^{-1}\partial/\partial x_n)$, as in Chapter 1. The highest-order part, or the principal part, of $A(x, D)$ is denoted by $\mathring{A}(x, D)$: $\mathring{A}(x, D) = \sum_{|\alpha| = m} a_\alpha(x) D^\alpha$. Assume that $A(x, D)$ is elliptic in $\bar{\Omega}$, by which it is meant that $\mathring{A}(x, \xi) \neq 0$ for any $x \in \bar{\Omega}$ and any non-zero real vector $\xi = (\xi_1, \ldots, \xi_n)$. In addition, we assume $A(x, D)$ to be *properly elliptic*. That is to say, let m be assumed even and let ξ, η be real vectors which are linearly independent. Then, the roots of the polynomial $\mathring{A}(x, \xi + \tau\eta)$ of one variable τ are never real, since $A(x, D)$ is elliptic. In this situation, we suppose that, from amongst the roots, just $m/2$ have positive imaginary parts and the remaining $m/2$ have negative imaginary parts. Every elliptic differential operator for $n \geq 3$ is properly elliptic ([22], p. 625), but this is not the case for $n = 2$, as is shown by the counter-example of the Cauchy–Riemann operator $\partial/\partial x + i\partial/\partial y$.

Next, let $B_j(x, D) = \sum_{|\beta| \leq m_j} b_{j\beta}(x) D^\beta$ for each $j = 1, \ldots, m/2$ be a linear differential operator of order m_j whose coefficients are defined in $\partial\Omega$. Assume that each coefficient $b_{j\beta}$ belongs to $C^{m-m_j}(\partial\Omega)$. In what follows, we will be concerned only with the case $m_j < m$. (This restriction is not assumed in [22].)

In this way, we are going to study the boundary-value problem

$$A(x, D)u(x) = f(x), \qquad x \in \Omega, \tag{3.65}$$

$$B_j(x, D)u(x) = g_j(x), \qquad x \in \partial\Omega, \qquad j = 1, \ldots, m/2, \tag{3.66}$$

where f, g_j are known functions belonging to $L^p(\Omega)$ and $W_p^{m-m_j-1/p}(\partial\Omega)$ with $1 < p < \infty$, respectively. We will seek a solution u in $W_p^m(\Omega)$. One of the main results of [22] is that, apart from its existence, the solution of (3.65), (3.66) has the *a priori* estimate

$$\|u\|_{m,p} \leq C\{\|f\|_p + \sum_{j=1}^{m/2} [g_j]_{m-m_j-1/p} + \|u\|_p\}$$

if and only if the following condition is satisfied:

Complementing condition Let x be an arbitrary point of $\partial\Omega$, ν the normal vector at x, and ξ an arbitrary non-zero real vector tangent to $\partial\Omega$ at x. Since $A(x, D)$ is assumed to be properly elliptic, just $m/2$ roots of the polynomial $\mathring{A}(x, \xi + \tau\nu)$ of one variable τ have positive imaginary parts. They will be denoted by $\tau_1^+(\xi), \ldots, \tau_{m/2}^+(\xi)$. Then $m/2$ polynomials $\mathring{B}_j(x, \xi + \tau\nu)$ are linearly independent modulo $\prod_{k=1}^{m/2} (\tau - \tau_k^+(\xi))$. In other words, a polynomial $\sum_{j=1}^{m/2} c_j \mathring{B}_j(x, \xi + \tau\nu)$ of τ is algebraically divisable by $\prod_{k=1}^{m/2}(\tau - \tau_k^+(\xi))$ only if $c_1 = \cdots = c_{m/2}$, where $\mathring{B}_j(x, D) = \sum_{|\beta|=m_j} b_{j\beta}(x)D^\beta$ is the principal part of $B_j(x, D)$.

The proof of this statement is so lengthy that we will omit it. It can be simplified to some extent for $p = 2$, because then we are allowed to use the Fourier transformation (Schecter [153]). Here we content ourselves with a brief explanation of the meaning of the complementing condition. We do not claim that the following argument is strict. Suppose that Ω is the half-space $\mathbb{R}_+^n = \{x \in \mathbb{R}^n : x_n > 0\}$ and that $A(x, D)$ and $\{B_j(x, D)\}$ have constant coefficients and consist only of the principal part, so that

$$A(x, D) = A(D) = \sum_{|\alpha|=m} a_\alpha D^\alpha, \qquad B_j(x, D) = B_j(D) = \sum_{|\beta|=m_j} b_{j\beta} D^\beta.$$

With the notations $x' = (x_1, \ldots, x_{n-1})$, $D' = (-i\partial/\partial x_1, \ldots, -i\partial/\partial x_{n-1})$, $D_n = -i\partial/\partial x_n$, $A(D) = A(D', D_n)$ and $B_j(D) = B_j(D', D_n)$, we consider the boundary-value problem in $\mathbb{R}_+^n$

$$A(D', D_n)u(x) = 0, \qquad x_n > 0, \tag{3.67}$$

$$B_j(D', D_n)u(x', 0) = g_j(x'), \qquad j = 1, \ldots, m/2. \tag{3.68}$$

Let us denote the partial Fourier transform of u with respect to x' by $\hat{u}(\xi', x_n)$ and the Fourier transform of g_j by $\hat{g}_j$:

$$\hat{u}(\xi', x_n) = \int e^{ix'\xi'} u(x', x_n) \, dx',$$

$$\hat{g}_j(\xi') = \int e^{ix'\xi'} g(x') \, dx',$$

where $\xi' = (\xi_1, \ldots, \xi_{n-1})$ is a real vector. If u is a solution of (3.67), (3.68), then $\hat{u}$ is a solution of the initial-value problem of the following ordinary differential equation containing ξ' as a parameter:

$$A(\xi', D_n)\hat{u}(\xi', x_n) = 0, \qquad x_n > 0, \tag{3.69}$$

$$B_j(\xi', D_n)\hat{u}(\xi', 0) = \hat{g}_j(\xi'), \qquad j = 1, \ldots, m/2. \tag{3.70}$$

Since $A(x, D)$ is properly elliptic, the roots of $A(\xi', \eta) = 0$ for $\xi' \neq 0$ are all complex; just $m/2$ roots have positive imaginary parts and the remaining $m/2$ have negative imaginary parts. Now let $p(t) = a_0 t^m + a_1 t^{m-1} + \cdots + a_m$ be a polynomial of order m such that its $m/2$ roots have positive imaginary parts, while the remaining $m/2$ have negative imaginary parts. Consider the ordinary differential equation

$$p(-i\, d/dx)u(x) = 0. \tag{3.71}$$

We can choose m linearly independent solutions $\varphi_1, \ldots, \varphi_m$ in such a way that each φ_j can be represented as $\varphi_j(x) = x^k e^{\lambda_j x}$, where λ_j is a root of $p(t)$ and k is a non-negative integer less than the multiplicity of λ_j. As $x \to \infty$, $|\varphi_j(x)|$ decreases exponentially if $\operatorname{Im} \lambda_j > 0$, while it increases exponentially if $\operatorname{Im} \lambda_j < 0$. Therefore, if the numbering of the roots is fixed so that $\operatorname{Im} \lambda_j > 0$ for $j = 1, \ldots, m/2$ and $\operatorname{Im} \lambda_j < 0$ for others, a solution of (3.71) belonging to $L^p(0, \infty)$ $(1 < p < \infty)$ can be expressed as

$$u(x) = \sum_{j=1}^{m/2} c_j \varphi_j(x). \tag{3.72}$$

Consider the initial-value problem on the half-line $0 \leqslant x < \infty$:

$$p(-i\, d/dx)u(x) = 0, \qquad x > 0, \tag{3.73}$$

$$b_j(-i\, d/dx)u(0) = a_j, \qquad j = 1, \ldots, m/2, \tag{3.74}$$

where b_j is a polynomial of order m_j. The substitution of (3.72) into (3.74) yields a system of linear equations with unknowns $c_1, \ldots, c_{m/2}$. If the matrix consisting of the coefficients of these equations is regular, then, for $a_1, \ldots, a_{m/2}$ arbitrarily specified, there exists one and only one solution of (3.73), (3.74) which belongs to $L^p(0, \infty)$. That $(A(D), \{B_j(D)\}, \mathbb{R}^n_+)$ satisfies the complementing condition means that the initial-value problem of ordinary differential equations, (3.69), (3.70), satisfies this condition for each $\xi' \neq 0$. In the case of $\{B_j(x, D)\}_{j=1}^{m/2} = \{(\partial/\partial\nu)^{j-1}\}_{j=1}^{m/2}$, that is, if the boundary condition is of Dirichlet type, the complementing condition is satisfied by every properly elliptic operator.

Definition 3.7.1 $\{B_j(x, D)\}_{j=1}^{m/2}$ *is said to be* normal *if it satisfies the following conditions.*

(i) *$j \neq k$ implies that $m_j \neq m_k$, in other words, the orders of the $B_j(x, D)$ differ from each other.*

(ii) *For each $j = 1, \ldots, m/2$ the boundary $\partial\Omega$ is not characteristic with respect to $B_j(x, D)$, i.e., $\overset{\circ}{B}_j(x, \nu) = 0$, where $x \in \partial\Omega$ and ν is the normal vector of $\partial\Omega$ at x.*

Definition 3.7.2 Let $B_j(x, D)$ $(j = 1, \ldots, m)$ be a linear differential operator of order m_j with coefficients defined on $\partial\Omega$. $\{B_j(x, D)\}_{j=1}^m$ is called a Dirichlet system *of order* m *if it satisfies the following conditions.*

(i) $\{B_j(x, D)\}_{j=1}^m$ *is normal.*
(ii) $m_j < m$ *for each* $j = 1, \ldots, m$.

If $\{B_j(x, D)\}_{j=1}^m$ is a Dirichlet system, then $\{m_1, \ldots, m_m\} = \{0, \ldots, m\}$. A normal system $\{B_j(x, D)\}_{j=1}^{m/2}$ is a subset of some Dirichlet system of order m. Evidently, $\{(\partial/\partial\nu)^{j-1}\}_{j=1}^m$ is a Dirichlet system of order m. Let $\{B_j(x, D)\}_{j=1}^m$ and $\{C_j(x, D)\}_{j=1}^m$ be two Dirichlet systems such that their orders are m_j and l_j, and their coefficients belong to $C^{m-m_j}(\partial\Omega)$ and $C^{m-l_j}(\partial\Omega)$, respectively. Then, for each $j = 1, \ldots, m$, we have a unique representation

$$B_j(x, D) = \sum_{k=1}^m T_{jk}(x, D) C_k(x, D), \tag{3.75}$$

where $T_{jk}(x, D)$ is a linear differential operator containing only tangential differentiations and its order is equal to $m_j - l_k$; $T_{jk}(x, D)$ is a non-vanishing function if $l_k = m_j$ and $T_{jk}(x, D) = 0$ if $l_k > m_j$. Let g_j $(j = 1, \ldots, m)$ be an arbitrary function belonging to $W_p^{m-m_j-1/p}(\partial\Omega)$, then there exists a $u \in W_p^m(\Omega)$ such that $B_j(x, D)u = g_j$ $(j = 1, \ldots, m)$ on $\partial\Omega$. The proof goes as follows: first prove the statement for $g_j \in C^m(\partial\Omega)$ and $\{(\partial/\partial\nu)^{j-1}\}_{j=1}^m$, then the general case will be obtained by using (3.75) with $\{C_j(x, D)\}_{j=1}^m = \{(\partial/\partial\nu)^{j-1}\}_{j=1}^m$ and by approximating g_j by a smooth function.

Theorem 3.7.1 Let $A(x, D)$ be an elliptic operator of order m and $\{B_j(x, D)\}_{j=1}^{m/2}$ a system of normal boundary differential operators such that their coefficients are sufficiently smooth and the order of $B_j(x, D)$ is m_j. The formal adjoint of $A(x, D)$ is denoted by $A'(x, D)$. Choose arbitrary boundary differential operators with smooth coefficients, $\{S_j(x, D)\}_{j=1}^{m/2}$, in such a way that $\{B_1, \ldots, B_{m/2}, S_1, \ldots, S_{m/2}\}$ forms a Dirichlet system of order m. Then there exists another unique Dirichlet system $\{B_1', \ldots, B_{m/2}', S_1', \ldots, S_{m/2}'\}$, for which the following equality holds:

$$\int_\Omega A(x, D)u \cdot \bar{v} \, dx - \int_\Omega u \cdot \overline{A'(x, D)v} \, dx$$

$$= \sum_{j=1}^{m/2} \int_{\partial\Omega} S_j(x, D)u \cdot \overline{B_j'(x, D)v} \, d\sigma$$

$$- \sum_{j=1}^{m/2} \int_{\partial\Omega} B_j(x, D)u \cdot \overline{S_j'(x, D)v} \, d\sigma. \tag{3.76}$$

Let m'_j be the order of $B'_j(x, D)$, then S'_j and S_j are of order $m - m_j - 1$ and $m - m'_j - 1$, respectively. Though $\{B'_j(x, D)\}_{j=1}^{m/2}$ depends on the choice of $\{S_j(x, D)\}_{j=1}^{m/2}$, different $\{B'_j\}$ obtained through different $\{S_j\}$ are equivalent to each other in the following sense. Suppose $\{S_j^{(1)}\}_{j=1}^{m/2}$ and $\{S_j^{(2)}\}_{j=1}^{m/2}$ provide $\{B_j'^{(1)}\}_{j=1}^{m/2}\{B_j'^{(2)}\}_{j=1}^{m/2}$, respectively. Then $B_j'^{(1)}(x, D)u|_{\partial\Omega} = 0$ $(j = 1, \ldots, m/2)$ and $B_j'^{(2)}(x, D)u|_{\partial\Omega} = 0$ $(j = 1, \ldots, m/2)$ for $u \in C^m(\bar{\Omega})$ are equivalent. In case $u \in C^m(\bar{\Omega})$, the equality

$$\int_\Omega A(x, D)u \cdot \bar{v}\, dx = \int_\Omega u \cdot \overline{A'(x, D)v}\, dx \tag{3.77}$$

holds for all $v \in C^m(\bar{\Omega})$ satisfying $B'_j(x, D)v = 0$ $(j = 1, \ldots, m/2)$ in $\partial\Omega$ if and only if $B_j(x, D)u = 0$ $(j = 1, \ldots, m/2)$ is satisfied in $\partial\Omega$. On the other hand, if we consider the case $v \in C^m(\bar{\Omega})$, the equality (3.77) holds for all $u \in C^m(\Omega)$ satisfying $B_j(x, D)u = 0$ $(j = 1, \ldots, m/2)$ in $\partial\Omega$ if and only if $B'_j(x, D)v = 0$ $(j = 1, \ldots, m/2)$ is satisfied in $\partial\Omega$.

The proof of the theorem is obtained by mapping the intersection of a neighbourhood of a boundary point and Ω into a semi-ball and by integrating by parts. For the details see Schechter [154] or Lions and Magenes [15], pp. 115–121.

Definition 3.7.3 *$(A'(x, D), \{B'_j(x, D)\}_{j=1}^{m/2}, \Omega)$ is called the boundary-value problem adjoint to $(A(x, D), \{B_j(x, D)\}_{j=1}^{m/2}, \Omega)$. $B'_j(x, D)u|_{\partial\Omega} = 0$ $(j = 1, \ldots, m/2)$ is called the boundary condition adjoint to $B_j(x, D)u|_{\partial\Omega} = 0$ $(j = 1, \ldots, m/2)$ with respect to $A(x, D)$.*

If $\{B_j(x, D)\}_{j=1}^{m/2} = \{(\partial/\partial\nu)^{j-1}\}_{j=1}^{m/2}$, i.e., if $B_j(x, D)u|_{\partial\Omega} = 0$ $(j = 1, \ldots, m/2)$ is a Dirichlet boundary condition, then its adjoint boundary condition is necessarily of Dirichlet type. If $A(x, D)$ is properly elliptic, so also is $A'(x, D)$, and if $(A(x, D), \{B_j(x, D)\}_{j=1}^{m/2}, \Omega)$ satisfies the complementing condition, then so does its adjoint boundary problem, and vice versa.

Here is one of the results obtained by Schechter [154].

Theorem 3.7.2 *Let Ω be a bounded region of class C^∞ and $A(x, D)$ a properly elliptic operator of order m, defined on $\bar{\Omega}$, with coefficients belonging to $C^\infty(\bar{\Omega})$. Further, let $\{B_j(x, D)\}_{j=1}^{m/2}$ be a set of normal boundary differential operators such that $B_j(x, D)$ is of order m_j, $m_j < m$, and its coefficients belong to $C^\infty(\partial\Omega)$. Assume that $(A(x, D), \{B_j(x, D)\}_{j=1}^{m/2}, \Omega)$ satisfies the complementing condition. Its adjoint boundary-value problem* *is denoted by* $(A'(x, D), \{B'_j(x, D)\}_{j=1}^{m/2}, \Omega)$. *Then the boundary-value problem*

$$A(x, D)u(x) = f(x), \qquad x \in \Omega,$$

$$B_j(x, D)u(x) = g_j(x), \qquad x \in \partial\Omega, \qquad j = 1, \ldots, m/2,$$

for arbitrary $f \in C^\infty(\bar{\Omega})$ *and* $g_j \in C^\infty(\partial\Omega)$ $(j = 1, \ldots, m/2)$ *has a solution* $u \in C^\infty(\bar{\Omega})$ *if and only if* $v = 0$ *is the only solution, belonging to* $C^\infty(\bar{\Omega})$, *of the problem*

$$A'(x, D)v(x) = 0, \qquad x \in \Omega,$$

$$B'_j(x, D)v(x) = 0, \qquad x \in \partial\Omega, \qquad j = 1, \cdots, m/2\},$$

The same is also true if the roles of $(A(x, D), \{B_j(x, D)\})$ *and* $(A'(x, D), \{B'_j(x, D)\})$ *are interchanged.*

The technique of L^2 estimates was used for the proof of this theorem. However, the result itself is correct independently of such a function space.

3.8 Application 3. Parabolic mixed problems

Let $(A(x, D), \{B_j(x, D)\}_{j=1}^{m/2}, \Omega)$ be an elliptic boundary-value problem, as in the preceding section, and assume $\{B_j(x, D)\}_{j=1}^{m/2}$ to be normal. The operator A is defined as follows:

$$D(A) = \{u \in W_p^m(\Omega) : B_j(x, D)u(x) = 0$$

$$\text{on} \quad \partial\Omega, \qquad j = 1, \ldots, m/2\}, \tag{3.78}$$

$$(Au)(x) = A(x, D)u(x) \quad \text{for} \quad u \in D(A),$$

where $1 < p < \infty$. A is a closed operator in $L^p(\Omega)$ and the domain $D(A)$ is dense in $L^p(\Omega)$ since it contains $C_0^\infty(\Omega)$. A is defined for each p, so it is written as A_p when p is to be specified.

Definition 3.8.1 *Let A be a closed operator in a Banach space and θ a certain angle. Suppose there exists a constant $C > 0$ such that the half-line* $\{\lambda : \arg \lambda = \theta, |\lambda| > C\}$ *is contained in $\rho(A)$ and* $\|(A - \lambda)^{-1}\| \leq C|\lambda|^{-1}$ *holds for all λ on the half-line. Then the half-line $\arg \lambda = \theta$ is called the* ray of minimal growth *of the resolvent of A.*

Theorem 3.8.1 (Agmon [19]) *Let A be the operator defined by (3.78) and $\theta \in [0, 2\pi)$. The following two conditions are sufficient for the half-line $\arg \lambda = \theta$ to be the ray of minimal growth of the resolvent of A, and they are also necessary if $p = 2$.*

(i) $\arg \mathring{A}(x, \xi) = 0$ for all $x \in \bar{\Omega}$ and all real $\xi \neq 0$.

(ii) Let x be an arbitrary point of $\partial\Omega$, ξ any real vector tangent to $\partial\Omega$ at x, ν the normal vector to $\partial\Omega$ at x, and $\arg \lambda = \theta$. The polynomial $\mathring{A}(x, \xi + \tau\nu) - \lambda$ of one variable τ has just $m/2$ roots, $\tau_1^+(\xi, \lambda), \ldots, \tau_{m/2}^+(\xi, \lambda)$, with positive imaginary part and the remaining $m/2$ with negative imaginary part. $m/2$ polynomials $B_j(x, \xi + \tau\nu)$ $(j = 1, \ldots, m/2)$ are linearly independent modulo $\prod_{k=1}^{m/2} (\tau - \tau_k^+(\xi, \lambda))$.

Remark 3.8.1 If the boundary condition is of Dirichlet type, then (i) implies (ii).

Proof of Theorem 3.8.1 Let t be a real parameter and denote $D_t = -i\partial/\partial t$, then the condition (i) is equivalent to saying that $L(x, D_x, D_t) = A(x, D_x) - e^{i\theta} D_t^m$ is elliptic in $Q = \Omega \times (-\infty, \infty)$. Also, (ii) is equivalent to the statement that $(L(x, D_x, D_t), \{B_j(x, D_x)\}, Q)$ satisfies the complementing condition. Because of later requirements, we will prove the following lemma by slightly generalizing Agmon's proof.

Lemma 3.8.1 Let u be an arbitrary element of $W_p^m(\Omega)$ and g_j any function of $W_p^{m-m_i}(\Omega)$ which coincides with $B_j(x, D)u$ in $\partial\Omega$. Further, let $\arg \lambda = \theta$. Then, under the conditions of the theorem, there exists a constant C_0 such that the inequality

$$\sum_{j=0}^m |\lambda|^{(m-j)/m} \|u\|_{j,p}$$

$$\leq C_0 \left\{ \|(A(x, D) - \lambda)u\|_p + \sum_{j=1}^{m/2} |\lambda|^{(m-m_j)/m} \|g_j\|_p + \sum_{j=1}^{m/2} \|g_j\|_{m-m_j,p} \right\} \quad (3.79)$$

holds for all λ satisfying $|\lambda| > C_0$.

Proof Let ζ be a function, belonging to $C^\infty(-\infty, \infty)$, such that $\zeta(t) = 0$ for $|t| > 1$ and $\zeta(t) = 1$ for $|t| < \frac{1}{2}$. If $r > 0$ and $u \in W_p^m(\Omega)$, the function $v(x, t) = \zeta(t)e^{irt}u(x)$ belongs to $W_p^m(Q)$ and its support is contained in $\bar{\Omega} \times [-1, 1]$. Therefore, by the a priori estimate, we have

$$\|v\|_{m,p,Q} \leq C \left\{ \|L(x, D_x, D_t)v\|_{p,Q} + \sum_{j=1}^{m/2} [B_j(x, D_x)v]_{m-m_j-1/p,\partial Q} \right.$$
$$\left. + \|v\|_{p,Q} \right\}. \quad (3.80)$$

Since

$$L(x, D_x, D_t)v(x, t) = \zeta(t)e^{i\mu t}(A(x, D_x) - r^m e^{i\theta})u(x)$$
$$+ e^{i\theta} \sum_{k=0}^{m-1} \binom{m}{k} D_t^{m-k}\zeta(t) \cdot r^k e^{irt}u(x)$$

by Leibnitz's formula, we get the estimate

$$\|L(x, D_x, D_t)v\|_{p,Q} \leq \|(A(x, D) - r^m e^{i\theta})u\|_p + C \sum_{k=0}^{m-1} r^k \|u\|_p. \tag{3.81}$$

Now, $(x, t) \in \partial Q$ is equivalent to $x \in \partial \Omega$ and we then have

$$B_j(x, D_x)v(x, t) = \zeta(t)e^{i\mu t}B_j(x, D)u(x) = \zeta(t)e^{i\mu t}g_j(x).$$

Thus, by the help of the interpolation inequality (1.16), the estimate

$$[B_j(x, D)v]_{m-m_j-1/p, \partial Q} \leq \|\zeta e^{i\mu t}g_j\|_{m-m_j, Q}$$

$$\leq C \sum_{k=0}^{m-m_j} r^{m-m_j-k} \|g_j\|_{k,p}$$

$$\leq C(r^{m-m_j} \|g_j\|_p + \|g_j\|_{m-m_j, p}) \tag{3.82}$$

holds for $r > 1$. On the other hand, it is found that

$$\|v\|_{m,p,Q}^p = \sum_{|\alpha|+k \leq m} \int_{-\infty}^{\infty} \int_{\Omega} |D_x^\alpha D_t^k v(x, t)|^p \, dx \, dt$$

$$\geq \sum_{|\alpha|+k \leq m} \int_{-1/2}^{1/2} \int_{\Omega} |D_t^k e^{irt} D^\alpha u(x)|^p \, dx \, dt$$

$$= \sum_{k=0}^{m} r^{pk} \sum_{|\alpha| \leq m-k} \int_{\Omega} |D^\alpha u(x)|^p \, dx = \sum_{j=0}^{m} r^{p(m-j)} \|u\|_{j,p}^p. \tag{3.83}$$

Collecting (3.80)–(3.83), we finally obtain

$$\sum_{j=0}^{m} r^{m-j} \|u\|_{j,p} \leq C \left\{ \|(A(x, D) - r^m e^{i\theta})u\|_p + \sum_{k=0}^{m-1} r^k \|u\|_p \right.$$

$$\left. + \sum_{j=1}^{m/2} (r^{m-m_j} \|g_j\|_p + \|g_j\|_{m-m_j, p}) + \|u\|_p \right\}.$$

For sufficiently large r, we may disregard the second and the fourth terms in the curly brackets on the right-hand side, since then they become much smaller than the terms on the left-hand side. By putting $\lambda = r^m e^{i\theta}$, the inequality (3.79) is established. □

If $u \in D(A)$, it is possible to take $g_j = 0$ and then (3.79) implies that

$$|\lambda| \|u\|_p + \|u\|_{m,p} \leq C_0 \|(A - \lambda)u\|_p, \tag{3.84}$$

which shows that the operator $A - \lambda$ has a continuous inverse for $\arg \lambda = \theta$ and $|\lambda| > C_0$. Therefore, $R(A - \lambda)$ is a closed subspace of $L^p(\Omega)$. Next, in order to see that $\lambda \in \rho(A)$, we must have $R(A - \lambda) = L^p(\Omega)$. First, we observe that, since $\{\mathring{A}(x, \xi): x \in \bar{\Omega}, |\xi| = 1\}$ is a compact set and has no intersection with the half-line $\{re^{i\theta}: 0 \leq r < \infty\}$, it also does not intersect with the half-line $\{re^{i\theta}: 0 \leqslant r < \infty, |\varphi - \theta| \leqslant \delta\}$ for some $\delta > 0$. To begin with, we assume that Ω is of class C^∞. Each coefficient a_α ($|\alpha| = m$) of the principal part of $A(x, D)$ is approximated uniformly by a sequence of functions $\{a_\alpha^{(k)}\}$ belonging to $C^\infty(\bar{\Omega})$. Also, the coefficient of lower order, a_α ($|\alpha| < m$), is approximated by a sequence of uniformly-bounded functions $\{a_\alpha^{(k)}\}$ belonging to $C^\infty(\bar{\Omega})$ in such a way that $a_\alpha^{(k)}(x) \to a_\alpha(x)$ almost everywhere. For instance, let a be one such coefficient and φ_δ a mollifier, then $\varphi_\delta * a \in C^\infty(\bar{\Omega})$, $\|\varphi_\delta * a\|_\infty \leqslant \|a\|_\infty$, and $\varphi_\delta * a(x) \to a(x)$ at each Lebesgue point x of a (see the end of Chapter 1). Put $A^{(k)}(x, D) = \sum_{|\alpha| \leqslant m} a_\alpha^{(k)}(x)D^\alpha$. Nothing is lost by supposing that $\{\mathring{A}^{(k)}(x, \xi): x \in \bar{\Omega}, |\xi| = 1\} \cap \{re^{i\theta}: 0 \leqslant r < \infty\}$ is void. Next, each $b_{j\beta}$ ($|\beta| \leqslant m_j, j = 1, \ldots, m/2$) is extended to Ω so that $b_{j\beta} \in C^{m-m_j}(\bar{\Omega})$. $b_{j\beta}$ is approximated by a sequence of functions $\{b_{j\beta}^{(k)}\}$ belonging to $C^\infty(\bar{\Omega})$ in the norm of $C^{m-m_j}(\bar{\Omega})$. Put $B_j^{(k)}(x, D) = \sum_{|\beta| \leqslant m_j} b_{j\beta}^{(k)}(x)D^\beta$. We need only consider the case of sufficiently large k, so that it may be supposed that the assumptions of the theorem about $A^{(k)}(x, D)$ and $\{B_j^{(k)}(x, D)\}$ are satisfied uniformly in k. Accordingly, by Lemma 3.8.1, there exists a constant C_0 independent of k such that, for $\arg \lambda = \theta$ and $|\lambda| \geqslant C_0$, the inequality

$$\sum_{j=0}^{m} |\lambda|^{(m-j)/m} \|u\|_{j,p} \leqslant C_0 \Big\{ \|(A^{(k)}(x, D) - \lambda)u\|_p$$

$$+ \sum_{j=1}^{m/2} |\lambda|^{(m-m_j)/m} \|g_j\|_p + \sum_{j=1}^{m/2} \|g_j\|_{m-m_j,p} \Big\} \qquad (3.85)$$

holds. Here g_j is any function, belonging to $W_p^{m-m_j}(\Omega)$, which satisfies $B_j^{(k)}(x, D)u = g_j$ in $\partial\Omega$. Let $(A'^{(k)}(x, D), \{B_j'^{(k)}(x, D)\})$ for each k be the adjoint boundary-value problem. Then, with $-\theta$ replacing θ, all the assumptions of the theorem are satisfied for each k, but this time the uniformity in k is obscure. Thus, for each k, there exists a constant $C_k \geqslant C_0$ such that, for $\arg \lambda = -\theta$ and $|\lambda| \geqslant C_k$, the inequality

$$\sum_{j=0}^{m} |\lambda|^{(m-j)/m} \|v\|_{j,q} \leqslant C_k \|(A'^{(k)}(x, D) - \lambda)v\|_q$$

holds. Here v is an arbitrary function, belonging to $W_q^m(\Omega)$, which satisfies $B_j'^{(k)}(x, D)v = 0$ ($j = 1, \ldots, m/2$) in $\partial\Omega$, and q is a certain real number satisfying $1 < q < \infty$. Therefore, for any λ_k which satisfies $\arg \lambda_k = \theta$ and $|\lambda_k| \geqslant C_k$, the adjoint boundary-value problem

$$(A'^{(k)}(x, D) - \overline{\lambda_k})v(x) = 0, \qquad x \in \Omega,$$

$$B_j'^{(k)}(x, D)v(x) = 0, \qquad x \in \partial\Omega, \qquad j = 1, \ldots, m/2,$$

admits the only solution $v = 0$ in $C^\infty(\bar{\Omega})$. Hence, by Theorem 3.7.2, for any $f \in C^\infty(\bar{\Omega})$, the boundary-value problem

$$(A^{(k)}(x, D) - \lambda_k)u(x) = f(x), \qquad x \in Q,$$
$$B_j^{(k)}(x, D)u(x) = 0, \qquad x \in \partial\Omega, \qquad j = 1, \ldots, m/2, \tag{3.86}$$

has a solution $u \in C^\infty(\bar{\Omega})$. From $(A(x, D), \{B_j(x, D)\}, \Omega)$, we have defined by (3.78) an operator A in $L^p(\Omega)$. In the same way, an operator $A^{(k)}$ can be defined from $(A^{(k)}(x, D), \{B_j^{(k)}(x, D)\}, \Omega)$. Then (3.86) is equivalent to $(A^{(k)} - \lambda_k)u = f$ with $u \in D(A^{(k)})$. Accordingly, $R(A^{(k)} - \lambda_k) \supset C^\infty(\bar{\Omega})$ and, hence, $R(A^{(k)} - \lambda_k)$ is dense in $L^p(\Omega)$. Moreover, since it is a closed subspace, we have $R(A^{(k)} - \lambda_k) = L^p(\Omega)$, which implies that $\lambda_k \in \rho(A^{(k)})$. Since $\|(A^{(k)} - \lambda_k)^{-1}\| \leq C_0/|\lambda_k|$ by (3.85), it is found by the power-series expansion of $(A^{(k)} - \lambda)^{-1}$ around λ_k that a circle of radius $|\lambda_k|/C_0$, centred at λ_k, is contained in $\rho(A^{(k)})$. The same manipulation works around any point of the intersection of this circle and the half-line $\{\arg \lambda = \theta\}$. By repeating these processes, we find that there exist positive constants $C_0' \geq C_0$ and $\delta' > 0$ such that

$$\{\lambda : |\arg \lambda - \theta| \leq \delta', \quad |\lambda| \geq C_0'\} \subset \rho(A^{(k)}). \tag{3.87}$$

In addition, at each point of the set in the left member of (3.87), the inequality $\|(A^{(k)} - \lambda)^{-1}\| \leq C_0'/|\lambda|$ is satisfied for all k. Next, let λ be an arbitrary complex number satisfying $\arg \lambda = \theta$ and $|\lambda| \geq C_0'$, and f any element of $L^p(\Omega)$, then, by (3.87), there exists $u_k \in D(A^{(k)})$ for every k such that $(A^{(k)} - \lambda)u_k = f$. Since $\|u_k\|_{m,p} \leq C_0\|f\|_p$ by (3.85), the sequence $\{u_k\}$ is bounded in $W_p^m(\Omega)$. Since $W_p^m(\Omega)$ is reflexive, we can pick out, by applying Lemma 1.2.2, a subsequence of $\{u_k\}$ which converges weakly in $W_p^m(\Omega)$ but strongly in $W_p^{m-1}(\Omega)$. For notational simplicity, we suppose that $u_k \to u$ weakly in $W_p^m(\Omega)$ and $u_k \to u$ strongly in $W_p^{m-1}(\Omega)$. For each k, l, we have

$$\begin{aligned}
(A^{(k)}(x, D) - \lambda)(u_k - u_l) &= f - (A^{(k)}(x, D) - \lambda)u_l \\
&= (A^{(l)}(x, D) - \lambda)u_l - (A^{(k)}(x, D) - \lambda)u_l \\
&= (A^{(l)}(x, D) - A^{(k)}(x, D))u_l \\
&= \sum_{|\alpha| \leq m} (a_\alpha^{(l)}(x) - a_\alpha^{(k)}(x))D^\alpha u_l.
\end{aligned}$$

Let us examine the convergence as $k, l \to \infty$. For $|\alpha| = m$,

$$\|(a_\alpha^{(l)} - a_\alpha^{(k)})D^\alpha u_l\|_p \leq \sup_\Omega |a_\alpha^{(l)}(x) - a_\alpha^{(k)}(x)| \|u_l\|_{m,p} \to 0, \qquad k, l \to \infty,$$

whereas, for $|\alpha| < m$,

$$\|(a_\alpha^{(l)} - a_\alpha^{(k)})D^\alpha u_l\|_p \leq \|(a_\alpha^{(l)} - a_\alpha^{(k)})(D^\alpha u_l - D^\alpha u)\|_p + \|(a_\alpha^{(l)} - a_\alpha^{(k)})D^\alpha u\|_p.$$

The first term on the right-hand side of the last inequality converges to 0 as $k, l \to \infty$, because it does not exceed

$$\sup_{\Omega} |a_\alpha^{(l)}(x) - a_\alpha^{(k)}(x)| \, \|u_l - u\|_{m-1,p}.$$

The second term also converges to 0, since it is given explicitly by

$$\left\{ \int_\Omega |a_\alpha^{(l)}(x) - a_\alpha^{(k)}(x)|^p \, |D^\alpha u(x)|^p \, dx \right\}^{1/p},$$

whose integrand is majorized by an absolutely integrable function independent of k, l and, moreover, converges to 0 almost everywhere as $k, l \to \infty$. Thus, we obtain

$$\lim_{k,l\to\infty} \|(A^{(k)}(x, D) - \lambda)(u_k - u_l)\|_p = 0. \tag{3.88}$$

On the other hand, on $\partial\Omega$ we have

$$B_j^{(k)}(x, D)(u_k - u_l) = -B_j^{(k)}(x, D)u_l$$
$$= (B_j^{(l)}(x, D) - B_j^{(k)}(x, D))u_l \equiv g_{j,k,l}.$$

It is easy to see that

$$\lim_{k,l\to\infty} \|g_{j,k,l}\|_{m-m_j,p} = 0. \tag{3.89}$$

By applying (3.85) to $u_k - u_l$, we find that

$$\sum_{j=0}^m |\lambda|^{(m-j)/m} \|u_k - u_l\|_{j,p} \leqslant C_0 \Big\{ \|(A^{(k)}(x, D) - \lambda)(u_k - u_l)\|_p$$
$$+ \sum_{j=1}^{m/2} |\lambda|^{(m-m_j)/m} \|g_{j,k,l}\|_p + \sum_{j=1}^{m/2} \|g_{j,k,l}\|_{m-m_j,p} \Big\}.$$

The right-hand side converges to 0 as $k, l \to \infty$ by (3.88) and (3.89). Thus $\{u_k\}$ converges strongly in $W_p^m(\Omega)$ to a function $u \in W_p^m(\Omega)$. It is readily seen that the function u is a solution of $(A - \lambda)u = f$. Therefore, $\lambda \in \rho(A)$, i.e.,

$$\{\lambda : \arg \lambda = \theta, |\lambda| \geqslant C_0'\} \subset \rho(A).$$

Moreover it has been found that $\|(A - \lambda)^{-1}\| \leqslant C_0/|\lambda|$ holds at each point of the set in the left-hand member. When Ω is not of class C^∞, we only need to map Ω into a region of class C^∞ by a transformation of class C^m, to solve there an equation transformed from $(A - \lambda)u = f$, and then to come back again to Ω. The proof for the rest of the theorem, that is, for the part pertaining to the case $p = 2$, will be omitted. This completes the proof of Theorem 3.8.1. $\square$

An operator $A(x, D)$ is said to be *strongly elliptic* if Re $\mathring{A}(x, \xi) \neq 0$ for all $x \in \bar{\Omega}$ and all non-zero real vectors ξ. A strongly elliptic operator is necessarily properly elliptic. Hereafter, by a strongly elliptic operator we always mean an operator $A(x, D)$ for which Re $\mathring{A}(x, \xi) > 0$ for all $x \in \bar{\Omega}$ and all $\xi \neq 0$. Then, as is easily seen, there exists an angle $\theta_0 \in [0, \pi/2)$ such that $\mathring{A}(x, \xi) \neq e^{i\theta}$ for each $\theta \in [\theta_0, 2\pi - \theta_0]$.

Theorem 3.8.2 *Let $A(x, D)$ be a strongly elliptic operator of order m, $\{B_j(x, D)\}_{j=1}^{m/2}$ a set of normal boundary differential operators, and θ_0 some angle as above. Then, if the assumptions of Theorem 3.8.1 are satisfied for each angle $\theta \in [\theta_0, 2\pi - \theta_0]$, the operator $-A$ defined by (3.78) generates an analytic semigroup in $L^p(\Omega)$. In particular, if $\{B_j(x, D)\}_{j=1}^{m/2}$ is of Dirichlet type, then $-A$ for any strongly elliptic operator $A(x, D)$ of order m generates an analytic semigroup in $L^p(\Omega)$.*

Remark 3.8.2 The boundedness of Ω is not an indispensable restriction. Assume that Ω is uniformly regular of class C^m, but of class C^{2m} locally. Further, let the following conditions all be satisfied. The coefficients of $A(x, D)$ belong to $L^\infty(\Omega)$ and particularly those of the principal part belong to $B^0(\bar{\Omega})$. The same is assumed for the coefficients of the operator $A'(x, D)$ which is formally adjoint to $A(x, D)$. The coefficients of $B_j'(x, D)$, as well as $B_j(x, D)$, belong to $B^{m-m_j}(\partial\Omega)$, where $\{B_j'(x, D)\}$ are adjoint boundary differential operators constructed by choosing the $\{S_j\}$ appropriately in Theorem 3.7.1. Then, if the conditions of Theorem 3.8.2 are satisfied by $(A(x, D), \{B_j(x, D)\})$ uniformly in $\bar{\Omega}$, the following proposition due to Browder [36] can be proved. Let, as above, an operator A_p be defined by $(A(x, D), \{B_j(x, D)\})$. Similarly, from $(A'(x, D), \{B_j'(x, D)\})$, we define an operator A_q' in $L^q(\Omega)$ with $1/p + 1/q = 1$. Then $(A_p)^* = A_q'$ and $-A_p$ generates an analytic semigroup in $L^p(\Omega)$.

Remark 3.8.3 It has been shown by Kielhöfer [94] and von Wahl [175] that, in the space of Hölder continuous functions, an operator corresponding to A in Theorem 3.8.2 generates a semigroup which is not of class C_0.

4

Temporally inhomogeneous equations

4.1 Fundamental solutions of temporally inhomogeneous equations

In the preceding chapter, we discussed an equation with an operator A as its coefficient, which is independent of t; in this chapter, we consider the initial-value problem

$$\mathrm{d}u(t)/\,\mathrm{d}t = A(t)u(t) + f(t), \qquad 0 \le t \le T, \tag{4.1}$$

$$u(0) = u_0, \tag{4.2}$$

where the operator A depends on t. We always assume that $A(t)$ for each t generates a semigroup. Similar to the semigroup $\exp(tA)$ for A independent of t, we can think of the following operator $U(t, s)$:

$$\begin{cases} U(t, s) \text{ is a strongly continuous function, defined on} \\ 0 \le s \le t \le T, \text{ which takes on values in } B(X), \end{cases} \tag{4.3}$$

$$U(t, r)U(r, s) = U(t, s) \quad \text{for} \quad 0 \le s \le r \le t \le T, \tag{4.4}$$

$$U(s, s) = I \quad \text{for each} \quad s \in [0, T], \tag{4.5}$$

$$(\partial/\partial t)U(t, s) = A(t)U(t, s), \tag{4.6}$$

$$(\partial/\partial s)U(t, s) = -U(t, s)A(s). \tag{4.7}$$

Since, in general, the equations (4.6) and (4.7) involve unbounded operators on both sides, we assume that they hold in a dense subspace which is to be determined for each equation. Such an operator-valued function $U(t, s)$ is called a *fundamental solution* of (4.1). If $A(t) = A$ is independent of t, $U(t, s) = \exp((t-s)A)$ is the fundamental solution, and (4.6) and (4.7) hold on $D(A)$. When the fundamental solution exists, one expects that the solution of (4.1), (4.2) can be written as

$$u(t) = U(t, 0)u_0 + \int_0^t U(t, s)f(s)\,\mathrm{d}s. \tag{4.8}$$

89

Kato [76, 77] was the first to succeed in constructing the fundamental solution of temporally inhomogeneous equations; the results were later revised by himself [87]. In the next section, the construction of the fundamental solution is explained principally following the arguments given in [87]. Reference [87] aims mainly at applications to the initial-value problem of 'hyperbolic' equations, as its title suggests, but applications are possible beyond the field of hyperbolic equations. The reader should refer to Yosida [18] in addition to [87].

4.2 Admissible subspace with respect to a generator

Let X be a Banach space and denote its norm by $\|\cdot\|$. Let Y be a dense subspace of X and assume Y to be by itself a Banach space with the norm denoted by $\|\cdot\|_Y$. Suppose the topology of Y is stronger than that of X, so that there exists a constant C such that $\|v\| \leqslant C \|v\|_Y$ for all $v \in Y$. We also denote the norms of $B(X)$ and $B(Y)$ by $\|\cdot\|$ and $\|\cdot\|_Y$, respectively. In this section, we consider as preliminaries the case when $A \in G(X)$ and it is independent of t.

Definition 4.2.1 *Let T be an operator from X into X. An operator $\tilde{T}$ defined by*

$$\begin{cases} D(\tilde{T}) = \{u \in D(T) \cap Y : Tu \in Y\}, \\ \tilde{T}u = Tu \quad \text{for} \quad u \in D(T) \end{cases}$$

is said to be the part *of T in Y. In particular, if T maps $D(T) \cap Y$ into Y, then $\tilde{T}$ coincides with the restriction of T to $D(T) \cap Y$.*

Definition 4.2.2 *Let $A \in G(X)$. Y is called* admissible with respect to A *or simply A-admissible if $\exp(tA)$ maps Y into Y and if the restriction of $\exp(tA)$ to Y forms a semigroup in Y. In this section, we assume $A \in G(X, M, \beta)$ throughout. Hence, $D(A)$ is dense, any $\lambda > \beta$ belongs to $\rho(A)$, and*

$$\|(A - \lambda)^{-n}\| \leqslant M(\lambda - \beta)^{-n} \tag{4.9}$$

for all $\lambda > \beta$ and $n = 1, 2, \ldots$. Furthermore, the semigroup generated by A satisfies $\|\exp(tA)\| \leqslant Me^{\beta t}$.

Proposition 4.2.1 *Y is A-admissible if and only if there exist numbers $\tilde{M}$ and $\tilde{\beta}$ satisfying the following conditions.*

(i) *For $\lambda > \max(\beta, \tilde{\beta})$, the resolvent $(A - \lambda)^{-1}$ maps Y into Y; hence, so does $(A - \lambda)^{-n}$, and the following inequality holds:*

$$\|(A - \lambda)^{-n}\|_Y \leq \tilde{M}(\lambda - \tilde{\beta})^{-n}, \qquad n = 1, 2, \ldots . \tag{4.10}$$

(ii) *$(A - \lambda)^{-1}Y$ is dense in Y for each $\lambda > \max(\beta, \tilde{\beta})$.*

In this case, if we denote the part of A in Y by $\tilde{A}$, then $\tilde{A} \in G(Y, \tilde{M}, \tilde{\beta})$ and $\exp(t\tilde{A})$ coincides with the restriction of $\exp(tA)$ to Y. Hence, we have $\|\exp(t\tilde{A})\|_Y = \|\exp(tA)\|_Y \leq \tilde{M}e^{\tilde{\beta}t}$. If Y is reflexive, the condition (ii) follows from (i).

Proof Let Y be A-admissible. Let $\tilde{A}$ denote a generator of the restriction of $\exp(tA)$ to Y. Then, there exist numbers $\tilde{M}$ and $\tilde{\beta}$ such that $\tilde{A} \in G(Y, \tilde{M}, \tilde{\beta})$. Since $\exp(tA)v = \exp(t\tilde{A})v$ for $t \geq 0$ and $v \in Y$, if $\lambda > \max(\beta, \tilde{\beta})$, we obtain from (3.6) that $(A - \lambda)^{-1}v = (\tilde{A} - \lambda)^{-1}v \in Y$, so that

$$\|(A - \lambda)^{-n}\|_Y = \|(\tilde{A} - \lambda)^{-n}\|_Y \leq \tilde{M}(\lambda - \tilde{\beta})^{-n},$$

which provides the proof of (i). Also, $(A - \lambda)^{-1}Y = (\tilde{A} - \lambda)^{-1}Y = D(\tilde{A})$ is dense in Y. Let $\tilde{D}$ be the domain of the part of A in Y. For $v \in D(\tilde{A})$, the limit

$$\lim_{t \to +0} t^{-1}(e^{tA}v - v) = \lim_{t \to +0} t^{-1}(e^{t\tilde{A}}v - v)$$

exists in the strong topology of Y and, hence, of X, and is equal to $\tilde{A}v$. Accordingly, $v \in D(A)$ and $Av = \tilde{A}v \in Y$, so that $v \in \tilde{D}$. Conversely, suppose $v \in \tilde{D}$. From

$$\exp(t\tilde{A})v - v = \exp(tA)v - v = \int_0^t \exp(sA)Av \, ds = \int_0^t \exp(s\tilde{A})Av \, ds,$$

we obtain $v \in D(\tilde{A})$. Hence, $\tilde{A}$ coincides with the part of A in Y.

Next, we assume (i) and (ii) hold. Let $\tilde{A}$ be the part of A in Y. Put $w = (A - \lambda)^{-1}v$, where $v \in Y$ and $\lambda > \max(\beta, \tilde{\beta})$. From the assumption, $w \in Y \cap D(A)$; since $Aw = v + \lambda w \in Y$, we obtain $w \in D(\tilde{A})$. $v = (A - \lambda)w = (\tilde{A} - \lambda)w$ and $v \in Y$ is arbitrary, so that $R(\tilde{A} - \lambda) = Y$. Clearly, the inverse of $\tilde{A} - \lambda$ exists, so that $\lambda \in \rho(\tilde{A})$. Let v and w be as above, then, since $(\tilde{A} - \lambda)^{-1}v = w = (A - \lambda)^{-1}v$, we have $(\tilde{A} - \lambda)^{-1} = (A - \lambda)^{-1}$ on Y. Hence, from (4.10) it follows that

$$\|(\tilde{A} - \lambda)^{-n}\|_Y = \|(A - \lambda)^{-n}\|_Y \leq \tilde{M}(\lambda - \tilde{\beta})^{-n}.$$

Also, by (ii), $D(\tilde{A}) = (\tilde{A} - \lambda)^{-1}Y = (A - \lambda)^{-1}Y$ is dense in Y. Therefore, $\tilde{A} \in G(Y, \tilde{M}, \tilde{\beta})$; we can observe from (3.16) that $\exp(tA)v = \exp(t\tilde{A})v$ for each $v \in Y$. Hence, Y is A-admissible.

The last part of the proof proceeds as follows. Suppose Y is reflexive and (i) is satisfied. It is easy to see that $D = (A - \lambda)^{-1} Y$ is independent of λ if $\lambda > \max(\beta, \tilde{\beta})$. From the inequality (4.10) with $n = 1$, it follows that $\|\lambda(\lambda - A)^{-1}\|_Y$ is bounded as $\lambda \to \infty$. Therefore, by Theorem 1.1.7, for any $v \in Y$ there exists a sequence $\{\lambda_j\}$, tending to infinity, such that $\lambda_j(\lambda_j - A)^{-1}v$ converges weakly to an element w of Y as $\lambda_j \to \infty$. On the other hand, because of $A \in G(X)$, we obtain from (3.12) that $\lambda_j(\lambda_j - A)^{-1}v \to v$ on X. Therefore, we have $w = v$, which, together with $\lambda_j(\lambda_j - A)^{-1}v \in D$ and Corollary 2 of Theorem 1.1.8, proves (ii). $\square$

Remark 4.2.1 In the above proof, we have used Remark 3.1.4.

Proposition 4.2.2 *Let S be an isomorphism from Y onto X. Then Y is A-admissible if and only if $A_1 = SAS^{-1} \in G(X)$. In this case, $S \exp(tA)S^{-1} = \exp(tA_1)$.*

Proof Let $\tilde{A}$ be the part of A in Y.

$$D(A_1) = \{u \in X: S^{-1}u \in D(A), AS^{-1}u \in Y\}$$

$$= \{u \in X: S^{-1}u \in D(\tilde{A})\} = SD(\tilde{A}). \tag{4.11}$$

Therefore, for all $u \in D(A_1)$,

$$(A_1 - \lambda)u = SAS^{-1}u - \lambda u = S\tilde{A}S^{-1}u - \lambda u = S(\tilde{A} - \lambda)S^{-1}u.$$

From this it follows that $\lambda \in \rho(A_1)$ and $\lambda \in \rho(\tilde{A})$ are equivalent and that we have

$$(A_1 - \lambda)^{-1} = S(\tilde{A} - \lambda)^{-1}S^{-1}. \tag{4.12}$$

By Proposition 4.2.1, if Y is A-admissible, there exist numbers $\tilde{M}$ and $\tilde{\beta}$ such that $\tilde{A} \in G(Y, \tilde{M}, \tilde{\beta})$, so that, for any $\lambda > \tilde{\beta}$, we have $\lambda \in \rho(A_1)$ and

$$\|(A_1 - \lambda)^{-n}\| = \|S(\tilde{A} - \lambda)^{-n}S^{-1}\| \leq \|S\| \|S^{-1}\| \tilde{M}(\lambda - \tilde{\beta})^{-n}$$

for all $n = 1, 2, \ldots$. Also, $A_1 \in G(X)$, since $D(A_1)$ is dense in X by (4.11). Conversely, if $A_1 \in G(X)$, similar arguments lead to $\tilde{A} \in G(Y)$. Let $v \in Y$ and λ be sufficiently large, and put $(A - \lambda)^{-1}v = w$ and $(\tilde{A} - \lambda)^{-1}v = \tilde{w}$. Then, from $v = (A - \lambda)w = (\tilde{A} - \lambda)\tilde{w} = (A - \lambda)\tilde{w}$, we obtain $w = \tilde{w} \in Y$. Thus, $(A - \lambda)^{-1} = (\tilde{A} - \lambda)^{-1}$ on Y, from which it is easy to see that the conditions (i) and (ii) of Proposition 4.2.1 are satisfied, and, hence, Y is A-admissible. The rest of the proof follows at once from (4.12) and (3.16). $\square$

4.3 Stability of a system of generators

We consider a family of operators $A(t)$ in $G(X)$, where $0 \leqslant t \leqslant T$.

Definition 4.3.1 $\{A(t)\}$ is called stable *if there exist real numbers* $M \geqslant 1$ *and* β *such that*

$$\left\| \prod_{j=1}^{k} (A(t_j) - \lambda)^{-1} \right\| \leqslant M(\lambda - \beta)^{-k} \tag{4.13}$$

for all $\lambda > \beta$, $0 \leqslant t_1 \leqslant t_2 \leqslant \cdots \leqslant t_k \leqslant T$, *and* $k = 1, 2, \ldots$. *Here,*

$$\prod_{j=1}^{k} (A(t_j) - \lambda)^{-1} = (A(t_k) - \lambda)^{-1} (A(t_{k-1}) - \lambda)^{-1} \cdots (A(t_1) - \lambda)^{-1}.$$

From now on, we always express a product of factors containing $\{t_j\}$ *in descending order of* t_j *as above. We call* M *and* β *the* stability constants *of* $\{A(t)\}$.

Remark 4.3.1 If $A(t) \in G(X, 1, \beta)$ for each t, the family $\{A(t)\}$ is clearly stable with stability constants 1 and β.

Proposition 4.3.1 *The condition* (4.13) *is equivalent to each of the following*:

$$\left\| \prod_{j=1}^{k} \exp(s_j A(t_j)) \right\| \leqslant M e^{\beta(s_1 + \cdots + s_k)}, \qquad s_j \geqslant 0, \tag{4.14}$$

$$\left\| \prod_{j=1}^{k} (A(t_j) - \lambda_j)^{-1} \right\| \leqslant M \prod_{j=1}^{k} (\lambda_j - \beta)^{-1}, \qquad \lambda_j > \beta, \tag{4.15}$$

where $\{t_j\}$ *is as defined in Definition* 4.3.1.

Proof Assume (4.13) holds. Apply it to the product of mk factors, of which m factors have the identical t_j for each j, $1 \leqslant j \leqslant k$, multiply the inequality thus obtained by λ^{mk}, set $\lambda = m/s$ ($s > 0$), and, finally, let $m \to \infty$. Then we have the inequality (4.14) with $s_j = s$. Next, by applying the inequality obtained in this way to the number m_j of t_j's for each j, we have (4.14) with $s_j = m_j s$. Thus, we can observe that (4.14) holds for each rational s_j. For the general case, we only have to choose an approximating sequence of s_j from rational numbers. The inequality (4.15) follows immediately from (4.14) and (3.6). The condition (4.15) contains (4.13). $\square$

To test the stability, the following two propositions are useful.

Proposition 4.3.2 *Assume that there can be defined for each t a norm $\|\cdot\|_t$ equivalent to the original norm in X and that there exists a positive constant c such that*

$$\|u\|_t / \|u\|_s \leq e^{c|t-s|}$$

holds for each $u \in X$ and each $s, t \in [0, T]$. Denote by X_t a Banach space obtained by equipping X with the norm $\|\cdot\|_t$. If $A(t) \in G(X_t, 1, \beta)$ for each $t \in [0, T]$, then $\{T(t)\}$ is stable with stability constants e^{2cT} and β.

Proof We also denote the norm in $B(X_t)$ by $\|\cdot\|_t$. For $\lambda > \beta$, we have $\|(A(t) - \lambda)^{-1}\|_t \leq (\lambda - \beta)^{-1}$ and

$$\left\| \prod_{j=1}^{k} (A(t_j) - \lambda)^{-1} u \right\|_T \leq e^{c(T - t_k)} \left\| \prod_{j=1}^{k} (A(t_j) - \lambda)^{-1} u \right\|_{t_k}$$

$$\leq e^{c(T - t_k)} (\lambda - \beta)^{-1} \left\| \prod_{j=1}^{k-1} (A(t_j) - \lambda)^{-1} u \right\|_{t_k}$$

$$\leq \cdots$$

$$\leq (\lambda - \beta)^{-k} e^{c(T - t_k)} e^{c(t_k - t_{k-1})} \cdots e^{ct_1} \|u\|_0$$

$$\leq (\lambda - \beta)^{-k} e^{2cT} \|u\|_T.$$

Thus, we have shown that (4.13) with $M = e^{2cT}$ and β is satisfied for the norm $\|\cdot\|_T$. It is easy to see that the same is true for the general norm $\|\cdot\|_t$. $\square$

Proposition 4.3.3 *Assume that $\{A(t)\}$ is stable with stability constants M and β. If $B(t)$ for each t is bounded and $\|B(t)\| \leq K < \infty$, then $A(t) + B(t)$ belongs to $G(X)$ and is stable with stability constants M and $\beta + MK$.*

Proof $A(t) + B(t) \in G(X, M, \beta + MK)$ follows from Theorem 3.4.1. Since the proof is the same as that of the theorem, it is omitted. $\square$

4.4 Construction of fundamental solutions

Let X and Y be the Banach spaces explained in Section 4.2.

Theorem 4.4.1 *Suppose $A(t) \in G(X)$ for each $t \in [0, T]$ and that the following conditions are satisfied.*

  (i) *$\{A(t)\}$ is stable with stability constants M and β.*
  (ii) *Y is A-admissible for each t. Let $\tilde{A}(t)$ be the part of $A(t)$ in Y. Then $\tilde{A}(t) \in G(Y)$ and $\{\tilde{A}(t)\}$ is also stable with stability constants $\tilde{M}$ and $\tilde{\beta}$.*

(iii) $Y \subset D(A(t))$ for each $t \in [0, T]$, so that $A(t) \in B(Y, X)$. The function $A(t)$ of t is continuous in the norm of $B(Y, X)$.

Under these conditions, there exists a unique bounded-operator-valued function $U(t, s) \in B(X)$ with $0 \leq s \leq t \leq T$ having the following four properties.

(a) $U(t, s)$ is strongly continuous in s and t, $U(s, s) = I$ and $\|U(t, s)\| \leq Me^{\beta(t-s)}$.

(b) $U(t, s) = U(t, r)U(r, s)$ for $s \leq r \leq t$.

(c) $D_t^+ U(t, s)v|_{t=s} = A(s)v$ for each $v \in Y$ and each $s \in [0, T]$.

(d) $(\partial/\partial s)U(t, s)v = -U(t, s)A(s)v$ on $0 \leq s \leq t \leq T$ for each $v \in Y$.

Here D^+ signifies right differentiation and, along with $\partial/\partial s$, it is taken in the sense of the strong topology of X.

Proof Divide the interval $[0, T]$ into n equal subintervals. Let $A_n(t) = A(T[nt/T]/n)$ be a step function approximating $A(t)$, i.e., $A_n(t) = A(kT/n)$ for $kT/n \leq t \lneq (k+1)T/n$. From (iii) it follows that, as $n \to \infty$,

$$\|A_n(t) - A(t)\|_{Y \to X} \to 0 \tag{4.16}$$

uniformly on $[0, T]$. Clearly, $\tilde{A}_n(t) = \tilde{A}(T[nt/T]/n)$ is the part of $A_n(t)$ in Y, and $\{A_n(t)\}$ and $\{\tilde{A}_n(t)\}$ are stable with stability constants M, β and $\tilde{M}, \tilde{\beta}$, respectively. Let

$$\begin{cases} U_n(t, s) = \exp\left((t-s)A(kT/n)\right), & kT/n \leq s \leq t \leq (k+1)T/n, \\ U_n(t, s) = \exp\left((t-lT/n)A(lT/n)\right)\exp\left(T/nA((l-1)T/n)\right)\ldots \\ \qquad \ldots \exp\left(T/nA((k+1)T/n)\right)\exp\left(((k+1)T/n-s)A(kT/n)\right), \\ kT/n \leq s < (k+1)T/n, \quad lT/n \leq t < (l+1)T/n, \quad k < l. \end{cases} \tag{4.17}$$

As is easily seen, $U_n(t, s)$ satisfies (a) and (b). It can be observed from (ii) that $U_n(t, s)$ maps Y into Y, and also, from (iii), that, for each $v \in Y$,

$$(\partial/\partial t)U_n(t, s)v = A_n(t)U_n(t, s)v \tag{4.18}$$

when t does not coincide with the division points and

$$(\partial/\partial s)U_n(t, s)v = -U_n(t, s)A_n(s)v \tag{4.19}$$

when s does not coincide with the division points. It is also obvious, from Proposition 4.3.1, that

$$\|U_n(t, s)\| \leq Me^{\beta(t-s)}, \qquad \|U_n(t, s)\|_Y \leq \tilde{M}e^{\tilde{\beta}(t-s)}. \tag{4.20}$$

Next, let $v \in Y$. From (4.18) and (4.19), it follows that

$$U_n(t, s)v - U_m(t, s)v = -\int_s^t (\partial/\partial r)\{U_n(t, r)U_m(r, s)v\} \, dr$$

$$= \int_s^t U_n(t, r)(A_n(r) - A_m(r))U_m(r, s)v \, dr.$$

Hence, by (4.20), we have

$$\|U_n(t, s)v - U_m(t, s)v\| \leqslant M\tilde{M}e^{\gamma(t-s)}\|v\|_Y \int_s^t \|A_n(r) - A_m(r)\|_{Y \to X} \, dr,$$

where $\gamma = \max(\beta, \tilde{\beta})$. By virtue of (4.1), $U_n(t, s)v$ converges as $n \to \infty$, strongly in X and uniformly in $0 \leqslant s \leqslant t \leqslant T$. From (4.5) and the fact that Y is dense in X, it follows that, for all $u \in X$,

$$U(t, s)u = \lim_{n \to \infty} U_n(t, s)u$$

exists in the strong topology of X uniformly in $0 \leqslant s \leqslant t \leqslant T$. Since $U_n(t, s)$ satisfies (a) and (b), so, obviously, does $U(t, s)$. Next, for $v \in Y$, from

$$\|U_n(t, s)v - \exp((t-s)A(s))v\|$$

$$= \left\| \int_s^t (\partial/\partial r)\{U_n(t, r) \exp((r-s)A(s))v\} \, dr \right\|$$

$$= \left\| \int_s^t U_n(t, r)(A_n(r) - A(s)) \exp((r-s)A(s))v \, dr \right\|$$

$$\leqslant M\tilde{M}e^{\gamma(t-s)}\|v\|_Y \int_s^t \|A_n(r) - A(s)\|_{Y \to X} \, dr,$$

we obtain

$$\|U(t, s)v - \exp((t-s)A(s))v\| \leqslant M\tilde{M}e^{\gamma(t-s)}\|v\|_Y \int_s^t \|A(r) - A(s)\|_{Y \to X} \, dr.$$

$$(4.21)$$

The order of diminishment of the right-hand side of (4.21) as $t \to s$ is $o(t-s)$. Hence, by dividing both sides of (4.21) by $t-s$ and letting $t \to s$, we get (c). Similarly, we have

$$D_s - U(t, s)v|_{s=t} = -A(t)v. \tag{4.22}$$

If $s < t$, the property (c) and the strong continuity of $U(t, s)$ imply that

$$h^{-1}\{U(t, s+h)v - U(t, s)v\}$$

$$= U(t, s+h)h^{-1}\{v - U(s+h, s)v\} \to -U(t, s)A(s)v$$

as $h \to +0$, whereas, if $s \le t$, we obtain

$$h^{-1}\{U(t, s)v - U(t, s-h)v\}$$
$$= U(t, s)h^{-1}\{v - U(s, s-h)v\} \to -U(t, s)A(s)v$$

from (4.22) as $h \to +0$. Thus the property (d) has been proved.

Suppose $\{V(t, s)\}$ is another family of operator-valued functions satisfying the properties (a) to (d). For each $v \in Y$, we integrate the derivative of $V(t, r)U_n(r, s)v$ with respect to r from s to t and find

$$V(t, s)v - U_n(t, s)v = \int_s^t V(t, r)(A(r) - A_n(r))U_n(r, s)v \, dr.$$

From (4.16) and (4.20), we immediately obtain $V(t, s)v = U(t, s)v$. Hence, $V(t, s) = U(t, s)$. $\square$

The following proposition is useful for testing the condition (ii) in Theorem 4.4.1.

Proposition 4.4.1 *The condition* (ii) *holds if the following is satisfied.*

(ii′) *There exists a family* $\{S(t)\}$ *of isomorphic mappings from* Y *onto* X *such that* $S(t)A(t)S(t)^{-1} = A_1(t) \in G(X)$ *for* $0 \le t \le T$. $\{A_1(t)\}$ *is stable;* $\|S(t)\|_{Y \to X}$ *and* $\|S(t)^{-1}\|_{X \to Y}$ *are bounded on* $0 \le t \le T$; $S(t)$ *as a function of* t *is of bounded variation in the norm of* $B(Y, X)$.

Proof If (ii′) is satisfied, by Proposition 4.2.2, Y is $A(t)$-admissible for each t. Let $\tilde{A}(t)$ be the part of $A(t)$ in Y, then it follows from the proof of Proposition 4.2.2 that, for sufficiently large λ, we have

$$(\tilde{A}(t) - \lambda)^{-1} = S(t)^{-1}(A_1(t) - \lambda)^{-1}S(t),$$

so that

$$\prod_{j=1}^{k} (\tilde{A}(t_j) - \lambda)^{-1} = \prod_{j=1}^{k} S(t_j)^{-1}(A_1(t_j) - \lambda)^{-1}S(t_j). \tag{4.23}$$

If we put $P_j = (S(t_j) - S(t_{j-1}))S(t_{j-1})^{-1}$, the right-hand side of (4.23) becomes

$$S(t_k)^{-1}\{(A_1(t_k) - \lambda)^{-1}(1 + P_k)(A_1(t_{k-1}) - \lambda)^{-1}(1 + P_{k-1}) \cdots$$
$$(1 + P_2)(A_1(t_1) - \lambda)^{-1}\}S(t_1).$$

We expand the expression in curly brackets to estimate its norm. Let M_1 and β_1 be the stability constants of $\{A_1(t)\}$. Note that we need the $m + 1$ factors of M_1 in order to estimate a term containing m factors out of $\{P_j\}$, $j = 2, 3, \ldots, k$. Then, the norm of the expression in curly brackets is seen not to exceed

$$M_1(\lambda - \beta_1)^{-k}(1 + M_1 \|P_k\|) \cdots (1 + M_1 \|P_2\|).$$

Suppose $\|S(t)\|_{Y \to X} \leq c$ and $\|S(t)^{-1}\|_{X \to Y} \leq c$, and let V be a total variation of $S(t)$. By the use of the facts that $\|P_j\| \leq c \|S(t_j) - S(t_{j-1})\|_{Y \to X}$ and $1 + a < e^a$ for $a > 0$, we obtain

$$\left\| \prod_{j=1}^{k} (\tilde{A}(t_j) - \lambda)^{-1} \right\|_Y \leq c^2 M_1 e^{cM_1 V} (\lambda - \beta_1)^{-k}. \quad \square$$

In Theorem 4.4.1, the differentiability of $U(t, s)$ is not guaranteed. Under stronger conditions, we obtain a satisfactory result.

Theorem 4.4.2 *Replace the condition* (ii) *in Theorem 4.4.1 by*

(ii″) *There exists a family* $\{S(t)\}$ *of isomorphic mappings from Y onto X. $S(t)$ is strongly continuously differentiable on $[0, T]$ as a function with values in $B(Y, X)$. There exists a strongly continuous function $B(t)$ with values in $B(X)$ such that $S(t)A(t)S(t)^{-1} = A(t) + B(t)$.*

Under these conditions, besides (a), (b), (c) *and* (d) *of Theorem 4.4.1, the following properties hold.*

(e) $U(t, s)Y \subset Y$; $U(t, s)$ *is strongly continuous in Y.*

(f) *For all $v \in Y$ and all $s \in [0, T]$, $U(t, s)v$ is strongly continuously differentiable with respect to t in $[s, T]$;*

$$(\partial/\partial t)U(t, s)v = A(t)U(t, s)v.$$

Proof Since $\|dS(t)/dt\|_{Y \to X}$ is bounded on $0 \leq t \leq T$ by nature of the assumption and Theorem 1.3.1, $S(t)$ satisfies the Lipschitz condition in the norm of $B(Y, X)$, so that it is of bounded variation. From $S(t) = \{I + (S(t) - S(s))S(s)^{-1}\}S(s)$ and Theorem 1.1.15, it follows that $S(t)^{-1} \to S(s)^{-1}$ as $t \to s$ in the norm of $B(X, Y)$, so that $\|S(t)^{-1}\|_{X \to Y}$ is bounded. Hence, by Proposition 4.3.3 and Proposition 4.4.1, the condition (ii) of Theorem 4.4.1 is satisfied; there exists a unique $U(t, s)$ satisfying (a) to (d).

If we can show that

$$\begin{cases} W(t, s) = S(t)U(r, s)(S)(s)^{-1} \text{ belongs to } B(X) \text{ and} \\ \text{is strongly continuous in } s \text{ and } t, \end{cases} \tag{4.24}$$

the property (e) immediately follows. Since $A(t)U(t, s)v = A(t)S(t)^{-1}W(t, s)S(s)v$ with $v \in Y$ is strongly continuous and since, by virtue of $U(t, s)v \in Y$ and (c),

$$h^{-1}\{U(t+h, s)v - U(t, s)v\}$$
$$= h^{-1}\{U(t+h, t) - I\}U(t, s)v \rightarrow A(t)U(t, s)v$$

as $h \rightarrow +0$, the property (f) follows. Now we prove (4.24). Put $C(t) = (dS(t)/dt)S(t)^{-1}$, and then $C(t) \in B(X)$ and it is a strongly continuous function of t. Construct operator-valued functions $V(t, s)$ and $W(t, s)$ as follows:

$$\begin{cases} V(t, s) = \sum_{m=0}^{\infty} V^{(m)}(t, s), \\ V^{(0)}(t, s) = U(t, s), \\ V^{(m)}(t, s) = \int_s^t U(t, r)B(r)V^{(m-1)}(r, s)\, dr, \end{cases} \tag{4.25}$$

$$\begin{cases} W(t, s) = \sum_{m=0}^{\infty} W^{(m)}(t, s), \\ W^{(0)}(t, s) = V(t, s), \\ W^{(m)}(t, s) = \int_s^t W^{(m-1)}(t, r)C(r)V(r, s)\, dr. \end{cases} \tag{4.26}$$

Let $\|B(t)\| \leqslant K$ on $[0, T]$. Then it is easy to see that, for each m,

$$\|V^{(m)}(t, s)\| \leqslant M^{m+1}K^m e^{\beta(t-s)}(t-s)^m/m!, \tag{4.27}$$

so that the series defining $V(t, s)$ strongly converges uniformly in $0 \leqslant s \leqslant t \leqslant T$; $V(t, s)$ is strongly continuous in two variables s and t. Then same is also true for $W(t, s)$. Hence, it is enough to show that

$$S(t)U(t, s)S(s)^{-1} = W(t, s). \tag{4.28}$$

An operator $\bar{A}(t)$ defined by $\bar{A}(t) = A(t) + B(t)$ is stable with stability constants M and $\alpha = \beta + MK$ by Proposition 4.3.3. As in the proof of Theorem 4.4.1, divide the interval $[0, T]$ into n equal subintervals and let $A_n(t) = A(T[nt/T]/n)$. Similarly, define $\bar{A}_n(t)$ and $B_n(t)$. Consider an operator obtained by replacing $A(jT/n)$, $j = k, \ldots, l$, on the right-hand side of (4.17) by $\bar{A}(jT/n)$ and denote it by $V_n(t, s)$. Similar to the first inequality in (4.20), there holds the result

$$\|V_n(t, s)\| \leqslant Me^{\alpha(t-s)}. \tag{4.29}$$

Define an operator $\bar{V}_n(t, s)$ as follows:

$$\begin{cases} \bar{V}_n(t, s) = \sum_{m=0}^{\infty} V_n^{(m)}(t, s), \\[2mm] V_n^{(0)}(t, s) = U_n(t, s), \\[2mm] V_n^{(m)}(t, s) = \int_s^t U_n(t, r) B_n(r) V_n^{(m-1)}(r, s) \, dr. \end{cases} \quad (4.30)$$

That is, $\bar{V}_n(t, s)$ is obtained from (4.25) by replacing $U(t, s)$ and $B(r)$ by $U_n(t, s)$ and $B_n(r)$, respectively. It is easy to show that $V_n^{(m)}(t, s)$ satisfies the same inequality as (4.27) and that, for each m, $V_n^{(m)}(t, s)$ strongly converges to $V^{(m)}(t, s)$ uniformly as $n \to \infty$. Hence, $\bar{V}_n(t, s)$ strongly converges to $V(t, s)$ uniformly in $0 \le s \le t \le T$. Next, $\bar{V}_n(t, s)$ also satisfies the property (b) in Theorem 4.4.1 as may be seen as follows. First, note that

$$\bar{V}_n(t, s) = U_n(t, s) + \int_s^t U_n(t, r) B_n(r) \bar{V}_n(r, s) \, dr, \quad (4.31)$$

$$\bar{V}_n(t, s) = U_n(t, s) + \int_s^t \bar{V}_n(t, r) B_n(r) U_n(r, s) \, dr. \quad (4.32)$$

Equation (4.31) is easily obtained from (4.30), and (4.32) from

$$V_n^{(m)}(t, s) = \int_s^t V_n^{(m-1)}(t, r) B_n(r) U_n(r, s) \, dr.$$

Equation (4.32) implies that

$$\bar{V}_n(t, r) \bar{V}_n(r, s) = U_n(t, r) \bar{V}_n(r, s) + \int_r^t \bar{V}_n(t, \sigma) B_n(\sigma) U_n(\sigma, r) \bar{V}_n(r, s) \, d\sigma. \quad (4.33)$$

By (4.31), the first term on the right-hand side of (4.33) is equal to

$$U_n(t, s) + \int_s^r U_n(t, \tau) B_n(\tau) \bar{V}_n(\tau, s) \, d\tau.$$

Similarly, from (4.31), it follows that

$$U_n(\sigma, r) \bar{V}_n(t, s) = U_n(\sigma, s) + \int_s^r U_n(\sigma, \tau) B_n(\tau) \bar{V}_n(\tau, s) \, d\tau.$$

Again, by (4.31), the right-hand side of the equation is equal to

$$\bar{V}_n(\sigma, s) - \int_r^\sigma U_n(\sigma, \tau) B_n(\tau) \bar{V}_n(\tau, s) \, d\tau.$$

Substitute this result into the second term on the right-hand side of (4.33), change the orders of integration, and then we find that the right-hand side of (4.33) is equal to $\bar{V}_n(t, s)$. $A_n(t)$ and $B_n(t)$ are independent of t on each subinterval obtained by dividing $[0, T]$, so that, by Theorem 3.4.2, $\bar{V}_n(t, s) = V_n(t, s)$ if t and s are on the same subinterval. This is the case for general t and s since both $\bar{V}_n$ and V_n satisfy the property (b) in Theorem 4.4.1, and, hence, $V_n(t, s)$ converges strongly to $V(t, s)$ as $n \to \infty$ uniformly in $0 \le s \le t \le T$. Put $W_n(t, s) = S(t)U_n(t, s)S(s)^{-1}$, then $W_n(t, s) \in B(X)$ since $U_n(t, s) \in B(Y)$. If we can show that $W_n(t, s) \to W(t, s)$ as $n \to \infty$ uniformly in $0 \le s \le t \le T$, we have $S(t)^{-1}W(t, s) = \lim_{n \to \infty} S(t)^{-1}W_n(t, s) = \lim_{n \to \infty} U_n(t, s)S(s)^{-1} = U(t, s)S(s)^{-1}$ in the strong topology of X, so that (4.24) is obtained. Now we will show that: there exists a number N such that

$$\|C(t, s)\| \le N \, |t - s|, \tag{4.34}$$

where $C(t, s) = (S(t) - S(s))S(s)^{-1}$. Since $S(t) \exp(\tau A(t))S(t)^{-1} = \exp(\tau \bar{A}(t))$, for $kT/n \le s < (k+1)T/n$ and $T/n \le t < (l+1)T/n$ we have

$$W_n(t, s) = S(t)S\left(\frac{l}{n}T\right)^{-1} \exp\left(\left(t - \frac{l}{n}T\right)\bar{A}\left(\frac{l}{n}T\right)\right)S\left(\frac{l}{n}T\right)$$

$$\times S\left(\frac{l-1}{n}T\right)^{-1} \exp\left(\frac{T}{n}\bar{A}\left(\frac{l-1}{n}T\right)\right)S\left(\frac{l-1}{n}T\right)$$

$$\cdots S\left(\frac{k+1}{n}T\right)^{-1} \exp\left(\frac{T}{n}\bar{A}\left(\frac{k+1}{n}T\right)\right)S\left(\frac{k+1}{n}T\right)$$

$$\times S\left(\frac{k}{n}T\right)^{-1} \exp\left(\left(\frac{k+1}{n}T - s\right)\bar{A}\left(\frac{k}{n}T\right)\right)S\left(\frac{k}{n}T\right)S(s)^{-1}$$

$$= \left(1 + C\left(t, \frac{l}{n}T\right)\right)\left\{\exp\left(\left(t - \frac{l}{n}T\right)\bar{A}\left(\frac{l}{n}T\right)\right)\right.$$

$$\times \left(1 + C\left(\frac{1}{n}T, \frac{l-1}{n}T\right)\right)\exp\left(\frac{T}{n}\bar{A}\left(\frac{l-1}{n}T\right)\right)\left(1 + C\left(\frac{l-1}{n}T, \frac{l-2}{n}T\right)\right)$$

$$\cdots \left(1 + C\left(\frac{k+1}{n}T, \frac{k}{n}T\right)\right)\exp\left(\left(\frac{k+1}{n}T - s\right)\bar{A}\left(\frac{k}{n}T\right)\right)\right\}$$

$$\times \left(1 + C\left(\frac{k}{n}T, s\right)\right).$$

We expand the expression in curly brackets on the right-hand side of the above equation and rearrange it as follows. Denote the sum of terms containing the number m of factors $C(iT/n, (i-1)T/n)$ by $W_n^{(m)}(t, s)$. Then we have

$$\{\cdots\} = \sum_{m=0}^{l-k} W_n^{(m)}(t, s),$$

where

$$W_n^{(0)}(t, s) = V_n(t, s),$$

$$W_n^{(m)}(t, s) = \sum_{i=k+1}^{l-m+1} W_n^{(m-1)}\left(t, \frac{i}{n}T\right) C\left(\frac{i}{n}T, \frac{i-1}{n}T\right) V_n\left(\frac{i}{n}T, s\right).$$

From (4.29) and (4.34), it follows, by induction, that

$$\|W_n^{(m)}(t, s)\| \leqslant M^{m+1}N^m e^{\alpha(t-s)}(t - s + T/n)^m/m!,$$

where T/n on the right-hand side should be omitted if s coincides with the dividing point kT/n. We obtain

$$W_n^{(1)}(t, s) = \sum_{i=k+1}^{l} \int_{(i-1)T/n}^{iT/n} V_n\left(t, \frac{i}{n}T\right) \frac{d}{dr}S(r)S\left(\frac{i-1}{n}T\right)^{-1} V_n\left(\frac{i}{n}T, s\right) dr,$$

the right-hand side of which can be considered as an integral from kT/n to lT/n; the range of integration approaches (s, t) as $n \to \infty$. The norm of the integrand is bounded by $M^2 N e^{\alpha(t-s)}$ and the integrand converges strongly and uniformly to $V(t, r)C(r)V(r, s)$ as $n \to \infty$. Hence, $W_n^{(1)}(t, s)$ converges strongly to $W^{(1)}(t, s)$ as $n \to \infty$ uniformly in $0 \leqslant s \leqslant t \leqslant T$. By induction, we find that, for each m, $W_n^{(m)}(t, s)$ converges strongly and uniformly to $W^{(m)}(t, s)$. Hence, we have

$$W_n(t, s) = \left(1 + C\left(t, \frac{l}{n}T\right)\right) \sum_{m=0}^{l-k} W_n^{(m)}(t, s)\left(1 + C\left(\frac{k}{n}T, s\right)\right)$$

$$\to \sum_{m=0}^{\infty} W^{(m)}(t, s) = W(t, s)$$

strongly and, moreover, the convergence is uniform in $0 \leqslant s \leqslant t \leqslant T$. $\square$

Corollary   Suppose that $\{A(t)\}$ is stable, its domain D is independent of t and $A(t)v$ for each $v \in D$ is strongly continuously differentiable on $[0, T]$. Then there exists a function $U(t, s) \in B(X)$, satisfying (a) and (b) of Theorem 4.4.1, such that $U(t, s)$ maps D into D, $U(t, s)v$ for each $v \in D$ is strongly continuously differentiable in t and s, and the following results hold:

$$(\partial/\partial t)U(t, s)v = A(t)U(t, s)v, \tag{4.35}$$

$$(\partial/\partial s)U(t, s)v = -U(t, s)A(s)v. \tag{4.36}$$

Both sides of these equations are strongly continuous on $0 \leqslant s \leqslant t \leqslant T$. Such a $U(t, s)$ is unique.

Proof Let $\|v\|_Y = \|A(0)v\| + \|v\|$ for each $v \in D$, then D becomes a Banach space Y with this norm. For sufficiently large real number λ_0, $S(t) = A(t) - \lambda_0$ is an isomorphism from Y onto X and the property (ii″) is satisfied with $B(t) = 0$. $\square$

Kato [76] constructed fundamental solutions under the assumption $A(t) \in G(X, 1, \beta)$, which is stronger than the stability of $\{A(t)\}$ assumed in the corollary, and initiated the research in temporally inhomogeneous equations of evolution.

Theorem 4.4.3 *Assume that $\{A(t)\}$ satisfies the conditions of the preceding corollary. Assume also that $B(t)$ is defined on $[0, T]$ as a strongly continuous function with values in $B(X)$ and that there exists a real number λ_0, satisfying $\lambda_0 \in \rho(A(t))$ for all $t \in [0, T]$, such that $A(t)B(t)(A(t) - \lambda_0)^{-1}$ belongs to $B(X)$ for each $t \in [0, T]$ and is strongly continuous in $[0, T]$. Under these conditions, if $\{A(t)\}$ is replaced by $\{A(t) + B(t)\}$ in the above corollary, the conclusion remains valid.*

Proof Let us denote by $U_0(t, s)$ the operator whose existence is proved by the above corollary. For simplicity, assume $\lambda_0 = 0$. By letting

$$U_m(t, s) = \int_s^t U_0(t, r)B(r)U_{m-1}(r, s)\, dr, \qquad m = 1, 2, \ldots, \tag{4.37}$$

$$U(t, s) = \sum_{m=0}^{\infty} U_m(t, s), \tag{4.38}$$

it is clear that the series on the right-hand side of (4.38) is strongly convergent uniformly in $0 \le s \le t \le T$. By the assumption and the above corollary, $W(t, s) = A(t)U_0(t, s)A(s)^{-1}$ and $B_1(t) = A(t)B(t)A(t)^{-1}$ are strongly continuous in t, so that there exist numbers L and N such that $\|W(t, s)\| \le L$ and $\|B_1(t)\| \le N$. Since

$$A(t)U_m(t, s)A(s)^{-1} = \int_s^t W(t, r)B_1(r)A(r)U_{m-1}(r, s)A(s)^{-1}\, dr,$$

we have, by induction,

$$\|A(t)U_m(t, s)A(s)^{-1}\| \le L^{m+1}N^m(t - s)^m/m!$$

for all m. Hence, $A(t)U(t, s)A(s)^{-1}$ is bounded and its norm does not exceed $Le^{LN(t-s)}$. It is not hard to see that

$$(\partial/\partial t)U_m(t, s)A(s)^{-1} = B(t)U_{m-1}(t, s)A(s)^{-1} + A(t)U_m(t, s)A(s)^{-1}$$

for $m = 1, 2, \ldots$, so that $U(t, s)A(s)^{-1}$ is strongly differentiable in t and

$$(\partial/\partial t)U(t, s)A(s)^{-1} = (A(t) + B(t))U(t, s)A(s)^{-1}.$$

Since we can show, similarly to (4.32), that

$$U(t, s) = U_0(t, s) + \int_s^t U(t, r)B(r)U_0(r, s) \, dr,$$

it is evident that $U(t, s)v$ is differentiable in s for each $v \in D$ and satisfies

$$(\partial/\partial s)U(t, s)v = -U(t, s)(A(s) + B(s))v. \quad \square$$

Under the assumptions of Theorem 4.4.3, we choose Y and $S(t)$ as in the proof of the preceding corollary. Then the conditions (i) and (ii″) of Theorem 4.4.2 are satisfied if $A(t)$ is replaced by $A(t) + B(t)$, but we cannot claim the condition (iii) to be satisfied, since it is not certain whether $B(t)$ is norm continuous or not. Hence, a perturbational approach has been adopted.

From the proof of Theorem 4.4.2, the following theorem can also be easily obtained.

Theorem 4.4.4 *In addition to the assumptions of Theorem 4.4.2, suppose the following condition is satisfied:*

(iv) *$A(t)$ for each $t \in [0, T]$ generates a group and*

$$\|(A(t_1) - \lambda)^{-1}(A(t_2) - \lambda)^{-1} \cdots (A(t_k) - \lambda)^{-1}\| \le M(-\lambda - \beta)^{-k}$$

for $\lambda < -\beta$ and $0 \le t_1 \le \cdots \le t_k \le T$.

Then the function $U(t, s)$ in Theorem 4.4.2 can be constructed on $0 \le t \le T$ and $0 \le s \le T$, and the conclusion of the theorem still holds when $0 \le s \le t \le T$ is replaced by $0 \le t \le T$ and $0 \le s \le T$. In particular, we have $U(s, t) = U(t, s)^{-1}$.

Similar conclusions corresponding to the corollary of Theorem 4.4.2 and to Theorem 4.4.3 also hold.

The above presentation is due to Kato [87]. For the proof of Theorem 4.4.2, the technique by Yosida [18], pp. 425–429, has been used. Kato [87] showed that even if the condition (ii″) is not satisfied, but if Y is reflexive or some other assumptions are satisfied, one can obtain stronger conclusions on the continuity and the differentiability of $U(t, s)$ than Theorem 4.4.1. In [88] he revised the results of [87] and used them in [89] to solve non-linear equations.

4.5 Inhomogeneous equations

Let us consider the initial-value problem for the inhomogeneous equations (4.1) and (4.2), where u_0 and f are given elements of Y and

$C([0, T]; X)$, respectively. If $u \in C^1([0, T]; X) \cap C([0, T]; Y)$ satisfies (4.1) and (4.2), we call it a *solution* of (4.1) and (4.2).

Theorem 4.5.1 *Under the assumptions of Theorem 4.4.2, the solution of (4.1) and (4.2) is unique.*

Proof We can show that a solution u of (4.1) and (4.2) is represented by (4.8) in terms of the fundamental solution $U(t, s)$ constructed in Theorem 4.4.2. Indeed, by Theorem 4.4.2, we have

$$(\partial/\partial s)(U(t, s)) = U(t, s)u'(s) - U(t, s)A(s)u(s)$$

$$= U(t, s)(u'(s) - A(s)u(s)) = U(t, s)f(s),$$

which, being integrated from 0 to t, yields (4.8). $\square$

Equation (4.8) really gives a solution of (4.1) and (4.2) if a slightly stronger condition is imposed on f.

Theorem 4.5.2 *Suppose the assumptions of Theorem 4.4.2 are satisfied. Then, for any $u_0 \in Y$ and any $f \in C([0, T]; Y)$, the function defined by (4.8) is a solution of (4.1) and (4.2).*

Proof The theorem is obvious from the fact that by hypothesis; $A(t)U(t, s)f(s)$ is strongly continuous in s. $\square$

Theorem 4.5.3 *Suppose the assumptions of the corollary of Theorem 4.4.2 are satisfied. For any $u_0 \in D$ and $f \in C^1([0, T]; X)$, the function u defined by (4.8) is a solution of (4.1) and (4.2).*

Proof We use the same notations as in the proof of the corollary of Theorem 4.4.2. Without loss of generality, we may assume $\lambda_0 = 0$, so that $S(t) = A(t)$. Put $g(t) = -(dA(t)/dt)A(t)^{-1}f(t) + f'(t)$, then $g \in C([0, T]; X)$ and

$$(d/dt)(A(t)^{-1}f(t)) = A(t)^{-1}g(t). \tag{4.39}$$

Hence, we have

$$(\partial/\partial s)(U(t, s)A(s)^{-1}f(s)) = -U(t, s)f(s) + U(t, s)A(s)^{-1}g(s).$$

By integrating both sides of this equation from 0 to t, we obtain

$$v(t) \equiv \int_0^t U(t, s)f(s)\, ds = U(t, 0)A(0)^{-1}f(0) - A(t)^{-1}f(t)$$

$$+ \int_0^t U(t, s)A(s)^{-1}g(s)\, ds.$$

It is readily seen that v is the solution of (4.1) which satisfies the initial condition $v(0) = 0$. □

Theorem 4.5.4 *Suppose the conditions of Theorem 4.4.3 are satisfied. If $u_0 \in Y$ and $f \in C^1([0, T]; X)$, the solution of*

$$du(t)/dt = (A(t) + B(t))u(t) + f(t), \qquad (4.40)$$

$$u(0) = u_0 \qquad (4.41)$$

exists uniquely.

Since the proof is analogous to that of the preceding theorem, it is omitted.

4.6   Application 1. Initial-value problem of the symmetric hyperbolic system

Consider the following initial-value problem for simultaneous partial differential equations

$$\partial u / \partial t + \sum_{j=1}^{n} a_j(x, t) \partial u / \partial x_j + b(x, t)u = f(x, t),$$

$$x \in \mathbb{R}^n, \qquad 0 \leqslant t \leqslant T, \qquad u(x, 0) = u_0(x).$$

where $u = {}^t(u_1, \ldots, u_n)$ is a set of unknowns, $a_j(x, t)$ and $b(x, t)$ are square matrices of order N for each x and t, and $a_j(x, t)$, $j = 1, \ldots, n$, is assumed to be Hermitian. The above equations correspond to those in Section 3.5; their coefficients, however, depend on t. We use the same notations for function spaces, norms and so on, as were used in that section. As for the smoothness of the coefficients a_j and b, we assume

$$\begin{cases} \text{each } a_j(\cdot, t) \text{ is a continuous function of } t \in [0, T] \text{ with} \\ \text{values in } B^1(\mathbb{R}^n), \end{cases} \qquad (4.42)$$

$$\begin{cases} b(\cdot, t) \text{ is a continuous function of } t \in [0, T] \text{ with values in} \\ B^0(\mathbb{R}^n), \text{ and } \partial b / \partial x_j \text{ is continuous and bounded on } \mathbb{R}^n \times \\ [0, T]. \end{cases}$$

$$(4.43)$$

For each t, let us put

$$\mathscr{A}(t)u = \sum_{j=1}^{n} a_j(x, t) \partial u / \partial x_j + b(x, t)u$$

and define an operator $A(t)$ as follows:

$$D(A(t)) = \{u \in X : \mathscr{A}(t)u \in X\},$$
$$A(t)u = \mathscr{A}(t)u \quad \text{for} \quad u \in D(A(t)). \tag{4.44}$$

By the proof of Theorem 3.5.1, there exists a number $\beta \geqslant 0$ such that both $A(t)$ and $-A(t)$ belong to $G(X, 1, \beta)$ for all $t \in [0, T]$. Therefore, $\{A(t)\}$ and $\{-A(t)\}$ are stable with stability constants 1 and β. Evidently, the condition (iii) of Theorem 4.4.1 is satisfied. Using the Fourier transform, an operator $S(t) \equiv S = (1 - \Delta)^{1/2}$ is defined by

$$(Su)(x) = (2\pi)^{-n/2} \int e^{ix\xi} (1 + |\xi|^2)^{1/2} \hat{u}(\xi) \, d\xi \quad \text{for} \quad u \in Y,$$

where $\hat{u}$ is the Fourier transform of u and the integration is taken in the sense of mean convergence in $L^2(\mathbb{R}^n)$. S is an isomorphism from Y onto X. Now we will show that there exists a strongly continuous function $B(t)$ with values in $B(X)$ such that

$$SA(t)S^{-1} = \mathscr{A}(t) + B(t).$$

Let $u \in Y$. Then we represent

$$SA(t)S^{-1}u = A(t)u + \sum_{j=1}^{m} [S, a_j(x, t)](\partial/\partial x_j)S^{-1}u + [S, b(x, t)]S^{-1}u$$

$$\equiv A(t)u + B_1(t)u + B_2(t)u,$$

where $[S, a_j(x, t)]u = Sa_j(x, t)u - a_j(x, t)Su$. It is seen by Theorem 1 of Calderon [45] that $[S, a_j]$ can be extended as an element of $B(X)$. The theorem states that if R is a singular integral operator and $a \in B^1(\mathbb{R}^n)$, then $[R, a]\partial/\partial x_j$ can be extended to a bounded operator from $L^2(\mathbb{R}^n)$ into itself. Similarly to $S = (1 - \Delta)^{1/2}$, we define $\Lambda = (-\Delta)^{1/2}$ by using the Fourier transform. It is not hard to see that $S - \Lambda$ can be extended to an element of $B(X)$. Hence, in order to show that $[S, a]$ has a bounded extension, it is enough to ascertain that $[\Lambda, a]$ has a bounded extension. Since it is well known [43] that there exist singular operators R_k ($k = 1, \ldots, n$) such that $\Lambda = \sum_{k=1}^{n} R_k(\partial/\partial x_k)$, we have

$$[\Lambda, a] = \sum_{k=1}^{n} [R_k, a]\partial/\partial x_k + \sum_{k=1}^{n} R_k \partial a/\partial x_k.$$

The right-hand side of this equation can be extended to an element of $B(X)$ by Calderon's theorem mentioned above. Furthermore, as the proof of the theorem tells us, there exists a constant c such that $\|[S, a]u\| \leqslant c \, |a|_1 \|u\|$, so that $B_1(t) \in B(X)$ is norm continuous in t because of (4.42) and $\|(\partial/\partial x_j)S^{-1}\| \leqslant 1$. Next, we can observe that $B_2(t) \in B(X)$ is strongly continuous in t by noting the representation

$$B_2(t)u = \left\{ S^{-1} - \sum_{j=1}^{n} (\partial/\partial x_j) S^{-1} (\partial/\partial x_j) \right\} b(x, t) S^{-1} u - b(x, t) u.$$

Here the result of [45] cannot be applied since it is not assumed that $b(\cdot, t)$ is a continuous $B^1(\mathbb{R}^n)$-valued function. In this way, we find that

$$SA(t)S^{-1}u = A(t)u + B(t)u \tag{4.46}$$

is valid for $u \in Y$, where $B(t) = B_1(t) + B_2(t)$ is a strongly continuous function with values in $B(X)$. Hence, to prove (4.45), it is enough to show that both sides of the equation (4.46) have the identical domain, i.e., $D(SA(t)S^{-1}) = D(A(t))$. Let $u \in D(A(t))$. By the use of a mollifier, we can approximate u by a sequence $\{u_j\}$, $u_j \in Y$, in such a way that $u_j \to u$ and $A(t)u_j \to A(t)u$ in X. Since $A(t)S^{-1}u_j = S^{-1}(A(t) + B(t))u_j \to S^{-1}(A(t) + B(t))u$ by (4.46) and, since $A(t)$ is a closed operator, we have $S^{-1}u \in D(A(t))$ and $A(t)S^{-1}u = S^{-1}(A(t) + B(t))u$, which proves $SA(t)S^{-1} \supset A(t) + B(t)$. Hence, for sufficiently large real number λ, $S(A(t) + \lambda)^{-1}S^{-1} \supset (A(t) + B(t) + \lambda)^{-1}$; the equality really holds since the domain of the right-hand side is X. Thus, we obtain (4.45). Since similar results are valid for $-A(t)$, we can apply Theorem 4.4.4 to $\{A(t)\}$.

4.7 A theorem on fractional powers of positive definite self-adjoint operators

In this section, we will explain a theorem on fractional powers of positive definite self-adjoint operators depending on a parameter. This provides a preparation for the next section, but the material is interesting in itself.

Lemma 4.7.1 Assume that $A(t)$ for each $t \in [0, T]$ is a closed operator in a Banach space X, it has a bounded inverse, its domain $D(A(t)) = D$ is independent of t, and $A(t)u$ for each $u \in D$ is strongly continuously differentiable on $[0, T]$. Then $A(t)A(s)^{-1}$ is strongly continuously differentiable in $(t, s) \in [0, T] \times [0, T]$.

Proof Since $(d/dt)A(t)A(0)^{-1}$ is strongly continuous by nature of the assumption, it follows from Theorem 1.3.1 that there exists a number M such that $\|(d/dt)A(t)A(0)^{-1}\| \leq M$ on $0 \leq t \leq T$. Therefore, we have

$$\|A(t)A(0)^{-1}u - A(s)A(0)^{-1}u\| \leq M |t - s| \|u\|$$

for every $u \in X$, so that $A(t)A(0)^{-1}$ is Lipschitz continuous in the norm of $B(X)$. Since $A(0)A(t)^{-1} = (A(t)A(0)^{-1})^{-1}$ is norm continuous by

Theorem 1.1.15, $A(0)A(t)^{-1}$ is uniformly bounded and

$$(\partial/\partial t)(A(t)A(s)^{-1}) = (d/dt)A(t)A(0)^{-1} \cdot A(0)A(s)^{-1},$$
$$h^{-1}(A(t)A(s+h)^{-1} - A(t)A(s)^{-1} = A(t)A(0)^{-1} \cdot A(0)A(s+h)^{-1}$$
$$\times h^{-1}(A(s) - A(s+h))A(0)^{-1} \cdot A(0)A(s)^{-1}$$
$$\rightarrow -A(t)A(s)^{-1}(d/ds)A(s)A(0)^{-1} \cdot A(0)A(s)^{-1}.$$

This completes the proof. □

From now on, in this section, we assume that X is a Hilbert space, $A(t)$ for all $t \in [0, T]$ is a self-adjoint operator in X, whose domain $D(A(t)) = D$ is independent of t, and it is uniformly positive definite on $[0, T]$, i.e., there exists a constant $\delta > 0$ such that $(A(t)u, u) \geq \delta \|u\|^2$ for all $t \in [0, T]$ and all $u \in D$. By Theorem 2.3.3, $D_\alpha = D(A(t)^\alpha)$ for each $\alpha \in (0, 1)$ is also independent of t.

Theorem 4.7.1 *In addition to the assumptions above, we assume that $A(t)u$ for each $u \in D$ is strongly continuous on $[0, T]$. Then $A(t)^\alpha u$ is strongly continuous on $[0, T]$ for all $\alpha \in (0, 1)$ and all $u \in D_\alpha$.*

Proof From the assumption and Theorem 2.3.1 it follows that

$$K = \sup_{0 \leq t \leq T} \|A(t)A(0)^{-1}\| < \infty. \tag{4.47}$$

Also, $\|A(t)^\alpha A(0)^{-\alpha}\| \leq K^\alpha$ by Theorem 2.3.3. By noting that

$$\|(A(t) + \mu)^{-1}\| \leq (\delta + \mu)^{-1},$$

we have, for $u \in X$ and $\mu \geq 0$,

$$\|(A(t) + \mu)^{-1}u - (A(s) + \mu)^{-1}u\|$$
$$= \|(A(t) + \mu)^{-1}(A(s) - A(t))(A(s) + \mu)^{-1}u\|$$
$$\leq (\delta + \mu)^{-1}\|(A(s) - A(t))(A(s) + \mu)^{-1}u\| \rightarrow 0$$

as $t \rightarrow s$, so that $(A(t) + \mu)^{-1}u$ is a strongly continuous function of t. Hence

$$A(t)^{\alpha-1}u = \frac{\sin \pi\alpha}{\pi} \int_0^\infty \mu^{\alpha-1}(A(t) + \mu)^{-1}u \, d\mu$$

is also a strongly continuous function of t. Therefore $A(t)^\alpha u = A(t)^{\alpha-1}A(t)u$ for $u \in D$ is strongly continuous in t. Let $w = A(0)^\alpha u$ with $u \in D_\alpha$ and let $w_n \in D_{1-\alpha}$ be so chosen that $\|w_n - w\| \rightarrow 0$. Put $u_n =$

$A(0)^{-\alpha}w_n$, then $A(t)u_n$ is strongly continuous in t since $u_n \in D$, and

$$\|A(t)^\alpha u_n - A(t)^\alpha u\| = \|A(t)^\alpha A(0)^{-\alpha}(w_n - w)\| \leqslant K^\alpha \|w_n - w\|.$$

Hence $A(t)^\alpha u_n$ converges strongly to $A(t)^\alpha u$ as $n \to \infty$ uniformly in $0 \leqslant t \leqslant T$. Consequently, $A(t)^\alpha u$ is strongly continuous on $0 \leqslant t \leqslant T$. □

When $A(t)u$ is strongly differentiable for each $u \in D$, we simply denote $(\mathrm{d}/\mathrm{d}t)A(t)u$ by $A'(t)u$.

Lemma 4.7.2 *besides the assumptions made above, we further assume that $A(t)u$ for each $u \in D$ is strongly continuously differentiable in $[0, T]$. Let t, s and r be arbitrary points in $[0, T]$ and let α be an arbitrary number belonging to $[0, 1]$. Then the operator $A(r)^{\alpha-1}A'(t)A(s)^{-\alpha}$ has a bounded extension which is defined on $D_{1-\alpha}$ and whose norm does not exceed $M = \sup_{0 \leqslant t \leqslant T}\|A'(t)A(s)^{-1}\|$.*

Proof $M < \infty$ by Lemma 4.7.1. Since the operator $A'(t)$ defined on D is symmetric, we have

$$|(A(r)^{-1}A'(t)u, v)| = |(A'(t)u, A(r)^{-1}v)|$$

$$= |(u, A'(t)A(r)^{-1}v)| \leqslant M \|u\| \|v\|$$

for each $u \in D$ and each $v \in X$. Therefore, there exists a bounded extension T of $A(r)^{-1}A'(t)$ which satisfies $\|T\| \leqslant M$. Clearly, T maps D into D, and

$$\|A(r)Tu\| = \|A'(t)u\| = \|A'(t)A(s)^{-1}A(s)u\| \leqslant M \|A(s)u\|$$

for each $u \in D$. Thus the assumptions of Theorem 2.3.3 are satisfied with $X_1 = X_2 = X$, $A = A(s)$ and $B = A(r)$, so that T maps D_α into D_α and

$$\|A(r)^\alpha Tu\| \leqslant M \|A(s)^\alpha u\|$$

for each $u \in D_\alpha$. Hence, for $u \in D$, we have

$$\|A(r)^{\alpha-1}A'(t)u\| \leqslant M \|A(s)^\alpha u\|. \tag{4.48}$$

Since $u \in D_{1-\alpha}$ implies that $A(s)^{-\alpha}u \in D$, from (4.48) it follows that

$$\|A(r)^{\alpha-1}A'(t)A(s)^{-\alpha}u\| \leqslant M \|u\|. □$$

Theorem 4.7.2 (Daleckiĭ [57]) *In addition to the assumptions above, we assume that $A(t)u$ for each $u \in D$ is strongly continuously differentiable on $[0, T]$. Then $A(t)^\alpha A(s)^{-\alpha}$ is strongly continuously differentiable on $[0, T] \times [0, T]$ if $0 < \alpha < 1$. Also, $A(t)^{\alpha-1}(\mathrm{d}/\mathrm{d}t)A(t)^\alpha \cdot A(t)^{-1}$ is a bounded operator and strongly continuous on $[0, T]$.*

Proof

$$A(t)^\alpha u = A(t)^{\alpha-1} A(t) u = \frac{\sin \pi\alpha}{\pi} \int_0^\infty \mu^{\alpha-1} (A(t) + \mu)^{-1} A(t) u \, d\mu$$

for $u \in D$. Now, since

$$\begin{aligned}
(\partial/\partial t)((A(t) + \mu)^{-1} A(t) u) &= (\partial/\partial t)(u - \mu(A(t) + \mu)^{-1} u) \\
&= -\mu(\partial/\partial t)(A(t) + \mu)^{-1} u \\
&= \mu(A(t) + \mu)^{-1} A'(t)(A(t) + \mu)^{-1} u \\
&= \mu(A(t) + \mu)^{-1} A'(t) A(t)^{-1} \\
&\quad \times (A(t) + \mu)^{-1} A(t) u,
\end{aligned}$$

the norm of the integrand satisfies

$$\| \mu^{\alpha-1} (\partial/\partial t)((A(t) + \mu)^{-1} A(t) u) \| \leqslant M \mu^\alpha (\mu + \delta)^{-2} \| A(t) u \|$$

and so it is integrable in $0 < \mu < \infty$. Therefore, we have

$$\frac{d}{dt} A(t)^\alpha u = \frac{\sin \pi\alpha}{\pi} \int_0^\infty \mu^\alpha (A(t) + \mu)^{-1} A'(t)(A(t) + \mu)^{-1} u \, d\mu.$$

Let $\alpha < \beta < 1$ and let v be an arbitrary element of X. From Lemma 4.7.2 there follows the result

$$\begin{aligned}
&|((A(t) + \mu)^{-1} A'(t)(A(t) + \mu)^{-1} u, v)| \\
&= |(A(t)^{\beta-1} A'(t) A(t)^{-\beta} \cdot A(t)^\beta (A(t) + \mu)^{-1} u, A(t)^{1-\beta}(A(t) + \mu)^{-1} v)| \\
&\leqslant M \| A(t)^\beta (A(t) + \mu)^{-1} u \| \| A(t)^{1-\beta}(A(t) + \mu)^{-1} v \|,
\end{aligned}$$

so that

$$\begin{aligned}
\left| \left(\frac{d}{dt} A(t)^\alpha u, v \right) \right| &= \left| \frac{\sin \pi\alpha}{\pi} \int_0^\infty \mu^\alpha ((A(t) + \mu)^{-1} A'(t)(A(t) + \mu)^{-1} u, v) \, d\mu \right| \\
&\leqslant \frac{\sin \pi\alpha}{\pi} M \int_0^\infty \mu^\alpha \| A(t)^\beta (A(t) + \mu)^{-1} u \| \\
&\quad \times \| A(t)^{1-\beta}(A(t) + \mu)^{-1} v \| \, d\mu,
\end{aligned}$$

which, by choosing $\beta = (1 + \alpha)/2$ and making use of the Schwarz inequality, turns out to be

$$\leqslant \frac{\sin \pi\alpha}{\pi} M \left\{ \int_0^\infty \mu^\alpha \| A(t)^{(1-\alpha)/2}(A(t) + \mu)^{-1} A(t)^\alpha u \|^2 \, d\mu \right\}^{1/2}$$

$$\times \left\{ \int_0^\infty \mu^\alpha \| A(t)^{(1-\alpha)/2}(A(t) + \mu)^{-1} v \|^2 \, d\mu \right\}^{1/2}.$$

Let $A(t) = \int_\delta^\infty \lambda \, dE(\lambda)$ be the spectral resolution of $A(t)$. Then we have

$$
\int_0^\infty \mu^\alpha \, \|A(t)^{(1-\alpha)/2}(A(t)+\mu)^{-1}v\|^2 \, d\mu
$$

$$
= \int_0^\infty \mu^\alpha \int_\delta^\infty \lambda^{1-\alpha}(\lambda+\mu)^{-2} \, d\|E(\lambda)v\|^2 \, d\mu
$$

$$
= \int_\delta^\infty \lambda^{1-\alpha} \int_0^\infty \mu^\alpha(\lambda+\mu)^{-2} \, d\mu \, d\|E(\lambda)v\|^2
$$

$$
= \int_0^\infty \mu^\alpha(1+\mu)^{-2} \, d\mu \, \|v\|^2 = \frac{\pi\alpha}{\sin \pi\alpha} \|v\|^2,
$$

so that

$$
|((d/dt)A(t)^\alpha u, v)| \leq M\alpha \, \|A(t)^\alpha u\| \, \|v\|.
$$

Hence,

$$
\|(d/dt)A(t)^\alpha u\| \leq M\alpha \, \|A(t)^\alpha u\| \tag{4.49}
$$

holds for all $u \in D$. Next, let u be an arbitrary element of D_α. As in the proof of Theorem 4.7.1, we choose sequences $\{u_n\} \subset D$ and $\{w_n\} \subset D_{1-\alpha}$ so that $A(t)^\alpha u_n \to A(t)^\alpha u$ strongly and uniformly in $[0, T]$. Since, from (4.47) and (4.49), we obtain

$$
\|(d/dt)A(t)^\alpha u_n - (d/dt)A(t)^\alpha u_m\| \leq M\alpha \, \|A(t)^\alpha (u_n - u_m)\|
$$

$$
\leq M\alpha \, \|A(t)^\alpha A(0)^{-\alpha}(w_n - w_m)\|
$$

$$
\leq M\alpha K^\alpha \, \|w_n - w_m\|,
$$

$(d/dt)A(t)^\alpha u_n$ is strongly convergent uniformly in $[0, T]$. Hence, $A(t)^\alpha u$ also is strongly continuously differentiable and $(d/dt)A(t)^\alpha u = $ s-$\lim_{n\to\infty} (d/dt)A(t)^\alpha u_n$. This, combined with Lemma 4.7.1, completes the proof of the first half of the theorem.

Let u be an arbitrary element of D. The right-hand side of

$$
A(t+h)^{1-\alpha}h^{-1}(A(t+h)^\alpha - A(t)^\alpha)u
$$

$$
= h^{-1}(A(t+h) - A(t))u - h^{-1}(A(t+h)^{1-\alpha} - A(t)^{1-\alpha})A(t)^\alpha u
$$

is strongly convergent to $A'(t)u - (d/dt)A(t)^{1-\alpha} \cdot A(t)^\alpha u$ as $h \to 0$. Since

$$
(A(t+h)^{1-\alpha}h^{-1}(A(t+h)^\alpha - A(t)^\alpha)u, v)
$$

$$
= (h^{-1}(A(t+h)^\alpha - A(t)^\alpha)u, A(t+h)^{1-\alpha}v)
$$

$$
\to ((d/dt)A(t)^\alpha u, A(t)^{1-\alpha}v)
$$

for all $v \in D_{1-\alpha}$, we have

$$
((d/dt)A(t)^\alpha u, A(t)^{1-\alpha}v) = (A'(t)u - (d/dt)A(t)^{1-\alpha} \cdot A(t)^\alpha u, v).
$$

Hence, $(d/dt)A(t)u \in D_{1-\alpha}$ and

$$A(t)^{1-\alpha}(d/dt)A(t)^{\alpha} \cdot A(t)^{-1} = A'(t)A(t)^{-1} - (d/dt)A(t)^{1-\alpha} \cdot A(t)^{\alpha-1},$$

the right-hand side of which is strongly continuous in t. $\square$

4.8   Application 2.   Mixed problem of hyperbolic equations

Consider the initial-value problem of a second-order hyperbolic equation

$$d^2u(t)/dt^2 + A(t)u(t) = f(t), \qquad 0 \leqslant t \leqslant T, \tag{4.50}$$

$$u(0) = u_0, \qquad (d/dt)u(0) = u_1 \tag{4.51}$$

in a Hilbert space X, where the operator $A(t)$ in X is bounded from below and self-adjoint. We will first explain why this equation is called hyperbolic. Let Ω be a region in $\mathbb{R}^n$ and, for each $t \in [0, T]$ and $u, v \in \overset{\circ}{H}_1(\Omega)$, consider

$$a(t; u, v) = \sum_{i,j=1}^{n} \int_{\Omega} a_{ij}(x, t) \frac{\partial u}{\partial x_i} \frac{\overline{\partial v}}{\partial x_j} \, dx - \int_{\Omega} c(x, t)u\bar{v} \, dx, \tag{4.52}$$

where a_{ij} and c are real-valued functions which are continuous and bounded on $\bar{\Omega} \times [0, T]$, and the matrix $(a_{ij}(x, t))$ is uniformly positive definite, i.e., there exists a positive constant δ such that

$$\sum_{i,j=1}^{n} a_{ij}(x, t)\xi_i\xi_j \geqslant \delta |\xi|^2 \tag{4.53}$$

for all $x \in \bar{\Omega}$, $t \in [0, T]$ and for all real vectors ξ. Denote the operator defined by this quadratic form for each t by $A(t)$, then the mixed problem for the hyperbolic equation

$$\frac{\partial^2 u}{\partial t^2} = \sum_{i,j=1}^{n} \frac{\partial}{\partial x_j} \left(a_{ij}(x, t) \frac{\partial u}{\partial x_i} \right) + c(x, t)u + f(x, t),$$
$$x \in \Omega, \qquad 0 \leqslant t \leqslant T, \tag{4.54}$$

$$u(x, t) = 0, \qquad x \in \partial\Omega, \qquad 0 \leqslant t \leqslant T, \tag{4.55}$$

$$u(x, 0) = u_0(x), \qquad (\partial/\partial t)u(x, 0) = u_1(x), \qquad x \in \Omega, \tag{4.56}$$

can be written in the abstract form of (4.50) and (4.51). From what was stated in Section 2.2, it follows that $A(t)$ is bounded from below and self-adjoint in $L^2(\Omega)$. All the discussions in this section can be applied to the case when $A(t)$ is an operator defined by the Dirichlet problem of higher-order elliptic operators, apart from the question whether (4.50) is then of hyperbolic type.

In this section, we assume that $A(t)$, for each $t \in [0, T]$, is self-adjoint and bounded, its domain $D(A(t)) = D$ is independent of t, and $A(t)u$ for each $u \in D$ is strongly continuously differentiable on $[0, T]$. For simplicity, we further assume that $A(t)$ is uniformly positive definite; otherwise, we may choose a sufficiently large positive number λ_0 and put $A_1(t) = A(t) + \lambda_0$ to express (4.50) in the form

$$\mathrm{d}^2 u(t)/\mathrm{d}t^2 + A_1(t)u(t) - \lambda_0 u(t) = f(t).$$

Then, as will be discussed below, the term $-\lambda_0 u(t)$ can be treated as a perturbation by Theorem 4.4.3. When $u \in C^2([0, T]; X)$, $u(t) \in D$, $\mathrm{d}u(t)/\mathrm{d}t \in D_{1/2} \equiv D(A(t)^{1/2})$ for each $t \in [0, T]$, Au, $A^{1/2}\,\mathrm{d}u/\mathrm{d}t \in C([0, T];$ $X)$, and (4.50) and (4.51) are satisfied, u is called a solution of (4.50) and (4.51). If Ω is locally of class C^4 and uniformly of class C^2, if

$$a_{ij}, \quad \frac{\partial a_{ij}}{\partial t}, \quad \frac{\partial a_{ij}}{\partial x_i}, \quad \frac{\partial^2 a_{ij}}{\partial x_i \partial x_j}, \quad \frac{\partial^2 a_{ij}}{\partial x_i \partial t}, \quad c, \quad \frac{\partial c}{\partial t}$$

are all continuous and bounded on $\Omega \times [0, T]$, and if a_{ij} is uniformly continuous, then the operator $A(t)$ given by the quadratic form (4.52) satisfies all the assumptions in this section (except that of uniform positive definiteness). It is due to Browder [36] that $D(A(t)) = H_2(\Omega) \cap \overset{\circ}{H}_1(\Omega)$.

Let $\mathfrak{X} = X \times X$ be a direct product of two copies of the Hilbert space X. That is, it is given by the set of all $U = {}^t(u, v)$ with $u, v \in X$ and its norm is defined by $\|U\|^2 = \|u\|^2 + \|v\|^2$. Put

$$\mathfrak{A}(t) = \begin{pmatrix} 0 & iA(t)^{1/2} \\ iA(t)^{1/2} & 0 \end{pmatrix}, \qquad \mathfrak{B}(t) = \begin{pmatrix} (\mathrm{d}/\mathrm{d}t)A(t)^{1/2} \cdot A(t)^{-1/2} & 0 \\ 0 & 0 \end{pmatrix}.$$

$\mathfrak{A}(t)$ has a domain $\mathfrak{D} = D_{1/2} \times D_{1/2}$ and is a skew-symmetric operator in $\mathfrak{X}$, and $\mathfrak{B}(t)$ is bounded on $\mathfrak{X}$ by Theorem 4.7.2; $\mathfrak{A}(t)U$ for each $U \in \mathfrak{D}$ is strongly continuously differentiable and $\mathfrak{B}(t)$ is strongly continuous in $\mathfrak{X}$. When $u_0 \in D$, $u_1 \in D_{1/2}$ and $u(t)$ is a solution of (4.50) and (4.51), the function $U(t)$ defined by $U(t) = {}^t(u_0(t), u_1(t))$, where $u_0(t) = iA(t)^{1/2}u(t)$ and $u_1(t) = \mathrm{d}u(t)/\mathrm{d}t$, is a solution of

$$\mathrm{d}U(t)/\mathrm{d}t = \mathfrak{A}(t)U(t) + \mathfrak{B}(t)U(t) + F(t), \tag{4.57}$$

$$U(0) = U_0 = {}^t(u_0, u_1). \tag{4.58}$$

Here we have put $F(t) = {}^t(0, f(t))$. Conversely, if $U(t) = {}^t(u_0(t), u_1(t))$ is a solution of (4.57) and (4.58), then, by tracing back the calculations, we can show that $u(t) = -iA(t)^{-1/2}u_0(t)$ is a solution of (4.50) and (4.51). Hence, the two problems, (4.50) and (4.51) on one hand, and (4.57) and

(4.58) on the other, are equivalent. Since $i\mathfrak{A}(t)$ is self-adjoint in $\mathfrak{X}$ for each t, $\mathfrak{A}(t)$ generates a contraction group.

$$\mathfrak{A}(t)^{-1} = \begin{pmatrix} 0 & -iA(t)^{-1/2} \\ -iA(t)^{-1/2} & 0 \end{pmatrix}$$

is bounded and a simple calculation gives

$$\mathfrak{A}(t)\mathfrak{B}(t)\mathfrak{A}(t)^{-1} = \begin{pmatrix} 0 & 0 \\ 0 & A(t)^{1/2}(d/dt)A(t)^{1/2} \cdot A(t)^{-1} \end{pmatrix}.$$

Therefore, by Theorem 4.7.2, $\mathfrak{A}(t)\mathfrak{B}(t)\mathfrak{A}(t)^{-1}$ is also bounded on $\mathfrak{X}$ and strongly continuous in t. Consequently, $\mathfrak{A}(t)$ and $\mathfrak{B}(t)$ satisfy the assumptions of Theorem 4.4.3. Thus we have shown that there exists a fundamental solution $\mathfrak{U}(t, s)$ of (4.57) and (4.58) and established the following:

Theorem 4.8.1 *If $u_0 \in D$, $u_1 \in D$ and $f \in C^1([0, T]; X)$, then a solution of* (4.50) *and* (4.51) *exists uniquely.*

Remark 4.8.1 The above result holds also for the following rather more general equation:

$$\frac{\partial^2 u}{\partial t^2} = \sum_{i=1}^{n} a_i \frac{\partial^2 u}{\partial x_i \partial t} + \sum_{i,j=1}^{n} a_{ij} \frac{\partial^2 u}{\partial x_i \partial x_j} + \sum_{i=1}^{n} b_i \frac{\partial u}{\partial x_i} + cu + f,$$

where the assumption on (a_{ij}) is the same as above, a_i is real, b_i and c may be complex-valued, and

$$a_i, \quad \frac{\partial a_i}{\partial x_i}, \quad \frac{\partial a_i}{\partial t}, \quad \frac{\partial^2 a_i}{\partial x_i \partial t}, \quad b_i, \quad \frac{\partial b_i}{\partial t}, \quad c, \quad \frac{\partial c}{\partial x_i}$$

are continuous and bounded on $\Omega \times [0, T]$. If we put

$$B_0(t) = \sum_{i=1}^{n} \left(a_i \frac{\partial}{\partial x_i} + \frac{1}{2}\frac{\partial a_i}{\partial x_i} \right), \qquad A_0(t) = -\sum_{i,j=1}^{n} \left(a_{ij}\frac{\partial^2}{\partial x_i \partial x_j} + \frac{\partial a_{ij}}{\partial x_j}\frac{\partial}{\partial x_i} \right),$$

$$B_1(t) = -\frac{1}{2}\sum_{i=1}^{n} \frac{\partial a_i}{\partial x_i}, \quad A_1(t) = \sum_{i,j=1}^{n} \left(b_i - \frac{\partial a_{ij}}{\partial x_j} \right)\frac{\partial}{\partial x_i}, \qquad C(t) = c,$$

$$\mathfrak{A}_0(t) = \begin{pmatrix} 0 & iA_0^{1/2} \\ iA_0^{1/2} & B_0 \end{pmatrix}, \qquad \mathfrak{A}_1(t) = \begin{pmatrix} 0 & 0 \\ -iA_1 \cdot A_0^{-1/2} & B_1 \end{pmatrix},$$

$$\mathfrak{B}(t) = \begin{pmatrix} (d/dt)A_0^{1/2} \cdot A_0^{-1/2} & 0 \\ -iCA_0^{-1/2} & 0 \end{pmatrix}, \qquad \mathfrak{A}(t) = \mathfrak{A}_0(t) + \mathfrak{A}_1(t),$$

the equation (4.57) reduces to

$$dU(t)/dt = (\mathfrak{A}(t) + \mathfrak{B}(t))U(t) + F(t).$$

Since $B_0(t)$ is skew-symmetric on $\mathfrak{D}$, the operator $i\mathfrak{A}_0(t)$ is self-adjoint with $\mathfrak{D}$ as its domain and strongly continuously differentiable on $\mathfrak{D}$, and $\mathfrak{A}_1(t)$ is bounded and strongly continuously differentiable on $\mathfrak{X}$. Hence, by Theorem 4.7.2, $\mathfrak{A}(t)$ satisfies the assumption of the corollary of Theorem 4.4.2. There exists a bounded inverse of $\mathfrak{A}_0(t)$ and it is given by

$$\mathfrak{A}_0(t)^{-1} = \begin{pmatrix} A_0^{-1/2}B_0A_0^{-1/2} & -iA_0^{-1/2} \\ -iA_0^{-1/2} & 0 \end{pmatrix}.$$

It is easy to see that $\mathfrak{A}_0(t)\mathfrak{B}(t)\mathfrak{A}_0(t)^{-1}$ is strongly continuous. From this we can observe that $\mathfrak{A}(t)$ and $\mathfrak{B}(t)$ satisfy the condition of Theorem 4.4.3.

References related to Sections 4.7 and 4.8 are Carroll and State [50], Pogorelenko and Sobolevskiĭ [150] and so on. The reader should also be referred to Chapter 8 of Lions [11] and Chapter 2 of Lions and Magenes [15].

5

Parabolic equations

5.1 Parabolic equations

In this chapter, we will write the equation with a coefficient depending on t as

$$\mathrm{d}u(t)/\mathrm{d}t + A(t)u(t) = f(t), \qquad 0 < t \leq T \tag{5.1}$$

by reversing the sign of $A(t)$, as compared with the preceding chapter. This is done for the purpose of avoiding the notational complexity due to the appearance of fractional powers of $A(t)$ and of quadratic forms satisfying Gårding's inequality.

Definition 5.1.1 (5.1) *is said to be a* parabolic equation *if* $-A(t)$ *for each* $t \in [0, T]$ *generates an analytic semigroup.*

As in the preceding chapter, we are going to study the existence and the uniqueness of a solution of (5.1) under the initial condition

$$u(0) = u_0. \tag{5.2}$$

The fundamental solution of the parabolic equation can be constructed by a method different from the one adopted in the preceding chapter. It is called *Levi's method*, and it has been used for a long time to construct the fundamental solutions of elliptic and parabolic partial differential equations. Throughout this chapter, (5.1) is always assumed to be parabolic and, if necessary, the unknown $u(t)$ is changed into $e^{-kt}u(t)$ with a sufficiently large positive number k so that $0 \in \rho(A(t))$. In other words, we make the following

Assumption 5.1.1 $A(t)$ *for each* $t \in [0, T]$ *is a closed operator defined densely in a Banach space* X. *Its resolvent set* $\rho(A(t))$ *contains the half-plane* $\operatorname{Re} \lambda \leq 0$, *and* $(1 + |\lambda|)(A(t) - \lambda)^{-1}$*is uniformly bounded in* $0 \leq t \leq T$ *and* $\operatorname{Re} \lambda \leq 0$.

Hence, by Remark 3.3.2 there exist a certain number M and an angle $\theta \in (0, \pi/2)$ such that $\rho(A(t))$ contains the closed sector $\Sigma = \{\lambda : |\arg \lambda| \geq \theta\} \cup \{0\}$ and the estimate

$$\|(A(t) - \lambda)^{-1}\| \leq M/(1 + |\lambda|) \tag{5.3}$$

holds for $0 \leq t \leq T$ and $\lambda \in \Sigma$. In this chapter, the letter C is commonly used to denote a constant determined only by the assumptions made on each occasion.

5.2   The case in which the domain of $A(t)$ is independent of t

An additional assumption in this section is

Assumption 5.2.1   The domain $D(A(t)) \equiv D$ of $A(t)$ is independent of t and, accordingly, $A(t)A(0)^{-1}$, being a bounded operator, is a Hölder continuous function of t in the norm of $B(X)$. In other words, there exist positive numbers $\alpha \leq 1$ and L such that

$$\|A(t)A(0)^{-1} - A(s)A(0)^{-1}\| \leq L|t - s|^{\alpha} \tag{5.4}$$

is satisfied for $0 \leq s \leq T$ and $0 \leq t \leq T$.

In this section, the fundamental solutions of (5.1) will be constructed under Assumptions 5.1.1 and 5.2.1. Assumption 5.2.1 means that $A(t)A(0)^{-1}$ is a norm continuous function of t and that $A(0)A(t)^{-1} = (A(t)A(0)^{-1})^{-1}$ (by Theorem 1.1.15). Thus, $A(0)A(t)^{-1}$ is uniformly bounded. Therefore, we may suppose that

$$\|A(t)A(r)^{-1} - A(s)A(r)^{-1}\| \leq L|t - s|^{\alpha} \tag{5.5}$$

holds for all $t, s, r \in [0, T]$, if necessary, by substituting L by another number. The fundamental solution $U(t, s)$ is constructed as follows. Put

$$U(t, s) = \exp\left(-(t - s)A(s)\right) + W(t, s), \tag{5.6}$$

$$W(t, s) = \int_{s}^{t} \exp\left(-(t - \tau)A(\tau)\right)R(\tau, s)\, d\tau. \tag{5.7}$$

A formal calculation gives

$$(\partial/\partial t)U(t, s) = -A(s)\exp\left(-(t - s)A(s)\right) + R(t, s)$$
$$- \int_{s}^{t} A(\tau)\exp\left(-(t - \tau)A(\tau)\right)R(\tau, s)\, d\tau,$$

$$A(t)U(t, s) = A(t)\exp\left(-(t - s)A(s)\right)$$
$$+ \int_{s}^{t} A(t)\exp\left(-(t - \tau)A(\tau)\right)R(\tau, s)\, d\tau.$$

Summing these two equalities side by side, we obtain

$$(\partial/\partial t)U(t, s) + A(t)U(t, s) = -R_1(t, s) + R(t, s)$$

$$-\int_s^t R_1(t, \tau)R(\tau, s)\,d\tau, \qquad (5.8)$$

where

$$R_1(t, s) = -(A(t) - A(s))\exp(-(t-s)A(s)). \qquad (5.9)$$

By Theorem 3.3.1, there exists a constant C_0 such that

$$\|\exp(-tA(s))\| \leq C_0, \qquad (5.10)$$

$$\|A(s)\exp(-tA(s))\| \leq C_0 t^{-1}. \qquad (5.11)$$

Thus, by (5.5) and (5.11), we can estimate the norm of $R_1(t, s)$ for $s < t$ as

$$\|R_1(t, s)\| \leq \|(A(t) - A(s))A(s)^{-1}\|\, \|A(s)\exp(-(t-s)A(s))\|$$

$$\leq LC_0(t-s)^{\alpha-1}. \qquad (5.12)$$

Since the right-hand side of (5.8) vanishes if $t > s$, we find $R(t, s)$ as a solution of the integral equation

$$R(t, s) - \int_s^t R_1(t, \tau)R(\tau, s)\,d\tau = R_1(t, s). \qquad (5.13)$$

It is easy to ascertain that $\exp(-(t-s)A(s))$ and $R_1(t, s)$ for $0 \leq s < t \leq T$ are continuous in the norm of $B(X)$. Because of (5.12), the integral equation (5.13) can be solved by successive iteration:

$$R(t, s) = \sum_{m=1}^{\infty} R_m(t, s), \qquad (5.14)$$

$$R_m(t, s) = \int_s^t R_1(t, \tau)R_{m-1}(\tau, s)\,d\tau. \qquad (5.15)$$

By induction, it is not difficult to see that

$$\|R_m(t, s)\| \leq (LC_0\Gamma(\alpha))^m (t-s)^{m\alpha-1}/\Gamma(m\alpha).$$

Thus, we have

$$\|R(t, s)\| \leq \sum_{m=1}^{\infty} (LC_0\Gamma(\alpha))^m (t-s)^{m\alpha-1}/\Gamma(m\alpha)$$

$$\leq \sum_{m=1}^{\infty} (LC_0\Gamma(\alpha))^m T^{(m-1)\alpha}\Gamma(m\alpha)^{-1}(t-s)^{\alpha-1} = C(t-s)^{\alpha-1},$$

$$\qquad (5.16)$$

$$\|W(t, s)\| \leq C(t-s)^{\alpha}, \qquad (5.17)$$

$$\|U(t, s)\| \leq C. \qquad (5.18)$$

Now we will make sure that $U(t, s)$ thus constructed is really the fundamental solution. To this end, let us prepare the following lemma.

Lemma 5.2.1 *Let β be an arbitrary positive number satisfying $0 < \beta < \alpha$. Then there exists a constant C_β such that the inequality*

$$\|R(t, s) - R(\tau, s)\| \leq C_\beta (t - \tau)^\beta (\tau - s)^{\alpha - \beta - 1} \tag{5.19}$$

holds for $0 \leq s_\tau < t \leq T$.

Proof   *Let us begin with the estimation of*

$$\begin{aligned}
R_1(t, s) - R_1(\tau, s) = &-(A(t) - A(\tau)) \exp\left(-(t - s)A(s)\right) \\
&- (A(\tau) - A(s))\{\exp\left(-(t - s)A(s)\right) \\
&- \exp\left(-(\tau - s)A(s)\right)\}.
\end{aligned} \tag{5.20}$$

The norm of the first term on the right-hand side does not exceed

$$\begin{aligned}
\|(A(t) - A(\tau))A(s)^{-1}\| \|A(s) \exp\left(-(t - s)A(s)\right)\| &\leq C(t - \tau)^\alpha (t - s)^{-1} \\
&\leq C(t - \tau)^\alpha (\tau - s)^{-1}.
\end{aligned}$$

The norm of the second term does not exceed

$$\begin{aligned}
\left\|(A(\tau) - A(s)) \int_\tau^t (d/dr) \exp\left(-(r - s)A(s)\right) dr\right\| & \\
= \left\|(A(\tau) - A(s))A(s)^{-1} \int_\tau^t A(s)^2 \exp\left(-(r - s)A(s)\right) dr\right\| & \\
\leq C(\tau - s)^\alpha \int_\tau^t (r - s)^{-2} \, dr = C(t - \tau)(t - s)^{-1}(\tau - s)^{\alpha - 1} & \\
\leq C(t - \tau)(\tau - s)^{\alpha - 2}, &
\end{aligned}$$

but it can also be estimated as

$$\begin{aligned}
\|(A(\tau) - A(s)) \exp\left(-(t - s)A(s)\right)\| + \|(A(\tau) - A(s)) \exp\left(-(\tau - s)A(s)\right)\| & \\
\leq C(\tau - s)^\alpha (t - s)^{-1} + C(\tau - s)^{\alpha - 1} \leq C(\tau - s)^{\alpha - 1}; &
\end{aligned}$$

thus, it does not exceed

$$C\{(t - \tau)(\tau - s)^{\alpha - 2}\}^\alpha \{(\tau - s)^{\alpha - 1}\}^{1 - \alpha} \leq C(t - \tau)^\alpha (\tau - s)^{-1}.$$

In this way, we have

$$\|R_1(t, s) - R_1(\tau, s)\| \leq C(t - \tau)^\alpha (\tau - s)^{-1}. \tag{5.21}$$

On the other hand, it follows from (5.12) that

$$\begin{aligned}
\|R_1(t, s) - R_1(\tau, s)\| &\leq \|R_1(t, s)\| + \|R_1(\tau, s)\| \\
&\leq C(t - s)^{\alpha - 1} + C(\tau - s)^{\alpha - 1} \leq C(\tau - s)^{\alpha - 1}. \tag{5.22}
\end{aligned}$$

The combination of (5.21) and (5.22) gives

$$\|R_1(t, s) - R_1(\tau, s)\| \le C\{(t-\tau)^\alpha (\tau-s)^{-1}\}^{\beta/\alpha} \{(\tau-s)^{\alpha-1}\}^{(\alpha-\beta)/\alpha}$$
$$= C(t-\tau)^\beta (\tau-s)^{\alpha-\beta-1}. \tag{5.23}$$

By the help of the relations

$$R(t, s) - R(\tau, s) = R_1(t, s) - R_1(\tau, s) + \int_\tau^t R_1(t, \sigma) R(\sigma, s) \, d\sigma$$

$$+ \int_s^\tau (R_1(t, \sigma) - R_1(\tau, \sigma)) R(\sigma, s) \, d\sigma,$$

$$\left\| \int_\tau^t R_1(t, \sigma) R(\sigma, s) \, d\sigma \right\| \le C \int_\tau^t (t-\sigma)^{\alpha-1} (\sigma-s)^{\alpha-1} \, d\sigma$$

$$\le C \int_\tau^t (t-\sigma)^{\alpha-1} \, d\sigma (\tau-s)^{\alpha-1}$$

$$= C(t-\tau)^\alpha (\tau-s)^{\alpha-1},$$

which are obtained from (5.13), (5.12), (5.16), the inequalities (5.23) and (5.16) lead immediately to the inequality (5.19). □

Lemma 5.2.2 *Let u be an arbitrary element of X. Then $\exp(-\varepsilon A(t))u \to u$ strongly and uniformly in $0 \le t \le T$ as $\varepsilon \to +0$.*

Proof If u belongs to D, it is easy to see that

$$\exp(-\varepsilon A(t))u - u = \int_0^\varepsilon (\partial/\partial\sigma) \exp(-\sigma A(t))u \, d\sigma$$

$$= -\int_0^\varepsilon \exp(-\sigma A(t)) A(t)u \, d\sigma.$$

Hence, by the use of (5.10) and the fact that $A(t)A(0)^{-1}$ is uniformly bounded, we have

$$\|\exp(-\varepsilon A(t))u - u\| \le \varepsilon C \|A(0)u\|. \tag{5.24}$$

Thus the lemma has been proved for $u \in D$. If u is a general element, then it may be approached by a sequence of elements of D. □

Lemma 5.2.3 *Define*

$$S(t, s) = A(t) \exp(-(t-s)A(t)) - A(s) \exp(-(t-s)A(s))$$

for $0 \le s < t \le T$. It has the estimate

$$\|S(t, s)\| \le C(t-s)^{\alpha-1}. \tag{5.25}$$

Proof $S(t, s)$ can be represented by the integral

$$S(t, s) = \frac{1}{2\pi i} \int_\Gamma \lambda e^{-\lambda(t-s)} \{(A(t) - \lambda)^{-1} - (A(s) - \lambda)^{-1}\} \, d\lambda, \tag{5.26}$$

where Γ is a smooth path in Σ connecting $\infty e^{-i\theta}$ and $\infty e^{i\theta}$. On the other hand, by (5.3) and (5.5), we have

$$\|(A(t) - \lambda)^{-1} - (A(s) - \lambda)^{-1}\| = \|(A(t) - \lambda)^{-1}(A(t) - A(s))(A(s) - \lambda)^{-1}\|$$
$$\leqslant \|(A(t) - \lambda)^{-1}\| \|(A(t) - A(s))A(s)^{-1}\| \|A(s)(A(s) - \lambda)^{-1}\|$$
$$\leqslant C(t - s)^\alpha / |\lambda|.$$

From these relations, we obtain (5.24) in a way similar to the proof of (3.27). $\square$

Put

$$W_\varepsilon(t, s) = \int_s^{t-\varepsilon} \exp\left(-(t - \tau)A(\tau)\right) R(\tau, s) \, d\tau$$

for $0 < \varepsilon < t - s$. $W_\varepsilon(t, s) \to W(t, s)$ as $\varepsilon \to 0$. By differentiation, we have

$$(\partial/\partial t) W_\varepsilon(t, s) = \exp\left(-\varepsilon A(t - \varepsilon)\right) R(t - \varepsilon, s)$$
$$- \int_s^{t-\varepsilon} A(\tau) \exp\left(-(t - \tau)A(\tau)\right) R(\tau, s) \, d\tau.$$

Upon observing the relation

$$A(t) \exp\left(-(t - \tau)A(t)\right) = (\partial/\partial \tau) \exp\left(-(t - \tau)A(t)\right),$$

the right-hand side can be rewritten as

$$(\partial/\partial t) W_\varepsilon(t, s) = \exp\left(-\varepsilon A(t - \varepsilon)\right) R(t - \varepsilon, s) + \int_s^{t-\varepsilon} S(t, \tau) R(\tau, s) \, d\tau$$
$$- \int_s^{t-\varepsilon} A(t) \exp\left(-(t - \tau)A(t)\right)(R(\tau, s) - R(t, s)) \, d\tau$$
$$- \{\exp\left(-\varepsilon A(t)\right) - \exp\left(-(t - s)A(t)\right)\} R(t, s). \tag{5.27}$$

By (5.10), (5.11), (5.16), and Lemmas 5.2.1 and 5.2.3, it is found that the norm of $(\partial/\partial t) W_\varepsilon(t, s)$ satisfies

$$\|(\partial/\partial t) W_\varepsilon(t, s)\| \leqslant C(t - s - \varepsilon)^{\alpha - 1}, \tag{5.28}$$

where C is a constant independent of ε as well. It is also easy to see, by Lemma 5.2.2, that each term on the right-hand side of (5.27) converges strongly as $\varepsilon \to 0$. Putting $W'(t, s) = \lim_{\varepsilon \to 0} (\partial/\partial t) W_\varepsilon(t, s)$, we obtain from (5.27) and (5.28) that

$$W'(t, s) = \int_s^t S(t, \tau) R(\tau, s) \, d\tau$$

$$- \int_s^t A(t) \exp\left(-(t-\tau)A(t)\right)(R(\tau, s) - R(t, s)) \, d\tau$$

$$+ \exp\left(-(t-s)A(t)\right)R(t, s), \tag{5.29}$$

$$\|W'(t, s)\| \le C(t-s)^{\alpha-1}. \tag{5.30}$$

Letting $\varepsilon \to 0$ in

$$W_\varepsilon(t', s) - W_\varepsilon(t, s) = \int_t^{t'} (\partial/\partial r) W_\varepsilon(r, s) \, dr$$

with $t' > t > s + \varepsilon$, we have, owing to (5.28),

$$W(t', s) - W(t, s) = \int_t^{t'} W'(r, s) \, dr.$$

Since $W'(t, s)$ is strongly continuous in $0 \le s < t \le T$, $W(t, s)$ is strongly continuously differentiable with respect to t, and, hence, it is found that

$$(\partial/\partial t) W(t, s) = W'(t, s). \tag{5.31}$$

Therefore, the derivative

$$(\partial/\partial t) U(t, s) = -A(s) \exp\left(-(t-s)A(s)\right) + (\partial/\partial t) W(t, s) \tag{5.32}$$

exists and satisfies

$$\|(\partial/\partial t) U(t, s)\| \le C(t-s)^{-1}.$$

For $0 < \varepsilon < t - s$, again, we put

$$U_\varepsilon(t, s) = \exp\left(-(t-s)A(s)\right) + W_\varepsilon(t, s),$$

then $R(U_\varepsilon(t, s)) \subset D$ and $Y_\varepsilon(t, s)$, defined by

$$Y_\varepsilon(t, s) = (\partial/\partial t) U_\varepsilon(t, s) + A(t) U_\varepsilon(t, s),$$

is evaluated as

$$Y_\varepsilon(t, s) = -R_1(t, s) + \exp\left(-\varepsilon A(t-\varepsilon)\right) R(t-\varepsilon, s)$$

$$- \int_s^{t-\varepsilon} R_1(t, \tau) R(\tau, s) \, d\tau.$$

It is found by (5.13) and Lemma 5.2.2 that

$$\text{s-}\lim_{\varepsilon \to 0} Y_\varepsilon(t, s) = 0.$$

On the other hand, since $U_\varepsilon(t, s) \to U(t, s)$ and $(\partial/\partial t) U_\varepsilon(t, s) \to (\partial/\partial t) U(t, s)$ as $\varepsilon \to 0$, the limit s-$\lim_{\varepsilon \to 0} A(t) U_\varepsilon(t, s)$ exists and, since the operator $A(t)$ is closed, $R(U(t, s)) \subset D$ and $A(t) U(t, s) = \lim_{\varepsilon \to 0} A(t) U_\varepsilon(t, s)$. Therefore, $U(t, s)$ for $s < t$ satisfies

$$(\partial/\partial t) U(t, s) + A(t) U(t, s) = 0.$$

Next we will show that $A(t) U(t, s) A(s)^{-1}$ is uniformly bounded.

Lemma 5.2.4 $\|R_1(t, s) A(s)^{-1}\| \le C(t - s)^\alpha$.

Proof   The lemma is evident from the identity $R_1(t, s) A(s)^{-1} = -(A(t) - A(s)) A(s)^{-1} \exp(-(t - s) A(s))$.   □

Lemma 5.2.5 $\|R(t, s) A(s)^{-1}\| \le C(t - s)^\alpha$.

Proof   This lemma is evident from the above lemma and the fact that $R(t, s)$ is also a solution of the equation

$$R(t, s) - \int_s^t R(t, \tau) R_1(\tau, s) \, d\tau = R_1(t, s).   □$$

By copying the proof of Lemma 5.2.1, we can prove

Lemma 5.2.6   *For each β satisfying $0 < \beta < \alpha$, there exists a constant C_β such that*

$$\|R(t, s) A(s)^{-1} - R(\tau, s) A(s)^{-1}\|$$
$$\le C_\beta \left\{ (t - \tau)^\alpha + (\tau - s)^\alpha \log \frac{t - s}{\tau - s} + (t - \tau)^\beta (\tau - s)^{2\alpha - \beta} \right\}.$$

Multiply (5.29) from the right by $A(s)^{-1}$. The equality thus obtained, (5.31), and Lemmas 5.2.3, 5.2.5, 5.2.6, imply the following lemma:

Lemma 5.2.7   *The inequality*

$$\|(\partial/\partial t) W(t, s) A(s)^{-1}\| \le C(t - s)^\alpha$$

holds for $0 \le s < t \le T$.

Lemma 5.2.8 $A(t) U(t, s) A(s)^{-1}$ *is uniformly bounded on $0 \le s \le t \le T$.*

Proof   The lemma is evident from (5.32), (5.11) and Lemma 5.2.7.   □

Next we will show, in a rather round-about way, that $U(t, s)u$ for each $u \in D$ is strongly differentiable with respect to $s \in [0, t]$ and that

$$(\partial/\partial s)U(t, s)u = U(t, s)A(s)u \qquad (5.33)$$

holds. First, suppose that $A(t)u$ for each $u \in D$ is strongly continuously differentiable in $[0, T]$. In this case, $(d/dt)A(t)A(0)^{-1} = A'(t)A(0)^{-1}$ is uniformly bounded. An operator-valued function $V(t, s)$ satisfying

$$(\partial/\partial s)V(t, s)u = V(t, s)A(s)u, \qquad 0 \leqslant s < t \leqslant T, \qquad V(t, t) = I$$

for each $u \in D$ is constructed as follows. Put

$$Q_1(t, s) = (\partial/\partial t + \partial/\partial s) \exp\left(-(t - s)A(s)\right).$$

Since $(A(s) - \lambda)^{-1}$ is differentiable with respect to s and

$$(\partial/\partial s)(A(s) - \lambda)^{-1} = -(A(s) - \lambda)^{-1}A'(s)(A(s) - \lambda)^{-1}.$$

there holds, by (5.3), the result

$$\|(\partial/\partial s)(A(s) - \lambda)^{-1}\| \leqslant C/|\lambda| \qquad (5.34)$$

for each $\lambda \in \Sigma$. In a manner analogous to the proof of (3.27), we obtain, from (5.34), that

$$\|Q_1(t, s)\| = \left\| \frac{1}{2\pi i} \int_\Gamma e^{-\lambda(t-s)}(\partial/\partial s)(A(s) - \lambda)^{-1} \, d\lambda \right\| \leqslant C,$$

$$\|(\partial/\partial s) \exp\left(-(t - s)A(s)\right)\| \leqslant C(t - s)^{-1},$$

where Γ is the path which appeared in the proof of Lemma 5.2.3. Put

$$V(t, s) = \exp\left(-(t - s)A(s)\right) + \int_s^t Q(t, \tau) \exp\left(-(\tau - s)A(s)\right) d\tau.$$

By a formal calculation, as in the construction of $U(t, s)$, we arrive at the integral equation

$$Q(t, s) - \int_s^t Q(t, \tau)Q_1(\tau, s) \, d\tau = Q_1(t, s). \qquad (5.35)$$

This equation can be solved by successive iteration and it is found that the solution $Q(t, s)$ is uniformly bounded and so is $V(t, s)$. Let u be an arbitrary element of X, then, for $s < r < t$, we have

$$(\partial/\partial r)\{V(t, r)U(r, s)u\} = V(t, r)A(r)U(r, s)u - V(t, r)A(r)U(r, s)u = 0,$$

which shows that $V(t, r)U(r, s)$ is independent of r in $s < r < t$. Therefore, by letting $r \to s$ and $r \to t$, we obtain $V(t, s) = U(t, s)$. Thus (5.33) has been proved and, along with it, $U(t, s)$ has turned out to be the desired fundamental solution. What we must do next is to remove the assumption that $A(t)A(0)^{-1}$ is continuously differentiable. For this purpose we use a mollifier which approximates $A(t)$ by a differentiable operator. Let $j(t)$ be a function, differentiable continuously in $-\infty < t < \infty$, such that $j(t) \geqslant 0$, $j(t) = 0$ for $|t| > 1$ and $\int_{-\infty}^{\infty} j(t) \, dt = 1$, and set $j_n(t) = nj(nt)$ for each natural number n. $A(t)$ is considered as defined in $-\infty < t < \infty$ by putting $A(t) = A(0)$ for $t < 0$ and $A(t) = A(T)$ for $t > T$. Define

$$A_n(t)u = \int_{-\infty}^{\infty} j_n(t-\tau)A(\tau)u \, d\tau \qquad (5.36)$$

for each $u \in D$. Then, for each $\lambda \in \Sigma$ and each $u \in X$, we have

$$u - (A_n(t) - \lambda)(A(t) - \lambda)^{-1}u = (A(t) - A_n(t))(A(t) - \lambda)^{-1}u$$

$$= \int_{-\infty}^{\infty} j_n(t-\tau)(A(t) - A(\tau))(A(t) - \lambda)^{-1}u \, d\tau.$$

Since $\|(A(t) - A(\tau))(A(t) - \lambda)^{-1}\| \leqslant Cn^{-\alpha}$ for t, τ such that $j_n(t-\tau) \neq 0$, it is found that

$$\|u - (A_n(t) - \lambda)(A(t) - \lambda)^{-1}u\| \leqslant Cn^{-\alpha}\|u\| \qquad (5.37)$$

and, hence, in particular, for $\lambda = 0$,

$$\|(A_n(t) - A(t))A(t)^{-1}\| \leqslant Cn^{-\alpha}. \qquad (5.38)$$

From (5.37) it follows that, for each $v \in D$,

$$(1 - Cn^{-\alpha})\|(A(t) - \lambda)v\| \leqslant \|(A_n(t) - \lambda)v\|$$

$$\leqslant (1 + Cn^{-\alpha})\|(A(t) - \lambda)v\|,$$

which shows that $A_n(t)$ is a closed operator if $Cn^{-\alpha} < 1$. In this case, it can be shown, similarly to (5.37), that the operator $(A_n(t) - \lambda) \times (A(t) - \lambda)^{-1}$ has a bounded inverse and, hence, $R(A_n(t) - \lambda) = X$ and $\rho(A_n(t)) \supset \Sigma$. It is also easy to see that there exists a constant C, independent of n, such that $\|(A_n(t) - \lambda)^{-1}\| \leqslant C/|\lambda|$ holds for $\lambda \in \Sigma$. Furthermore, since (5.38) implies the uniform boundedness of

$$A(0)A_n(0)^{-1} = \sum_{m=0}^{\infty} (1 - A_n(0)A(0)^{-1})^m,$$

it is also clear from (5.4) that

$$\|A_n(t)A_n(0)^{-1} - A_n(s)A_n(0)^{-1}\| \leqslant C|t - s|^\alpha.$$

From these facts, we can conclude that the equation

$$du(t)/dt + A_n(t)u(t) = 0$$

has a fundamental solution $U_n(t, s)$, for which $\|U_n(t, s)\|$ is bounded uniformly in n as well. In addition, since $A_n(t)u$ for each $u \in D$ is continuously differentiable with respect to t, the equality

$$(\partial/\partial s)U_n(t, s)u = U_n(t, s)A_n(s)u \qquad (5.39)$$

is valid for $u \in D$. The equation (5.39) can be used to derive

$$U(t, s)u - U_n(t, s)u = \int_s^t (\partial/\partial r)\{U_n(t, r)U(r, s)u\}\, dr$$

$$= \int_s^t U_n(t, r)(A_n(r) - A(r))U(r, s)A(s)^{-1}A(s)u\, dr.$$

On account of (5.38) and Lemma 5.2.8, we have

$$\|U(t, s)u - U_n(t, s)u\| \leq Cn^{-\alpha}(t-s)\|A(s)u\|,$$

so that $U_n(t, s)u \to U(t, s)u$ uniformly in t and s. Assuming $s < s' < t$, we consider

$$U_n(t, s')u - U_n(t, s)u = \int_s^{s'} (\partial/\partial r)U_n(t, r)u\, dr$$

$$= \int_s^{s'} U_n(t, r)A_n(r)u\, dr,$$

which, in the limit as $n \to \infty$, yields

$$U(t, s')u - U(t, s)u = \int_s^{s'} U(t, r)A(r)u\, dr,$$

so that (5.33) has been obtained.

Summing up, we have

Theorem 5.2.1 *Under Assumptions 5.1.1 and 5.2.1, a fundamental solution of (5.1) exists. For $0 \leq s < t \leq T$, the range $R(U(t, s)) \subset D$, the operator $(\partial/\partial t)U(t, s)$ exists as an element of $B(X)$, and the following inequalities hold:*

$$\|(\partial/\partial t)U(t, s)\| = \|A(t)U(t, s)\| \leq C(t-s)^{-1},$$

$$\|A(t)U(t, s)A(s)^{-1}\| \leq C.$$

$U(t, s)u$ for each $t \in (0, T]$ and each $u \in D$ is differentiable with respect to s in $0 \leq s \leq t$ and satisfies (5.33).

By this theorem if u_0 is an arbitrary element of X, then $u(t) = U(t, 0)u_0$ is continuous in $0 \leq t \leq T$, differentiable in $0 < t \leq T$, and is a solution of the homogeneous equation $du(t)/dt + A(t)u(t) = 0$, which coincides with u_0 at $t = 0$. Hence, as in Section 3.3, a function $u(t)$ is said to be a *solution* of the equations (5.1) and (5.2) if it satisfies these equations and if $u \in C([0, T]; X) \cap C^1((0, T]; X)$, $u(t) \in D$ for each $t \in (0, T]$ and $A(t)u(t) \in C((0, T]; X)$.

Theorem 5.2.2 *Let $u_0 \in X$ and $f \in C([0, T]; X)$. If u is a solution of (5.1) and (5.2), then it is expressible as*

$$u(t) = U(t, 0)u_0 + \int_0^t U(t, s)f(s)\, ds, \qquad (5.40)$$

so that the solution of (5.1) and (5.2) is unique.

Proof Assume $0 < \varepsilon < s < t$. Then, from (5.33), it follows that

$$(\partial/\partial s)(U(t, s)u(s)) = U(t, s)u'(s) + U(t, s)A(s)u(s)$$
$$= U(t, s)f(s). \qquad (5.41)$$

By integrating this equation from ε to t and letting $\varepsilon \to 0$, we obtain (5.40) immediately. $\square$

In conformity with Theorem 3.3.4, we have

Theorem 5.2.3 *Let u_0 be an arbitrary element of X and f an arbitrary function Hölder-continuous in $[0, T]$. Then the function u defined by (5.39) is the unique solution of (5.1) and (5.2).*

Proof The relations (5.30) and (5.31) imply that

$$(\partial/\partial t)\int_0^t W(t, s)f(s)\, ds = \int_0^t (\partial/\partial t)W(t, s)f(s)\, ds. \qquad (5.42)$$

Also, as in the proof of (5.29), we have

$$(\partial/\partial t)\int_0^t \exp\left(-(t-s)A(s)\right)f(s)\, ds = \int_0^t S(t, s)f(s)\, ds$$

$$- \int_0^t A(t)\exp\left(-(t-s)A(s)\right)(f(s) - f(t))\, ds + \exp\left(-tA(t)\right)f(t). \quad (5.43)$$

From these equations, and noting that

$$(\partial/\partial t)\int_0^{t-\varepsilon} U(t, s)f(s)\, ds - A(t)\int_0^{t-\varepsilon} U(t, s)f(s)\, ds - f(t)$$

$$= U(t, t-\varepsilon)f(t-\varepsilon) - f(t) \to 0,$$

as $\varepsilon \to +0$, we obtain the conclusion of the theorem. $\square$

Example Let Ω be a bounded region in $\mathbb{R}^n$ which is of class $\mathbb{C}^m$. The operator defined in Ω by

$$A(x, t, D) = \sum_{|\alpha| \leq m} a_\alpha(x, t) D^\alpha$$

is strongly elliptic, uniformly in $t \in [0, T]$. Now, it is assumed that for each t the coefficients of the highest-order derivatives are continuous in $\bar{\Omega}$ and the other coefficients are bounded and measurable in Ω. As for t, every coefficient is assumed to satisfy Hölder's condition of order h uniformly, namely

$$\max_{|\alpha| \leq m} \sup_{x \in \Omega} |a_\alpha(x, t) - a_\alpha(x, s)| \leq L|t - s|^h.$$

Put $D(A(t)) = W_p^m(\Omega) \cap \mathring{W}_p^{m/2}(\Omega)$ with $1 < p < \infty$ and $(A(t)u)(x) = A(x, t, D)u(x)$. Then, by Theorem 3.8.3, $-A(t)$ for each t generates an analytic semigroup in $L^p(\Omega)$ and satisfies Assumptions 5.1.1 and 5.1.2. Therefore, we can apply the results of the present section to this operator and the solution of (5.1) and (5.2) is a generalized solution of the following mixed problem:

$$\partial u(x, t)/\partial t + A(x, t, D)u(x, t) = f(x, t), \qquad x \in \Omega, \quad 0 < t \leq T,$$
$$u(x, 0) = u_0(x), \qquad x \in \Omega,$$
$$(\partial/\partial x)^\alpha u(x, t) = 0, \qquad |\alpha| \leq m/2 - 1, \qquad x \in \partial\Omega, \quad 0 < t \leq T.$$

5.3 The case in which the domain of A(t) varies with t

At the outset, we state the assumptions in the present section.

Assumption 5.3.1 $A(t)^{-1}$ *is continuously differentiable with respect to* $t \in [0, T]$ *in the norm of* $B(X)$.

Assumption 5.3.2 $dA(t)^{-1}/dt$ *is Hölder continuous with respect to* t *in the norm of* $B(X)$. *In other words, there exist a positive number* K *and* $\alpha \leq 1$ *such that, for each* $t, s \in [0, T]$,

$$\|dA(t)^{-1}/dt - \dot{d}A(s)^{-1}/ds\| \leq K|t - s|^\alpha. \tag{5.44}$$

Assumptions 5.1.1 and 5.3.1 imply that $(A(t) - \lambda)^{-1}$ is also a differentiable function of t and the relation

$$(\partial/\partial t)(A(t) - \lambda)^{-1} = A(t)(A(t) - \lambda)^{-1} dA(t)^{-1}/dt A(t)(A(t) - \lambda)^{-1} \tag{5.45}$$

holds. Since $A(t)(A(t)-\lambda)^{-1} = 1 + (A(t)-\lambda)^{-1}$ is uniformly bounded, the relation (5.43) gives a bound

$$\|(\partial/\partial t)(A(t)-\lambda)^{-1}\| \leq C, \tag{5.46}$$

but we will make a stronger assumption.

Assumption 5.3.3 *There exist two positive numbers N and $\rho \leq 1$ such that the inequality*

$$\|(\partial/\partial t)(A(t)-\lambda)^{-1}\| \leq N/|\lambda|^{\rho} \tag{5.47}$$

holds for every $\lambda \in \Sigma$ and $t \in [0, T]$.

In this section Γ again represents a smooth curve in Σ which connects $\infty e^{-i\theta}$ and $\infty e^{i\theta}$.

Lemma 5.3.1 *Under Assumptions 5.1.1 and 5.3.1, the operator $\exp(-(t-s)A(t))$ is differentiable with respect to t, s in $0 \leq s < t \leq T$ and its derivatives have the following estimates:*

$$\|(\partial/\partial t)\exp(-(t-s)A(t))\| \leq C(t-s)^{-1}, \tag{5.48}$$
$$\|(\partial/\partial s)\exp(-(t-s)A(t))\| \leq C(t-s)^{-1}. \tag{5.49}$$

Proof The latter is evident from $(\partial/\partial s)\exp(-(t-s)A(t)) = A(t)\exp(-(t-s)A(t))$. The former comes from

$$(\partial/\partial t)\exp(-(t-s)A(t))$$
$$= -A(t)\exp(-(t-s)A(t)) + \frac{1}{2\pi i}\int_{\Gamma} e^{-\lambda(t-s)}(\partial/\partial t)(A(t)-\lambda)^{-1}\,\mathrm{d}\lambda$$

combined with (5.49) and (5.46). $\square$

For the time being, we will use only Assumptions 5.1.1, 5.3.1 and 5.3.3. The fundamental solution $U(t, s)$ is constructed in the following form, slightly different from that in the preceding section. Putting

$$U(t, s) = \exp(-(t-s)A(t)) + W(t, s), \tag{5.50}$$

where

$$W(t, s) = \int_{s}^{t} \exp(-(t-\tau)A(t))R(\tau, s)\,\mathrm{d}\tau; \tag{5.51}$$

then, as in the preceding section, after a formal calculation, we arrive at the following integral equation of the same form as (5.13):

$$R(t, s) - \int_{s}^{t} R_1(t, \tau)R(\tau, s)\,\mathrm{d}\tau = R_1(t, s), \tag{5.52}$$

where we have put

$$R_1(t, s) = -(\partial/\partial t + \partial/\partial s) \exp(-(t-s)A(t)).$$ (5.53)

$R_1(t, s)$ is norm continuous in $0 \leq s < t \leq T$. Assumption 5.3.3 and

$$R_1(t, s) = \frac{-1}{2\pi i} \int_\Gamma e^{-\lambda(t-s)} \frac{\partial}{\partial t}(A(t) - \lambda)^{-1} d\lambda$$

imply immediately that

$$\|R_1(t, s)\| \leq C(t-s)^{\rho-1},$$ (5.54)

which enables us to solve (5.53) by successive iteration as follows:

$$R(t, s) = \sum_{m=1}^{\infty} R_m(t, s),$$ (5.55)

with

$$R_m(t, s) = \int_s^t R_1(t, \tau) R_{m-1}(\tau, s) \, d\tau.$$ (5.56)

Like (5.16), it is also found that the following estimate holds:

$$\|R(t, s)\| \leq C(t-s)^{\rho-1}.$$ (5.57)

Thus, $U(t, s)$ is constructed by (5.50) and (5.51). Hitherto, we have not used Assumption 5.3.2, but, without this assumption, it is not clear whether $U(t, s)$ thus constructed is really the desired fundamental solution or not. In particular, it is not clear whether or not the function given by

$$u(t) = U(t, 0)u_0 + \int_0^t U(t, s)f(s) \, ds$$ (5.58)

is a solution of (5.1) and (5.2), but it can be shown that (5.58) is a solution in the following weak sense.

Definition 5.3.1 *A function $u \in C([0, T]; X)$ is said to be a weak solution of (5.1) and (5.2) in $[0, T]$ if it satisfies*

$$\int_0^T (u(t), \varphi'(t) - A^*(t)\varphi(t)) \, dt + \int_0^t (f(t), \varphi(t)) \, dt + (u_0, \varphi(0)) = 0 \quad (5.59)$$

for every φ, subject to three conditions below:

(i) *$\varphi(t) \in D(A^*(t))$ for each t,*
(ii) *$\varphi, \varphi' \ (= d\varphi/dt)$ and $A^*\varphi$ all belong to $C([0, T]; X^*)$,*
(iii) *$\varphi(T) = 0$.*

It is not necessarily evident for us to conclude that we have sufficiently many functions φ satisfying the conditions (i), (ii), (iii), but the conclusion is supported by the result, derived in the sequel, that the weak solution is unique. Incidentally, let u^* be an element of X^*, q a complex-valued function, continuously differentiable in $[0, T]$, and assume that $q(T) = 0$, then $\varphi(t) = q(t)A^*(t)^{-1}u^*$ satisfies (i), (ii), (iii).

Theorem 5.3.1 *Suppose Assumptions 5.1.1, 5.3.1 and 5.3.3 are satisfied. The function $u(t)$ given by (5.58) with $u_0 \in X$ and $f \in C([0, T]; X)$ is a weak solution of (5.1) and (5.2) in $[0, T]$.*

Proof First, let ε be a sufficiently small positive number and put

$$U_\varepsilon(t, s) = \exp\left(-(t-s)A(t)\right) + \int_s^{t-\varepsilon} \exp\left(-(t-\tau)A(t)\right)R(\tau, s)\, d\tau.$$

$U_\varepsilon(t, s)$ is differentiable in t for $t > s + \varepsilon$ and its range $R(U_\varepsilon(t, s)) \subset D(A(t))$. The operator $Y_\varepsilon(t, s)$ defined by

$$Y_\varepsilon(t, s) = (\partial/\partial t)U_\varepsilon(t, s) + A(t)U_\varepsilon(t, s) \tag{5.60}$$

is easily shown to satisfy

$$\|Y_\varepsilon(t, s)\| \leqslant C(t - s - \varepsilon)^{\rho - 1} \tag{5.61}$$

and

$$\text{s-}\lim_{\varepsilon \to 0} Y_\varepsilon(t, s) = 0. \tag{5.62}$$

Let φ be a function satisfying the conditions (i), (ii), (iii) of Definition 5.3.1, then we find that

$$\int_0^T (U(t, 0)u_0, \varphi'(t))\, dt = \lim_{\eta \to +0} \lim_{\varepsilon \to +0} \int_\eta^T (U_\varepsilon(t, 0)u_0, \varphi'(t))\, dt. \tag{5.63}$$

Here the successive limits on the right-hand side are taken first with respect to ε and then with respect to η. By integration by parts and by (5.60), we obtain, for $0 < \varepsilon < \eta$,

$$\int_\eta^T (U_\varepsilon(t, 0)u_0, \varphi'(t))\, dt = -(U_\varepsilon(\eta, 0)u_0, \varphi(\eta)) - \int_\eta^T (Y_\varepsilon(t, 0)u_0, \varphi(t))\, dt$$

$$+ \int_\eta^T (U_\varepsilon(t, 0)u_0, A^*(t)\varphi(t))\, dt.$$

Now, letting $\varepsilon \to 0$, on account of (5.61) and (5.62), the right-hand side becomes

$$\rightarrow -(U(\eta, 0)u_0, \varphi(\eta)) + \int_\eta^T (U(t, 0)u_0, A^*(t)\varphi(t))\, \mathrm{d}t,$$

which, by letting $\eta \rightarrow 0$ next, turns into

$$\rightarrow -(u_0, \varphi(0)) + \int_0^T (U(t, 0)u_0, A^*(t)\varphi(t))\, \mathrm{d}t.$$

Thus, (5.63) yields

$$\int_0^T (U(t, 0)u_0, \varphi'(t) - A^*(t)\varphi(t))\, \mathrm{d}t + (u_0, \varphi(0)) = 0. \tag{5.64}$$

Further, we represent

$$\int_0^T \left(\int_0^t U(t, \sigma)f(\sigma)\, \mathrm{d}\sigma, \varphi'(t)\right) \mathrm{d}t = \int_0^T \int_\sigma^T (U(t, \sigma)f(\sigma), \varphi'(t))\, \mathrm{d}t\, \mathrm{d}\sigma$$

$$= \lim_{\eta \rightarrow +0} \lim_{\delta \rightarrow +0} \lim_{\varepsilon \rightarrow +0} \int_0^{T-\eta} \int_{\sigma+\delta}^T (U_\varepsilon(t, \sigma)f(\sigma), \varphi'(t))\, \mathrm{d}t\, \mathrm{d}\sigma.$$

For $0 < \varepsilon < \delta < \eta$, the integral on the right-hand side can be rewritten as

$$\int_0^{T-\eta} \int_{\sigma+\delta}^T (U_\varepsilon(t, \sigma)f(\sigma), \varphi'(t))\, \mathrm{d}t\, \mathrm{d}\sigma$$

$$= -\int_0^{T-\eta} (U_\varepsilon(\sigma+\delta, \sigma)f(\sigma), \varphi(\sigma+\delta))\, \mathrm{d}\sigma$$

$$- \int_0^{T-\eta} \int_{\sigma+\delta}^T (Y_\varepsilon(t, \sigma)f(\sigma), \varphi(t))\, \mathrm{d}t\, \mathrm{d}\sigma$$

$$+ \int_0^{T-\eta} \int_{\sigma+\delta}^T (U_\varepsilon(t, \sigma)f(\sigma), A^*(t)\varphi(t))\, \mathrm{d}t\, \mathrm{d}\sigma.$$

On account of (5.61) and (5.62), the last expression tends to

$$\rightarrow -\int_0^{T-\eta} (U(\sigma+\delta, \sigma)f(\sigma), \varphi(\sigma+\delta))\, \mathrm{d}\sigma$$

$$+ \int_0^{T-\eta} \int_{\sigma+\delta}^T (U(t, \sigma)f(\sigma), A^*(t)\varphi(t))\, \mathrm{d}t\, \mathrm{d}\sigma$$

as $\varepsilon \rightarrow 0$. Further, by letting $\delta \rightarrow 0$ and $\eta \rightarrow 0$ in this order, it turns into

$$\rightarrow -\int_0^T (f(\sigma), \varphi(\sigma))\, \mathrm{d}\sigma + \int_0^T \int_\sigma^T (U(t, \sigma)f(\sigma), A^*(t)\varphi(t))\, \mathrm{d}t\, \mathrm{d}\sigma.$$

In this way, we have

$$\int_0^T \left(\int_0^t U(t, \sigma) f(\sigma) \, d\sigma, \, \varphi'(t) - A^*(t)\varphi(t) \right) dt + \int_0^T (f(\sigma), \varphi(\sigma)) \, d\sigma = 0.$$

(5.65)

Result (5.59) follows from (5.64) and (5.65). □

Next, in order to prove the uniqueness of the weak solution, we construct an operator $V(t, s)$ having the following properties:

$$\begin{cases} V(t, s) \text{ is defined in } 0 \leq s \leq t \leq T \text{ as a} \\ \text{strongly continuous operator.} \end{cases}$$

(5.66)

$$V(t, t) = I.$$

(5.67)

$$\begin{cases} \text{For each } t, s \text{ in } 0 \leq s < t \leq T \text{ and each } u \in D(A(s)), \\ \text{the strong limit} \\ \quad \text{s-}\lim_{h \to 0} h^{-1}(V(t, s+h) - V(t, s))u \\ \text{exists and is equal to } V(t, s)A(s)u. \end{cases}$$

(5.68)

An operator satisfying these three conditions, like the operator $V(t, s)$ in the preceding section, can be constructed in the following form:

$$V(t, s) = \exp\left(-(t-s)A(s)\right) + Z(t, s),$$

(5.69)

$$Z(t, s) = \int_s^t Q(t, \tau) \exp\left(-(\tau-s)A(s)\right) d\tau,$$

(5.70)

$$Q(t, s) = \sum_{m=1}^{\infty} Q_m(t, s),$$

(5.71)

$$Q_1(t, s) = (\partial/\partial t + \partial/\partial s) \exp\left(-(t-s)A(s)\right),$$

(5.72)

$$Q_m(t, s) = \int_s^t Q_{m-1}(t, \tau)Q_1(\tau, s) \, d\tau.$$

(5.73)

Like (5.54), by Assumption 5.3.3, the estimate

$$\|Q_1(t, s)\| \leq C(t-s)^{\rho-1}$$

(5.74)

holds. From this, the inequality

$$\|Q(t, s)\| \leq C(t-s)^{\rho-1}$$

(5.75)

follows immediately. It is not difficult to ascertain that $Q(t, s)$ is a solution of an equation of the same form as (5.35) and that $V(t, s)$ satisfies the conditions (5.66)–(5.68).

Theorem 5.3.2 *The weak solution of* (5.1) *and* (5.2) *is unique.*

Proof It suffices to make sure that $u(t) = V(t, 0)u_0$ is the only weak solution of (5.1) and (5.2) when $f(t) \equiv 0$. Assume that $\varepsilon > 0$ and $s + \varepsilon < t$, and put

$$V_\varepsilon(t, s) = \exp\left(-(t - s)A(s)\right) + \int_{s+\varepsilon}^{t} Q(t, \tau) \exp\left(-(\tau - s)A(s)\right) d\tau.$$

Then there exist bounded extensions of $(\partial/\partial s)V_\varepsilon(t, s)$ and $V_\varepsilon(t, s)A(s)$ which are continuous for $0 \leqslant s \leqslant t - \varepsilon$ in the norm of $B(X)$. Accordingly, $A^*(s)V_\varepsilon^*(t, s) = (V_\varepsilon(t, s)A(s))^*$ is also bounded and norm continuous in s. Denote by $P_\varepsilon(t, s)$ the bounded extension of $(\partial/\partial s)V_\varepsilon(t, s) - V_\varepsilon(t, s)A(s)$. Then we have

$$\|P_\varepsilon(t, s)\| \leqslant C(t - s - \varepsilon)^{\rho - 1} \tag{5.76}$$

and

$$\text{s-}\lim_{\varepsilon \to 0} P_\varepsilon(t, s) = 0. \tag{5.77}$$

By definition, the equality

$$\int_0^T (u(t), \psi'(t) - A^*(t)\psi(t)) \, dt + (u_0, \psi(0)) = 0$$

holds for any function ψ satisfying the conditions (i), (ii), (iii) in Definition 5.3.1. Let t_0 be fixed arbitrarily in $0 < t_0 \leqslant T$ and let φ be any continuously differentiable function with values in X^* whose support is contained in $(0, t_0)$. We denote by ε a positive number smaller than the distance between the support of φ and t_0, and define a function $\psi_\varepsilon(t)$ as follows: $\psi_\varepsilon(t) = V_\varepsilon^*(t_0, t)\varphi(t)$ if t belongs to the support of φ and $\psi_\varepsilon(t) = 0$ otherwise. Then ψ_ε satisfies the conditions (i), (ii), (iii) in Definition 5.3.1 and we have

$$\int_0^{t_0} (V(t_0, t)u(t), \varphi'(t)) \, dt = \lim_{\varepsilon \to 0} \int_0^{t_0} (V_\varepsilon(t_0, t)u(t), \varphi'(t)) \, dt$$

$$= \lim_{\varepsilon \to 0} \int_0^{t_0} (u(t), V_\varepsilon^*(t_0, t)\varphi'(t)) \, dt$$

$$= \lim_{\varepsilon \to 0} \int_0^{t_0} (u(t), \psi_\varepsilon'(t) - (\partial/\partial t)V_\varepsilon^*(t_0, t) \cdot \varphi(t)) \, dt$$

$$= \lim_{\varepsilon \to 0} \left\{ \int_0^T (u(t), \psi_\varepsilon'(t) - A^*(t)\psi_\varepsilon(t)) \, dt \right.$$

$$\left. - \int_0^{t_0} (P_\varepsilon(t_0, t)u(t), \varphi(t)) \, dt \right\}.$$

The first term on the right-hand side is equal to 0 because $u(t)$ is a weak solution, and the second term also vanishes by (5.76) and (5.77). This shows that the derivative of $V(t_0, t)u(t)$ in the sense of a distribution vanishes in $(0, t_0)$, and, hence, $V(t_0, t)u(t)$ does not depend on t there. Therefore, $u(t_0) = V(t_0, 0)u_0$ is obtained by letting $t \to 0$ and $t \to t_0$. This completes the proof since t_0 is arbitrary in $(0, T]$.   □

It has already been proved that $U(t, 0)u_0$ is a weak solution of (5.1) and (5.2) when $f \equiv 0$, and, hence, by combining the result with the proof of the above theorem, we find that $V(t, 0)u_0 = U(t, 0)u_0$. Since a similar argument is applicable also for any initial time $s \in (0, T)$, we get

$$V(t, s) = U(t, s) \tag{5.78}$$

for $0 \le s \le t \le T$. From the uniqueness of the weak solution, it follows that

$$U(t, r)U(r, s) = U(t, s) \tag{5.79}$$

for $s \le r \le t$. Further, the limit

$$\lim_{h \to 0} h^{-1}(U(t, s+h) - U(t, s))u = U(t, s)A(s)u \tag{5.80}$$

is obtained for $u \in D(A(s))$.

As in the preceding section, a function $u(t)$ is said to be a *solution* of (5.1) and (5.2) if it satisfies these equations and if $u \in C([0, T]; X) \cap C^1((0, T]; X)$, $u(t) \in D(A(t))$ for all $t \in (0, T]$ and $Au \in C((0, T]; X)$. It is evident that a solution in this sense is a weak solution. Therefore, the uniqueness of the solution is guaranteed by Theorem 5.3.2.

Next, we suppose Assumption 5.3.2 is also satisfied. We will then show that the $U(t, s)$ constructed above is the desired fundamental solution of (5.1) and that (5.58) is a solution of (5.1) and (5.2).

Lemma 5.3.2   *With δ and β being arbitrary numbers satisfying $0 < \delta < \rho$ and $0 < \beta < \alpha$, the following inequality holds for $0 \le s < \tau < t \le T$:*

$$\|R(t, s) - R(\tau, s)\| \le C_{\alpha\beta}\{(t-\tau)^\delta(\tau-s)^{\rho-\delta-1} + (t-\tau)^\beta(\tau-s)^{\alpha-\beta-1}\}.$$

Proof   First, consider

$$R_1(t, s) - R_1(\tau, s) = \frac{-1}{2\pi i}\int_\Gamma e^{-\lambda(t-s)}\left\{\frac{\partial}{\partial t}(A(t) - \lambda)^{-1} - \frac{\partial}{\partial \tau}(A(\tau) - \lambda)^{-1}\right\}d\lambda$$

$$- \frac{1}{2\pi i}\int_\Gamma (e^{-\lambda(t-s)} - e^{-\lambda(\tau-s)})\frac{\partial}{\partial \tau}(A(\tau) - \lambda)^{-1}\,d\lambda = \text{I} + \text{II}.$$

By (5.45), we can rewrite the first term I on the right-hand side:

$$(\partial/\partial t)(A(t)-\lambda)^{-1}-(\partial/\partial\tau)(A(\tau)-\lambda)^{-1}$$
$$= -\{A(t)(A(t)-\lambda^{-1}-A(\tau)(A(\tau)-\lambda)^{-1}\}\,dA(t)^{-1}/dtA(t)(A(t)-\lambda)^{-1}$$
$$-A(\tau)(A(\tau)-\lambda)^{-1}\{dA(t)^{-1}/dt-dA(\tau)^{-1}/d\tau\}A(t)(A(t)-\lambda)^{-1}$$
$$-A(\tau)(A(\tau)-\lambda)^{-1}\,dA(\tau)^{-1}/d\tau\{A(t)(A(t)-\lambda)^{-1}-A(\tau)(A(\tau)-\lambda)^{-1}\}.$$

Since

$$A(t)(A(t)-\lambda)^{-1}-A(\tau)(A(\tau)-\lambda)^{-1}=\lambda\{(A(t)-\lambda)^{-1}-(A(\tau)-\lambda)^{-1}\}$$
$$=\lambda\int_\tau^t (\partial/\partial\sigma)(A(\sigma)-\lambda)^{-1}\,d\sigma,$$

we have, by (5.47),

$$\|A(t)(A(t)-\lambda)^{-1}-A(\tau)(A(\tau)-\lambda)^{-1}\|\le C(t-\tau)|\lambda|^{1-\rho},$$

which, combined with (5.44), yields

$$\|(\partial/\partial t)(A(t)-\lambda)^{-1}-(\partial/\partial\tau)(A(\tau)-\lambda)^{-1}\|\le C\{(t-\tau)|\lambda|^{1-\rho}+(t-\tau)^\alpha\}.$$

Thus, we obtain

$$\|\mathrm{I}\|\le C\{(t-\tau)(t-s)^{\rho-2}+(t-\tau)^\alpha(t-s)^{-1}\}$$
$$\le C\{(t-\tau)^\delta(t-s)^{-\delta}(\tau-s)^{\rho-1}+(t-\tau)^\beta(t-s)^{\alpha-\beta-1}\}. \tag{5.81}$$

Next, since

$$\mathrm{II}= -\int_{\tau-s}^{t-s}\frac{\partial}{\partial\sigma}\left\{\frac{1}{2\pi i}\int_\Gamma e^{-\lambda\sigma}\frac{\partial}{\partial\tau}(A(\tau)-\lambda)^{-1}\,d\lambda\right\}d\sigma$$
$$= -\frac{1}{2\pi i}\int_{\tau-s}^{t-s}\left\{\int_\Gamma \lambda e^{-\lambda\sigma}\frac{\partial}{\partial\tau}(A(\tau)-\lambda)^{-1}\,d\lambda\right\}d\sigma,$$
$$\left\|\int_\Gamma \lambda e^{-\lambda\sigma}\frac{\partial}{\partial\tau}(A(\tau)-\lambda)^{-1}\,d\lambda\right\|\le C\int_\Gamma e^{-\sigma\mathrm{Re}\lambda}|\lambda|^{1-\rho}|d\lambda|\le C\sigma^{\rho-2},$$

we have the estimate

$$\|\mathrm{II}\|\le C\int_{\tau-s}^{t-s}\sigma^{\rho-2}\,d\sigma = C\{(\tau-s)^{\rho-1}-(t-s)^{\rho-1}\}$$
$$= C(\tau-s)^{\rho-1}\{1-(\tau-s)^{1-\rho}/(t-s)^{1-\rho}\}$$
$$\le C(\tau-s)^{\rho-1}\{1-(\tau-s)/(t-s)\}$$
$$= C(t-\tau)(t-s)^{-1}(\tau-s)^{\rho-1}\le C(t-\tau)^\delta(t-s)^{-\delta}(\tau-s)^{\rho-1}. \tag{5.82}$$

From (5.81) and (5.82), it follows that

$$\|R_1(t, s) - R_1(\tau, s)\| \leq C\{(t-\tau)^\delta (t-s)^{-\delta}(\tau-s)^{\rho-1} + (t-\tau)^\beta (t-s)^{\alpha-\beta-1}\}.$$
(5.83)

The proof of the remainder of the lemma is easy. $\square$

For $0 < \varepsilon < t - s$, we define

$$W_\varepsilon(t, s) = \int_s^{t-s} \exp\left(-(t-\tau)A(t)\right)R(\tau, s)\,d\tau.$$

With a suitable deformation, we obtain

$$(\partial/\partial t)W_\varepsilon(t, s) = \exp\left(-\varepsilon A(t)\right)R(t-\varepsilon, s)$$
$$+ \int_s^{t-\varepsilon} (\partial/\partial t)\exp\left(-(t-\tau)A(t)\right)(R(\tau, s) - R(t, s))\,d\tau$$
$$- \int_s^{t-\varepsilon} R_1(t, \tau)\,d\tau R(t, s) - \exp\left(-\varepsilon A(t)\right)R(t, s)$$
$$+ \exp\left(-(t-s)A(t)\right)R(t, s),$$
(5.84)

which, by Lemmas 5.3.1, 5.3.2, (5.54) and (5.57), converges as $\varepsilon \to 0$ to

$$(\partial/\partial t)W(t, s) = \int_s^t (\partial/\partial t)\exp\left(-(t-\tau)A(t)\right)(R(\tau, s) - R(t, s))\,d\tau$$
$$- \int_s^t R_1(t, \tau)\,d\tau R(t, s) + \exp\left(-(t-s)A(t)\right)R(t, s).$$
(5.85)

Again, by Lemmas 5.3.1, 5.3.2, (5.54) and (5.57), we get

$$\|(\partial/\partial t)U(t, s)\| \leq C(t-s)^{-1}$$
(5.86)

and

$$\|(\partial/\partial t)W(t, s)\| \leq C\{(t-s)^{\rho-1} + (t-s)^{\alpha-1}\}.$$
(5.87)

Equations (5.60) and (5.62) imply that $R(U(t, s)) \subset D(A(t))$ and

$$(\partial/\partial t)U(t, s) + A(t)U(t, s) = 0$$
(5.88)

for $s < t$, so that

$$\|A(t)U(t, s)\| \leq C(t-s)^{-1}.$$
(5.89)

We can find analogues of (5.86) (5.88), (5.89) for $V(t, s)$ constructed by means of (3.69)–(3.73). Since $V(t, s) = U(t, s)$ by (3.78), it is shown that

$$(\partial/\partial s)U(t, s) = \text{a bounded extension of } U(t, s)A(s) \qquad (5.90)$$

and

$$\|(\partial/\partial s)U(t, s)\| \leq C(t - s)^{-1}. \qquad (5.91)$$

Summing up, we have

Theorem 5.3.3 *Under Assumptions 5.1.1 and 5.3.1–5.3.3, there exists a fundamental solution of (5.1) such that $R(U(t, s)) \subset D(A(t))$ for $0 \leq s < t \leq T$, $U(t, s)$ is differentiable with respect to t, s in the norm of $B(X)$, and the relations (5.86) and (5.89)–(5.91) hold.*

The inhomogeneous equation can be treated similarly, as in the preceding section, and the following theorem is obtained.

Theorem 5.3.4 *Under the same assumptions as above, let u_0 be an arbitrary element of X and f a function which is Hölder continuous in $[0, T]$. Then the function u defined by (5.58) is the unique solution of (5.1) and (5.2).*

This can be proved in the same way as Theorem 5.2.3, and so we will omit the proof.

Remark 5.3.1 In addition to Assumptions 5.1.1 and 5.3.1, we suppose that there exists a number ρ in $0 < \rho \leq 1$ such that $R(dA(t)^{-1}/dt) \subset D(A(t)^\rho)$ and $A(t)^\rho \, dA(t)^{-1}/dt$ is uniformly bounded, i.e.,

$$\|A(t)^\rho \, dA(t)^{-1}/dt\| \leq C. \qquad (5.92)$$

Then, since $\|A(t)^{1-\rho}(A(t) - \lambda)^{-1}\| \leq C|\lambda|^{-\rho}$ by (5.45), (5.92) and Proposition 2.3.3, we have

$$\|(\partial/\partial t)(A(t) - \lambda)^{-1}\| = \|A(t)^{1-\rho}(A(t) - \lambda)^{-1}A(t)^\rho \, dA(t)^{-1}/dtA(t)$$
$$\times (A(t) - \lambda)^{-1}\| \leq C|\lambda|^{-\rho},$$

showing that Assumption 5.3.3 is satisfied. In this case, even if Assumption 5.3.2 is not satisfied, it can be shown that the $U(t, s)$ constructed by means of (3.50)–(3.53) is the desired fundamental solution and that the conclusion of Theorem 5.3.3 is correct except for (5.90) and (5.91) (*see* [168]). A little more generally, Yagi [178] has shown that a similar result can be obtained under the bound

$$\|A(t)(A(t) - \lambda)^{-1} \, dA(t)^{-1}/dt\| \leq C|\lambda|^{-\rho} \qquad (5.93)$$

for $\lambda \in \Sigma$. It is easy to see that (5.93) comes out of (5.92).

Remark 5.3.2 Sobolevskiĭ [159, 160] has constructed the fundamental solution under the assumption that, for some $\rho \in (0, 1)$, $D(A(t)^\rho)$ is independent of t. His method is very complicated. Kato [79] has obtained a similar result in the case where ρ^{-1} is a positive integer. His method is concise. As both of them assumed that $A(t)^\rho A(0)^{-\rho}$ satisfies Hölder's condition of order α with some $\alpha \in (1 - \rho, 1)$, their results are considered as intermediate to those in the preceding and the present sections. The result of [79] is applicable to the case in which $A(t)$ is a regularly accretive operator in a Hilbert space. In such a case, however, the fundamental solution can be constructed differently, as will be described in the next section. In general, it is difficult to examine $D(A(t)^\rho)$ and the aim of the present section is to exclude entirely the assumption that $D(A(t)^\rho) = \text{constant}$. In this connection, we refer to Chapter 7 of Lions' book [11].

Now we are going to apply the result of the present section to the mixed problem for a parabolic equation:

$$\begin{cases} \partial u(x, t)/\partial t + A(x, t, D)u(x, t) = f(x, t) & x \in \Omega, \quad 0 < t \leq T, \\ u(x, 0) = u_0(x), & x \in \Omega, \\ B_j(x, t, D)u(x, t) = 0, & x \in \partial\Omega, \quad 0 < t \leq T, \quad j = 1, \ldots, m/2. \end{cases} \quad (5.94)$$

As in the example of the preceding section, we consider the problem in the scope of $L^p(\Omega)$ with $1 < p < \infty$. Let Ω and $A(x, t, p)$ be the same as in the example of the preceding section. Then, as stated in Section 3.8, there exists a certain angle $\theta_0 \in [0, \pi/2)$ such that $\arg \mathring{A}(x, t, \xi) \neq \theta$ for all t, x and for $\xi \neq 0$ and $\theta \in [\theta_0, 2\pi - \theta_0]$. For each t, let

$$B_j(x, t, D) = \sum_{|\beta| \leq m_j} b_{j\beta}(x, t)D^\beta, \quad j = 1, \ldots, m/2,$$

be a set of normal boundary differential operators. Assume that each $B_j(x, t, D)$ is of order m_j independently of t and that its coefficient belongs to $C^{m-m_j}(\partial\Omega)$. Assume, moreover, that the conditions of Theorem 3.8.1 are satisfied for each $t \in [0, T]$ and each $\theta \in [\theta_0, 2\pi - \theta_0]$. Then, by Theorem 3.8.2, the operator $-A(t)$, defined by

$$D(A(t)) = \{u \in W_p^m(\Omega): B_j(x, t, D)u(x) = 0,$$

$$x \in \partial\Omega, \quad j = 1, \ldots, m/2\}$$

and

$$(A(t)u)(x) = A(x, t, D)u(x) \quad \text{for} \quad u \in D(A(t)),$$

generates an analytic semigroup in $L^p(\Omega)$. In addition, if each of the coefficients of $A(x, t, D)$ and $\{B_j(x, t, D)\}$ is differentiable in t and if its derivative with respect to t is Hölder continuous uniformly in t, we can show that the results of this section are applicable. As usual, we may suppose that $0 \in \rho(A(t))$ by adding, if necessary, a sufficiently large positive number to $A(t)$. Also, we may extend all the coefficients of $B_j(x, t, D)$ into Ω while preserving their smoothness. First, assuming that coefficients are independent of t and writing $A(t) = A$, we consider the inhomogeneous boundary-value problem

$$A(x, D)u(x) = f(x), \qquad x \in \Omega, \tag{5.95}$$

$$B_j(x, D)u(x) = g_j(x), \qquad x \in \partial\Omega, \qquad j = 1, \ldots, m/2. \tag{5.96}$$

Lemma 5.3.3 For every $f \in L^p(\Omega)$ and $g_j \in W_p^{m-m_j-1/p}(\partial\Omega)$, $j = 1, \ldots, m/2$, the equations (5.95) and (5.96) admit a uniquely determined solution $u \in W_p^m(\Omega)$.

Proof $0 \in \rho(A)$ implies, by Lemma 1.2.2, that

$$\|u\|_{m,p} \leq C\{\|A(x, D)u\|_p + \sum_{j=1}^{m/2} [B_j(x, D)u]_{m-m_j-1/p}\}. \tag{5.97}$$

If $g_j \in C^{m-m_j}(\partial\Omega)$, $j = 1, \ldots, m/2$, we can construct a function $v \in C^m(\bar{\Omega})$ so that $B_j(x, D)v(x) = g_j(x)$ $(j = 1, \ldots, m/2)$ is satisfied in $\partial\Omega$. In this case, since (5.95) and (5.96) are equivalent to

$$A(x, D)(u(x) - v(x)) = f(x) - A(x, D)v(x), \qquad x \in \Omega,$$
$$B_j(x, D)(u(x) - v(x)) = 0, \qquad x \in \partial\Omega, \qquad j = 1, \ldots, m/2,$$

$u = v + A^{-1}(f - A(x, D)v)$ is the desired solution. In the more general case, we take a function $w_j \in W_p^{m-m_j}(\Omega)$ for each $j = 1, \ldots, m/2$ which coincides in $\partial\Omega$ with g_j and approximate it by a sequence of functions, $\{w_{j\nu}\}$, which belong to $C^{m-m_j}(\bar{\Omega})$ in such a way that $\|w_{j\nu} - w_j\|_{m-m_j,p} \to 0$. Denote by $g_{j\nu}$ the boundary value of $w_{j\nu}$. Since $g_{j\nu} \in C^{m-m_j}(\partial\Omega)$, there exists a solution u_ν of

$$A(x, D)u_\nu(x) = f(x), \qquad x \in \Omega,$$
$$B_j(x, D)u_\nu(x) = g_{j\nu}(x), \qquad x \in \partial\Omega, \qquad j = 1, \ldots, m/2.$$

By applying (5.97) to $u_\nu - u_\mu$, we have

$$\|u_\nu - u_\mu\|_{m,p} \leq C \sum_{j=1}^{m/2} [g_{j\nu} - g_{j\mu}]_{m-m_j-1/p}$$

$$\leq C \sum_{j=1}^{m/2} \|w_{j\nu} - w_{j\mu}\|_{m-m_j,p},$$

so that $\{u_\nu\}$ is a Cauchy sequence in $W_p^m(\Omega)$ and its limit, denoted by u, is clearly the desired solution. $\square$

Now we return to the case in which the coefficients depend on t. The following estimate corresponding to (5.97) holds for each t:

$$\|u\|_{m,p} \leq C\{\|A(x, t, D)u\|_p + \sum_{j=1}^{m/2} [B_j(x, t, D)u]_{m-m_j-1/p}\}. \tag{5.98}$$

Lemma 5.3.4 *If* $f \in L^p(\Omega)$, *then*

$$\|A(t)^{-1}f - A(s)^{-1}f\|_{m,p} \leq C|t-s| \|f\|_p. \tag{5.99}$$

Proof Putting $u(t) = A(t)^{-1}f$, we have $A(t)u(t) = f$, which means that

$$A(x, t, D)u(x, t) = f(x), \qquad x \in \Omega, \qquad 0 \leq t \leq T \tag{5.100}$$

and

$$B_j(x, t, D)u(x, t) = 0, \qquad x \in \partial\Omega, \qquad 0 \leq t \leq T, \qquad j = 1, \ldots, m/2. \tag{5.101}$$

Hence, we have

$$A(x, t, D)(u(x, t) - u(x, s))$$
$$= -(A(x, t, D) - A(x, s, D))u(x, s) \qquad x \in \Omega,$$
$$B_j(x, t, D)(u(x, t) - u(x, s)) = -(B_j(x, t, D) - B_j(x, s, D))u(x, s),$$
$$x \in \partial\Omega, \qquad j = 1, \ldots, m/2,$$

for which

$$\|(A(x, t, D) - A(x, s, D))u(s)\|_p \leq C|t-s| \|u(s)\|_{m,p},$$
$$[(B_j(x, t, D) - B_j(x, s, D))u(s)]_{\hat{m}-m_j-1/p}$$
$$\leq \|(B_j(x, t, D) - B_j(x, s, D))u(s)\|_{m-m_j,p} \leq C|t-s| \|u(s)\|_{m,p},$$
$$\|u(s)\|_{m,p} \leq C\|f\|_p$$

are easily found. Therefore, by applying (5.98) to $u(t) - u(s)$, we readily get (5.99). $\square$

Lemma 5.3.5 $A(t)^{-1}f$ *with* $f \in L^p(\Omega)$ *is strongly continuously differentiable in* $W_p^m(\Omega)$.

Proof Again we put $u(t) = A(t)^{-1}f$. By formally differentiating both sides of (5.100) and (5.101), regarded as identities of t, we obtain

$$A(x, t, D)\dot{u}(x, t) = -\dot{A}(x, t, D)u(x, t) \equiv f(x, t), \qquad x \in \Omega,$$
$$B_j(x, t, D)\dot{u}(x, t) = -\dot{B}_j(x, t, D)u(x, t) \equiv g_j(x, t), \qquad x \in \partial\Omega.$$

Evidently, $f(t) \in L^p(\Omega)$ and $g_j(t) \in W_p^{m-m_j-1/p}(\partial\Omega)$. With these functions, we consider the boundary-value problem

$$A(x, t, D)w(x, t) = f(x, t), \qquad x \in \Omega, \tag{5.102}$$

$$B_j(x, t, D)w(x, t) = g_j(x, t), \qquad x \in \partial\Omega, \qquad j = 1, \ldots, m/2. \tag{5.103}$$

By Lemma 5.3.3, a solution $w(t) \in W_p^m(\Omega)$ exists. By applying (5.98) to

$$v_{\Delta t}(x, t) = (\Delta t)^{-1}(u(x, t + \Delta t) - u(x, t)) - w(x, t)$$

and using Lemma 5.3.4, we easily find that

$$\lim_{\Delta t \to 0} \|v_{\Delta t}(t)\|_{m,p} = 0.$$

It is also easy to see that $w(t)$ is a continuous function with values in $W_p^m(\Omega)$. $\square$

By Lemma 3.8.1, the inequality

$$|\lambda| \, \|u\|_p + \|u\|_{m,p} \leq C\Big\{ \|(A(x, t, D) - \lambda)u\|_p + \sum_{j=1}^{m/2} |\lambda|^{(m-m_j)/m} \|g_j\|_p$$
$$+ \sum_{j=1}^{m/2} \|g_j\|_{m-m_j,p} \Big\} \tag{5.104}$$

holds for every $u \in W_p^m(\Omega)$ and every λ satisfying $\theta_0 \leq \arg \lambda \leq 2\pi - \theta_0$. Here g_j is an arbitrary function, belonging to $W_p^{m-m_j}(\Omega)$, which coincides with $B_j(x, t, D)u$ in $\partial\Omega$.

Lemma 5.3.6 *For any λ satisfying $\theta_0 \leq \arg \lambda \leq 2\pi - \theta_0$, the resolvent $(A(t) - \lambda)^{-1}$ is continuously norm differentiable with respect to t in the norm of $B(L^p(\Omega))$ and the following inequality holds:*

$$\|(\partial/\partial t)(A(t) - \lambda)^{-1}\| \leq C/|\lambda|.$$

Proof Put $u(t) = (A(t) - \lambda)^{-1}f$ with $f \in L^p(\Omega)$. Then we have

$$(A(x, t, D) - \lambda)u(x, t) = f(x), \qquad x \in \Omega, \qquad 0 \leq t \leq T,$$

$$B_j(x, t, D)u(x, t) = 0, \qquad x \in \partial\Omega, \qquad 0 \leq t \leq T, \qquad j = 1, \ldots, m/2.$$

As in the preceding lemma, $u(t)$ is strongly continuously differentiable in $W_p^m(\Omega)$, and the following results hold:

$$(A(x, t, D) - \lambda)\dot{u}(x, t) = -\dot{A}(x, t, D)u(x, t), \qquad x \in \Omega \tag{5.105}$$

and

$$B_j(x, t, D)\dot{u}(x, t) = -\dot{B}_j(x, t, D)u(x, t), \qquad x \in \partial\Omega, \qquad j = 1, \ldots, m/2. \tag{5.106}$$

Apply (5.104) to $\dot{u}(t)$. Then we may take the right-hand side of (5.106), $-\dot{B}_j(x, t, D)u \equiv g_j$, as g_j in (5.104). Here it is to be remarked that the coefficients of $B_j(x, t, D)$ have been extended into the interior of Ω. In this way, we obtain

$$|\lambda|\,\|\dot{u}(t)\|_p \le C\Big\{\|\dot{A}(x, t, D)u(t)\|_p + \sum_{j=1}^{m/2} |\lambda|^{(m-m_j)/m}\,\|g_j(t)\|_p$$
$$+ \sum_{j=1}^{m/2} \|g_j(t)\|_{m-m_j,p}\Big\},$$

from which it easily follows that

$$\|(\partial/\partial_t)(A(t)-\lambda)^{-1}f\|_p = \|\dot{u}(t)\|_p \le C\,\|f\|_p/|\lambda|. \quad \square$$

This completes the proof that Assumption 5.3.1 and 5.3.3 are satisfied. The proof for Assumption 5.3.2 is similar.

5.4   The case in which A(t) is a regularly accretive operator

Let X, V be the Hilbert spaces mentioned in Section 2.2, and $a(t; u, v)$ a quadratic form defined on $V \times V$. Assume that there exists a constant M such that

$$|a(t; u, v)| \le M\,\|u\|\,\|v\| \tag{5.107}$$

holds for all $u, v \in V$ and $0 \le t \le T$. Assume also that there exist a positive number $\delta > 0$ and a real number k such that

$$\text{Re } a(t; u, u) \ge \delta\|u\|^2 - k|u|^2 \tag{5.108}$$

is satisfied for all $t \in [0, T]$ and $u \in V$. Furthermore, $a(t; u, v)$ is assumed to be Hölder continuous in t in the following sense: there exists a certain number $\alpha \in (0, 1]$ such that

$$|a(t; u, v) - a(s; u, v)| \le K\,|t-s|^\alpha\,\|u\|\,\|v\| \tag{5.109}$$

for all $t \in [0, T]$ and $u, v \in V$. Denote by $A(t)$ the operator determined by $a(t; u, v)$, i.e., we set

$$a(t; u, v) = (A(t)u, v).$$

The purpose of the present section is to solve the equation of evolution in X, (5.1) and (5.2), under these assumptions. By Theorem 3.6.1, $-A(t)$ generates an analytic semigroup both in X and in V^*, and, when it is regarded as an operator in V^*, its domain coincides with V and, hence, is independent of t. Thus, if (5.1) and (5.2) are considered as equations in V^*, we can construct its fundamental solution $U(t, s)$ by applying the result of Section 5.2. Next, if $\alpha > 1/2$, the restriction of $U(t, s)$ to X is found to be the desired fundamental solution of (5.1) and (5.2). To carry this programme into effect, we first make (5.108) valid for $k = 0$ by considering $e^{-kt}u(t)$ anew as an unknown. Then, by (2.13), a series of inequalities

$$\delta\|u\| \leqslant \|A(t)u\|_* \leqslant M\|u\| \tag{5.110}$$

hold for all $t \in [0, T]$ and $u \in V$. It is also clear that the inequalities in Section 3.6 with A replaced by $A(s)$ are valid uniformly in s. In other words, we have

$$|(A(s) - \lambda)^{-1}f| \leqslant C|\lambda|^{-1}|f|, \tag{5.111}$$

$$|(A(s) - \lambda)^{-1}f| \leqslant C|\lambda|^{-1/2}\|f\|_*, \tag{5.112}$$

$$\|(A(s) - \lambda)^{-1}f\| \leqslant C|\lambda|^{-1/2}|f|, \tag{5.113}$$

$$\|(A(s) - \lambda)^{-1}f\| \leqslant C\|f\|_*, \tag{5.114}$$

$$\|(A(s) - \lambda)^{-1}f\|_* \leqslant C|\lambda|^{-1}\|f\|_*, \tag{5.115}$$

$$|\exp(-tA(s))| \leqslant 1, \tag{5.116}$$

$$\|\exp(-tA(s))\|_* \leqslant C, \tag{5.117}$$

$$|\exp(-tA(s))f| \leqslant Ct^{-1/2}\|f\|_*, \tag{5.118}$$

$$\|\exp(-tA(s))f\| \leqslant Ct^{-1/2}|f|, \tag{5.119}$$

$$\|\exp(-tA(s))f\| \leqslant Ct^{-1}\|f\|_*, \tag{5.120}$$

$$|A(s)\exp(-tA(s))f| \leqslant Ct^{-3/2}\|f\|_*, \tag{5.121}$$

$$\|A(s)\exp(-tA(s))f\| \leqslant Ct^{-3/2}|f|. \tag{5.122}$$

(5.109) immediately implies that

$$\|A(t)u - A(s)u\|_* \leqslant K|t - s|^\alpha \|u\|. \tag{5.123}$$

From this and (5.110) it follows that

$$\|A(t)A(0)^{-1} - A(s)A(0)^{-1}\|_* \leqslant C|t - s|^\alpha. \tag{5.124}$$

Therefore, the method in Section 5.2 makes it possible to construct the fundamental solution of (5.1), regarded as an equation in V^*, by means of (5.6), (5.7), (5.9), (5.14) and (5.15). $U(t, s)$ is a bounded operator in V^*. We next consider its restriction to X. As was shown in Section 5.2, the inequalities

$$\|R_1(t, s)\|_* \leqslant C(t - s)^{\alpha - 1}, \qquad 0 \leqslant s < t \leqslant T \tag{5.125}$$

and

$$\|R_1(t, s) - R_1(\tau, s)\|_* \leqslant C_\beta(t - \tau)^\beta(\tau - s)^{\alpha - \beta - 1},$$

$$0 \leqslant s < \tau < t \leqslant T \tag{5.126}$$

hold, where β is an arbitrary positive number smaller than α. Moreover, if $f \in X$, we have, from (5.119) and (5.123),

$$\|R_1(t, s)f\|_* \leq C(t-s)^{\alpha-1/2} |f|. \tag{5.127}$$

Lemma 5.4.1 *If $f \in X$ and $0 \leq s < t \leq T$, then*

$$\|R(t, s)f\|_* \leq C(t-s)^{\alpha-1/2} |f|. \tag{5.128}$$

Proof By induction, it can be shown that there exist some constants C_0 and C_1 such that the following inequality holds for all m:

$$\|R_m(t, s)f\|_* \leq C_0 \Gamma(\rho+1/2) C_1^{m-1} \Gamma(\rho)^{m-1} (t-s)^{m\rho-1/2} |f|/\Gamma(m\rho+1/2),$$

from which (5.128) follows immediately. □

Lemma 5.4.2 *For each $f \in X$, $0 \leq s < \tau < t \leq T$ and $0 < \beta < \alpha$, we have*

$$\|R(t, s)f - R(\tau, s)f\|_* \leq C_\beta \left\{ (t-\tau)^\alpha (\tau-s)^{-1/2} \right.$$

$$\left. + \int_\tau^t (t-\sigma)^{\alpha-1} (\sigma-s)^{\alpha-1/2} \, d\sigma + (t-\tau)^\beta (\tau-s)^{2\alpha-\beta-1/2} \right\} |f|. \tag{5.129}$$

Proof By the definition (5.9) of $R_1(t, s)$, we have

$$\|R_1(t, s)f - R_1(\tau, s)f\|_* \leq \|(A(t) - A(\tau)) \exp(-(t-s)A(s))f\|_*$$
$$+ \|(A(\tau) - A(s))$$
$$\times \{\exp(-(t-s)A(s)) - \exp(-(\tau-s)A(s))\}f\|_*. \tag{5.130}$$

Because of (9.119) and (5.123), the first term on the right-hand side does not exceed $C(t-\tau)^\alpha (t-s)^{-1/2} |f|$. The inequality (5.122) provides

$$\|\{\exp(-(t-s)A(s)) - \exp(-(\tau-s)A(s))\}f\|$$
$$= \left\| \int_\tau^t A(s) \exp(-(r-s)A(s))f \, dr \right\|$$
$$\leq C \int_\tau^t (r-s)^{-3/2} |f| \, dr$$
$$= C\{(\tau-s)^{-1/2} - (t-s)^{-1/2}\} |f|$$
$$= C(\tau-s)^{-1/2} \{1 - (\tau-s)^{1/2}(t-s)^{-1/2}\} |f|$$
$$\leq C(\tau-s)^{-1/2} \{1 - (\tau-s)/(t-s)^{-1}\} |f|$$
$$= C(t-\tau)(t-s)^{-1}(\tau-s)^{-1/2}|f| \leq C(t-\tau)(\tau-s)^{-3/2} |f|,$$

which, combining with (5.123), shows that the second term on the right-hand side does not exceed $C(t-\tau)(\tau-s)^{\alpha-3/2} |f|$. On the other hand, by (5.119) and (5.123), the same term does not exceed

$$\|(A(\tau) - A(s)) \exp(-(t-s)A(s))f\|_*$$
$$+ \|(A(\tau) - A(s)) \exp(-(\tau-s)A(s))f\|_*$$
$$\leqslant C(t-s)^\alpha (t-s)^{-1/2} |f| + C(\tau-s)^{\alpha-1/2} |f| \leqslant C(\tau-s)^{\alpha-1/2} |f|.$$

Thus, it is bounded by

$$C\{(t-\tau)(\tau-s)^{\alpha-3/2}\}^\alpha \{(\tau-s)^{\alpha-1/2}\}^{1-\alpha} |f| = C(t-\tau)^\alpha (\tau-s)^{-1/2} |f|.$$

Therefore, we have

$$\|R_1(t,s)f - R_1(\tau,s)f\|_* \leqslant C(t-\tau)^\alpha (\tau-s)^{-1/2} |f|. \tag{5.131}$$

The estimate (5.129) is easily obtained by the use of (5.13), (5.125), (5.126), (5.128) and (5.131). $\square$

Lemma 5.4.3 *Define*

$$Y(t,s) = \exp(-(t-s)A(s)) - \exp(-(t-s)A(t))$$

for $0 \leqslant s < t \leqslant T$. It has the estimate

$$\|Y(t,s)f\| \leqslant C(t-s)^{\alpha-1} \|f\|_*. \tag{5.132}$$

Proof $Y(t,s)$ can be expressed as

$$Y(t,s) = \frac{1}{2\pi i} \int_\Gamma e^{-\lambda(t-s)} (A(s) - \lambda)^{-1} (A(t) - A(s))(A(t) - \lambda)^{-1} \, d\lambda.$$

By (5.114) and (5.123), we have

$$\|(A(s) - \lambda)^{-1} (A(s) - A(t))(A(t) - \lambda)^{-1} f\| \leqslant C(t-s)^\alpha \|f\|_*,$$

from which (5.132) follows immediately. $\square$

Lemma 5.4.4

$$|W(t,s)f| \leqslant C(t-s)^\alpha |f|, \tag{5.133}$$
$$\|W(t,s)f\| \leqslant C(t-s)^{\alpha-1/2} |f|. \tag{5.134}$$

for $f \in X$ and $0 \leqslant s < t \leqslant T$.

Proof (5.133) is a direct consequence of Lemma 5.4.1 and (5.118). Next, (5.134) is obtained by deforming $W(t,s)$ as

$$W(t,s) = \int_s^t Y(t,\tau) R(\tau,s) \, d\tau + \int_s^t \exp(-(t-\tau)A(t))$$
$$\times (R(\tau,s) - R(t,s)) \, d\tau$$
$$+ A(t)^{-1}\{1 - \exp(-(t-s)A(t))\}R(t,s)$$

and by applying Lemmas 5.4.1 and 5.4.3 to the first term on the right-hand side, (5.120) and Lemma 5.4.2 to the second term, and (5.110), (5.117) and Lemma 5.4.1 to the third term, respectively. □

Theorem 5.4.1 $U(t, s)$ for $0 \leq s \leq t \leq T$ is a strongly continuous function with values in $B(X)$ and satisfies

$$|U(t, s)f| \leq C |f|$$
(5.135)

and

$$\|U(t, s)f\| \leq C(t-s)^{-1/2} |f|,$$
(5.136)

where $f \in X$ and $0 \leq s < t \leq T$. Moreover, for $f \in V^*$, it satisfies

$$|U(t, s)f| \leq C(t-s)^{-1/2} \|f\|_*.$$
(5.137)

Proof (5.135) and (5.136) are evident from Lemma 5.4.4 and (5.6). (5.137) follows from (5.118) and (5.125). □

This theorem implies that

$$u(t) = U(t, 0)u_0 + \int_0^t U(t, s)f(s) \, ds$$
(5.138)

belongs to $C([0, T]; X)$ if $u_0 \in X$ and $f \in C([0, T]; X)$, but, even if f is Hölder continuous, it is not certain whether (5.138) is a solution of (5.1) and (5.2) or not. However, Theorem 5.2.3 ensures that (5.138) forms a solution of the equations (5.1) and (5.2) in V^*.

Let $S(t, s)$ be the operator defined in Lemma 5.2.3:

$$S(t, s) = A(t) \exp\left(-(t-s)A(t)\right) - A(s) \exp\left(-(t-s)A(s)\right).$$

Lemma 5.4.5 For each $f \in V^*$ and $0 \leq s < t \leq T$, it holds that

$$|S(t, s)f| \leq C(t-s)^{\alpha - 3/2} \|f\|_*.$$
(5.139)

Proof By (5.112), (5.114) and (5.123), we have

$$|\{(A(t) - \lambda)^{-1} - (A(s) - \lambda)^{-1}\}f|$$
$$= |(A(t) - \lambda)^{-1}(A(s) - A(t))(A(s) - \lambda)^{-1}f| \leq C(t-s)^\alpha |\lambda|^{-1/2} \|f\|_*.$$

This, together with the expression (5.26) for $S(t, s)$, gives (5.139) at once. □

In the sequel we will assume that $\alpha > \frac{1}{2}$.

Lemma 5.4.6 The following inequality holds for $0 \leq s < t \leq T$:

$$|(\partial/\partial t)W(t, s)| \leq C(t-s)^{\alpha - 1}.$$
(5.140)

Proof Since $(\partial/\partial t)W(t, s)$ is expressed by the right-hand side of (5.29), it is easy to deduce (5.139) by the use of (5.118), (5.121), Lemmas 5.4.1, 5.4.2, 5.4.5 and the assumption $\alpha > \frac{1}{2}$. $\square$

Lemma 5.4.6 implies that $U(t, s)$ is differentiable with respect to t in $B(X)$ and that, for $0 \leq s < t \leq T$, the following results hold: $R(U(t, s)) \subset D(A(t))$, $(\partial/\partial t)U(t, s) + A(t)U(t, s) = 0$, and

$$|(\partial/\partial t)U(t, s)| = |A(t)U(t, s)| \leq C(t-s)^{-1}. \tag{5.141}$$

A quadratic form adjoint to $a(t; u, x)$ is denoted by $a^*(t; u, v)$, that is, $a^*(t; u, v) = \overline{a(t; v, u)}$. Let $A^*(t)$ be the operator determined by $a^*(t; u, v)$. Then, as in the above, we can construct an operator-valued function $V(t, s)$ $(0 \leq s < t \leq T)$ satisfying

$$-(\partial/\partial s)V(t, s) + A^*(s)V(t, s) = 0, \qquad 0 \leq s < t \leq T,$$
$$V(t, t) = I, \qquad 0 \leq t \leq T.$$

For $f, g \in X$ and $s < r < t$, we have

$$(\partial/\partial r)(U(r, s)f, V(t, r)g)$$
$$= -(A(r)U(r, s)f, V(t, r)g) + (U(r, s)f, A^*(r)V(t, r)g)$$
$$= -a(r; U(r, s)f, V(t, r)g) + a(r; U(r, s)f, V(t, r)g) = 0,$$

so that $(U(r, s)f, V(t, r)g)$ is independent of r in the open interval (s, t). Therefore, by letting $r \to s$ and $r \to t$, we get

$$U(t, s) = V^*(t, s). \tag{5.142}$$

As in (5.141), $V(t, s)$ is differentiable with respect to s and satisfies

$$|(\partial/\partial s)V(t, s)| = |A^*(s)V(t, s)| \leq C(t-s)^{-1}, \tag{5.143}$$

which, combining with (5.142), implies that $U(t, s)$ is also differentiable with respect to s in the interval $0 \leq s < t$ and that

$$(\partial/\partial s)U(t, s) = \text{a bounded extension of } U(t, s)A(s) \text{ in } X \tag{5.144}$$

and

$$|(\partial/\partial s)U(t, s)| \leq C(t-s)^{-1} \tag{5.145}$$

Theorem 5.4.2 *If the inequalities (5.107)–(5.109) with $\alpha > \frac{1}{2}$ are satisfied, there exists a fundamental solution $U(t, s)$ of the equation (5.1) in X. $U(t, s)$ is differentiable with respect to t, s in $0 \leq s < t \leq T$, its range $R(U(t, s)) \subset D(A(T))$, $U(t, s)A(s)$ has a bounded extension in X, and we have (5.141), (5.144) and (5.145).*

As in the preceding section, we agree to say that a function $u(t)$ is a *solution* of (5.1) and (5.2) if it satisfies these equations and if $u \in C([0, T]; X) \cap C^1((0, T]; X)$, $u(t) \in D(A(t))$ for each t and $Au \in C((0, T]; X)$.

Theorem 5.4.3 *Suppose that the assumptions of the above theorem hold. Let $u_0 \in X$ and let f be a Hölder continuous function with values in X. Then (5.138) is the unique solution of (5.1) and (5.2).*

The proof is similar to that of Theorem 5.1.3.

5.5 An alternative proof of a theorem due to J. L. Lions

Let X, V be the Hilbert spaces occurring in the preceding section, with an additional condition that they are separable. $a(t; u, v)$ is a quadratic form on $V \times V$ which satisfies (5.107) and (5.108) for each $t \in [0, T]$, and, considered as a function of t,

$$a(t; u, v) \text{ is only assumed to be measurable for each } u, v \in V.$$
$$(5.146)$$

As in the preceding section, $A(t)$ stands for the operator determined by $a(t; u, v)$. Put $(Au)(t) = A(t)u(t)$ with $u \in L^2(0, T; V)$. Then we have that $Au \in L^2(0, T; V^*)$. Under the assumptions above, Lions proved the following theorem.

Theorem 5.5.1 *Let u_0 and f be arbitrary elements of X and $L^2(0, T; V^*)$, respectively, then there exists a function $u \in L^2(0, T; V)$ satisfying*

$$u' \in L^2(0, T; V^*), \tag{5.147}$$

$$u' + Au = f, \tag{5.148}$$

$$u(0) = u_0. \tag{5.149}$$

Such a u is uniquely determined.

Since $u \in C([0, T]; X)$ by Lemma 5.5.1 (to be proved below), (5.149) makes sense as an element of X. It is also noted that in (5.148) every one of u', Au and f is an element of $L^2(0, T; V^*)$.

Lions proved Theorem 5.5.1 in [11] by an ingenious representation theorem of linear functionals in terms of some kind of quadratic forms and also in [12] by Galerkin's method. In this section, we will describe an alternative proof of it by using the results of the preceding section. As before, we assume that (5.108) is satisfied for $k = 0$.

Lemma 5.5.1 *If* $u \in L^2$ $(0, T; V)$ *and* $u' \in L^2(0, T; V^*)$, *then* $u \in C([0, T]; X)$. *With* v *being another function which satisfies the same conditions,* $(u(t), v(t))$ *is absolutely continuous and the following equality holds*:

$$(d/dt)(u(t)v(t)) = (u'(t), v(t)) + (u(t), v'(t)). \tag{5.150}$$

Proof With $0 < a < T$, we extend u to $(-a, T+a)$ by putting $u(t) = u(-t)$ for $-a < t < 0$ and $u(t) = u(2T-t)$ for $T < t < T+a$, so that $u \in L^2(-a, T+a; V)$ and $u' \in L^2(-a, T+a; V^*)$. Define $w(t) = \theta(t)u(t)$, where $\theta(t)$ is a real-valued, continuously differentiable function such that $\theta = 1$ in $[0, T]$ and $\theta = 0$ in some neighbourhoods of $-a$ and $T+a$. By means of the mollifier j_n, which appeared in Section 5.2, we construct

$$w_n(t) = \int_{-\alpha}^{T+\alpha} j_n(t-s)w(s)\, ds.$$

Then, as $n \to \infty$, $w_n \to w$ in $L^2(-a, T+a; V)$ and $w'_n \to w'$ in $L^2(-a, T+a; V^*)$. Moreover, by noting that $w'_n(t) \in V$, we have

$$|w_n(t) - w_m(t)|^2 = \int_{-a}^{t} (d/ds)|w_n(s) - w_m(s)|^2\, ds$$

$$= \int_{-a}^{t} 2\,\mathrm{Re}\,(w'_n(s) - w'_m(s), w_n(s) - w_m(s))\, ds$$

$$\leq \int_{-a}^{T+a} \|w'_n(s) - w'_m(s)\|_*^2\, ds + \int_{-a}^{T+a} \|w_n(s) - w_m(s)\|^2\, ds,$$

so that $\{w_n\}$ is a Cauchy sequence in $C([-a, T+a]; X)$. Therefore, on modifying values on the null set, it is found that $w(t) \in X$ and $|w_n(t) - w(t)| \to 0$ uniformly in $[-a, T+a]$ for all t. From this, it follows that $u \in C([0, T]; X)$. Next, in order to prove (5.150), we consider the integral

$$\int_0^T (w_n(t), v(t))\varphi'(t)\, dt$$

$$= \int_0^T ((d/dt)(\varphi(t)w_n(t)), v(t))\, dt - \int_0^T (\varphi(t)w'_n(t), v(t))\, dt$$

$$= -\int_0^T (\varphi(t)w_n(t), v'(t))\, dt - \int_0^T (\varphi(t)w'_n(t), v(t))\, dt$$

for $\varphi \in C_0^\infty(0, T)$. In the limit as $n \to \infty$, we have

$$\int_0^T (u(t), v(t))\varphi'(t)\, dt = -\int_0^T \{(u(t), v'(t)) + (u'(t), v(t))\}\varphi(t)\, dt,$$

which shows that (5.150) is valid in the sense of a distribution. Since the right-hand side of (5.150) is an absolutely measurable function of t, $(u(t), v(t))$ is absolutely continuous. $\square$

Corollary *The function u satisfying the conclusion of Theorem 5.5.1 is unique.*

Proof We only need to prove that $u = 0$ if $u_0 = 0$ and $f = 0$. By Lemma 5.5.1, $|u(t)|^2$ is absolutely continuous and

$$(d/dt)|u(t)|^2 + 2 \operatorname{Re} a(t; u(t), u(t))$$
$$= 2 \operatorname{Re} (u'(t), u(t)) + 2 \operatorname{Re} (A(t)u(t), u(t))$$
$$= 2 \operatorname{Re} (u'(t) + A(t)u(t), u(t)) = 0,$$

so that $(d/dt)|u(t)|^2 \leq 0$ and, hence, $|u(t)|$ is a decreasing function. Since $|u(0)| = 0$, we have $u(t) \equiv 0$. $\square$

The theorem will be proved first in the case in which (5.109) is satisfied. In this case, a fundamental solution $U(t, s)$ of (5.1) regarded as an equation in V^* exists.

Lemma 5.5.2 *If $u_0 \in X$ and $f \in L^2(0, T; V^*)$, the function u defined by (5.138) belongs to $C([0, T]; X) \cap L^2(0, T; V)$ and the following relations hold in $0 \leq t \leq T$:*

$$\tfrac{1}{2}|u(t)|^2 + \int_0^t \operatorname{Re} a(s; u(s), u(s)) \, ds = \tfrac{1}{2}|u_0|^2 + \int_0^t \operatorname{Re} (f(s), u(s)) \, ds \tag{5.151}$$

and

$$|u(t)|^2 + \delta \int_0^t \|u(s)\|^2 \, ds \leq |u_0|^2 + \frac{1}{\delta} \int_0^t \|f(s)\|_*^2 \, ds. \tag{5.152}$$

Proof First, let f belong to $C^1([0, T]; V^*)$. In this case, since $u \in C((0, T]; V) \cap C^1((0, T]; V^*)$, $Au \in C((0, T]; V^*)$ and

$$u'(t) + A(t)u(t) = f(t), \qquad 0 < t \leq T, \tag{5.153}$$

we have

$$(d/dt)|u(t)|^2 = 2 \operatorname{Re} (u'(t), u(t))$$
$$= 2 \operatorname{Re} (-A(t)u(t) + f(t), u(t))$$
$$= -2 \operatorname{Re} a(t; u(t), u(t)) + \operatorname{Re} (f(t), u(t)).$$

We integrate this equality from ε to T, where $0 < \varepsilon < T$, obtaining

$$\tfrac{1}{2}|u(t)|^2 + \int_\varepsilon^t \mathrm{Re}\, a(s; u(s), u(s))\, ds = \tfrac{1}{2}|u(\varepsilon)|^2 + \int_\varepsilon^t \mathrm{Re}\,(f(s), u(s))\, ds.$$

(5.154)

Hence, by (5.108) with $k = 0$ and by Schwarz's inequality, we have

$$|u(t)|^2 + \delta \int_\varepsilon^t \|u(s)\|^2\, ds \leq |u(\varepsilon)|^2 + \frac{1}{\delta} \int_\varepsilon^t \|f(s)\|_*^2\, ds.$$

(5.155)

Since $\lim_{\varepsilon \to 0} u(\varepsilon) = u_0$ by Theorem 5.4.1, the limiting procedure $\varepsilon \to 0$ in (5.154) and (5.155) gives $u \in L^2(0, T; V)$, (5.151) and (5.152). For a general function $f \in L^2(0, T; V^*)$, we may approach f in terms of functions f_n belonging to $C^1([0, T]; V^*)$ in the strong topology of $L^2(0, T; V^*)$. By putting

$$u_n(t) = U(t, 0)u_0 + \int_0^t U(t, s)f_n(s)\, ds$$

(5.156)

for each n, we get

$$u_n(t) - u_m(t) = \int_0^t U(t, s)(f_n(s) - f_m(s))\, ds,$$

which, on account of (5.152), implies that, for every $t \in [0, T]$,

$$|u_n(t) - u_m(t)|^2 + \delta \int_0^t \|u_n(s) - u_m(s)\|^2\, ds \leq \frac{1}{\delta} \int_0^t \|f_n(s) - f_m(s)\|_*^2\, ds.$$

Therefore $\{u_n\}$ is a Cauchy sequence both in $C([0, T]; X)$ and in $L^2(0, T; V)$. On the other hand, it is easy to see that $u_n(t) \to u(t)$ in V^* for each t. From this, (5.151) and (5.152) follow immediately. $\square$

Proposition 5.5.1 *Suppose that (5.109) is satisfied. Then the function u defined by (5.138) for $u_0 \in X$ and $f \in L^2(0, T; V^*)$ is the only function which satisfies the conclusion of Theorem 5.5.1.*

Proof This is obvious if $f \in C^1([0, T]; V^*)$. When f is a general element of $L^2(0, T; V^*)$, we choose functions $f_n \in C^1([0, T]; V^*)$ such that $f_n \to f$ strongly in $L^2(0, T; V^*)$. Then it is found that the function u_n defined by (5.156) satisfies $u_n \in C((0, T]; V) \cap C^1((0, T]; V^*)$, $Au_n \in C((0, T]; V^*)$ and

$$u_n'(t) + A(t)u_n(t) = f_n(t), \qquad u_n(0) = u_0.$$

(5.157)

Since, by the proof of Lemma 5.5.2, $u_n \to u$ in $L^2(0, T; V)$, we have $Au_n \to Au$ in $L^2(0, T; V^*)$ and so $u_n' \to u'$ in $L^2(0, T; V^*)$ by (5.157). Thus u satisfies (5.147)–(5.149). $\square$

Now let us turn to the general case. We extend $a(t; u, v)$ beyond $[0, T]$ so as to satisfy (5.107), (5.108) and (5.146) in $(-\infty, \infty)$. However, (5.108) is to be satisfied for $k = 0$. With the mollifier j_n given in Section 5.2, we set

$$a_n(t; u, v) = \int_{-\infty}^{\infty} j_n(t - s)a(s; u, v) \, ds.$$

$a_n(t; u, v)$ thus defined satisfies (5.107)–(5.109). An operator determined by this quadratic form will be denoted by $A_n(t)$. It is evident that

$$\|A_n^*(t)u\|_* \leq M \|u\|, \qquad \|A^*(t)u\|_* \leq M \|u\| \tag{5.158}$$

hold for every $t \in [0, T]$ and $u \in V$. We denote by $U(t, s)$ a fundamental solution of the equation $u' + A_n(t)u = 0$. For $u_0 \in X$ and $f \in L^2(0, T; V^*)$, we get

$$u_n(t) = U_n(t, 0)u_0 + \int_0^t U_n(t, s)f(s) \, ds. \tag{5.159}$$

By Lemma 5.5.2, the inequality (5.152) with u replaced by u_n holds, so that $\{u_n\}$ is bounded in $L^2(0, T; V^*)$ and, hence, we may consider, on replacing $\{u_n\}$ by its suitable subsequence, that $u_n \to u$ weakly in $L^2(0, T; V)$.

Lemma 5.5.3 *There exists a null set N of $[0, T]$ such that for all $t \in [0, T] \backslash N$ and $v \in V$ we have*

$$\lim_{n \to \infty} \|A_n^*(t)v - A^*(t)v\|_* = 0. \tag{5.160}$$

Proof For an arbitrary element w of V, we have

$$(w, A_n^*(t)v - A^*(t)v) = a_n(t; w, v) - a(t; w, v)$$

$$= \int_{-\infty}^{\infty} j_n(t - s)(a(s; w, v) - a(t; w, v)) \, ds$$

$$= \int_{-\infty}^{\infty} j_n(s)(w, A^*(t - s)v - A^*(t)v) \, ds,$$

so that

$$\|A_n^*(t)v - A^*(t)v\|_* \leq \int_{-\infty}^{\infty} j_n(s)\|A^*(t - s)v - A^*(t)v\|_* \, ds$$

$$\leq Cn \int_{-n^{-1}}^{n^{-1}} \|A^*(t - s)v - A^*(t)v\|_* \, ds. \tag{5.161}$$

$A^*(t)v$ is strongly measurable because it is clearly weakly measurable as a function of t, and V^* is separable. Therefore, by Lemma 1.3.2, there exists a null set N_v such that for every $t \in [0, T] \backslash N_v$ the right-hand side of (5.161) tends to 0 as $n \to \infty$, and so $\|A_n^*(t)v - A^*(t)v\|_* \to 0$. Let $\{v_j\}$ be a countable set dense in V and put $N = \cup_j N_{v_j}$, then N is also a null set in $[0, T]$. If v is an arbitrary element of V and $t \in [0, T] \backslash N$, then, on account of (5.158), for each j we have

$$\|A_n^*(t)v - A^*(t)v\|_* \leqslant \|A_n^*(t)(v - v_j)\|_*$$
$$+ \|A_n^*(t)v_j - A^*(t)v_j\|_* + \|A^*(t)(v_j - v)\|_*$$
$$\leqslant 2M\|v - v_j\| + \|A_n^*(t)v_j - A^*(t)v_j\|_*,$$

from which (5.160) then follows. $\square$

By Proposition 5.5.1, the function u_n defined by (5.159) satisfies $u_n' \in L^2(0, T; V^*)$ and

$$u_n' + A_n u_n = f, \qquad u_n(0) = u_0. \tag{5.162}$$

By (5.158) and Lemma 5.5.3, $A_n^* g$ converges strongly in $L^2(0, T; V^*)$ to $A^* g$ for any $g \in L^2(0, T; V)$. Since $u_n \to u$ weakly in $L^2(0, T; V)$, as we have seen above, it is found that

$$\int_0^T (A_n(t)u_n(t), g(t)) \, dt = \int_0^T (u_n(t), A_n^*(t)g(t)) \, dt$$
$$\to \int_0^T (u(t), A^*(t)g(t)) \, dt = \int_0^T (A(t)u(t), g(t)) \, dt$$

and thus $A_n u_n \to Au$ weakly in $L^2(0, T; V^*)$. Accordingly, $\{u_n'\}$ also converges weakly in $L^2(0, T; V^*)$ and $u_n(0) \to u(0)$ weakly, so that u satisfies (5.147)–(5.149). This completes the proof of Theorem 5.5.1. $\square$

5.6 Behaviour of the solution as $t \to \infty$

In this section, we consider the behaviour as $t \to \infty$ of a solution of the equation (5.1) given in $0 \leqslant t < \infty$. Assumptions 5.1.1 and 5.1.2 are supposed to be satisfied uniformly in $0 \leqslant t < \infty$, so that, in particular, there exist $\delta > 0$, $\theta \in (0, \pi/2)$ and $K > 0$ such that $\Sigma = \{\lambda : |\arg(\lambda - \delta)| \geqslant \theta\} \subset \rho(A)$ and the following inequality holds for every $\lambda \in \Sigma$ and $0 \leqslant t < \infty$:

$$\|(A(t) - \lambda)^{-1}\| \leqslant K(1 + |\lambda|)^{-1}.$$

We supplement this by the four assumptions below:

$$\begin{cases} A(t)A(s)^{-1} \text{ is uniformly bounded, i.e.,} \\ \quad \sup_{0 \leqslant t, s < \infty} \|A(t)A(s)^{-1}\| < \infty. \end{cases} \tag{5.163}$$

$$\begin{cases} \text{There exists a closed operator } A(\infty), \text{ having} \\ D \text{ as its domain, such that } \|A(t)-A(\infty)A(0)^{-1}\|\leftarrow 0. \end{cases} \qquad (5.164)$$

$$\begin{cases} f(t) \text{ is Hölder continuous uniformly in } 0\leqslant t<\infty, \\ \text{that is, there exist } F>0 \text{ and } \gamma\in(0,1) \text{ such} \\ \text{that, for all } t, s \text{ with } 0\leqslant t, s<\infty, \text{ we have} \\ \quad \|f(t)-f(s)\|\leqslant F|t-s|^{\gamma}. \end{cases} \qquad (5.165)$$

$$\begin{cases} \text{There exists an element } f(\infty) \text{ of } X \text{ such that} \\ \|f(t)-f(\infty)\|\to 0 \text{ as } t\to\infty. \end{cases} \qquad (5.166)$$

Theorem 5.6.1 *Under the assumptions stated above, the solution $u(t)$ of (5.1) converges strongly to a certain element $u(\infty)$ of X as $t\to\infty$. Moreover, we have that*

$$u(\infty)\in D, \qquad A(\infty)u(\infty)=f(\infty), \qquad (5.167)$$

$$\|du(t)/dt\|\to 0 \quad \text{as} \quad t\to\infty, \qquad (5.168)$$

$$\|A(t)u(t)-A(\infty)u(\infty)\|\to 0 \quad \text{as} \quad t\to\infty. \qquad (5.169)$$

First, by virtue of our assumption, $A(t)-\delta$ also generates a uniformly bounded parabolic semigroup, so that we have the estimates

$$\|\exp(-tA(s))\|\leqslant Ce^{-\delta t} \qquad (5.170)$$

and

$$\|A(s)\exp(-tA(s))\|\leqslant Ct^{-1}e^{-\delta t} \qquad (5.171)$$

for all t and s. Second, let us put

$$\eta(\mu)=\sup_{\mu\leqslant s<t, 0<\tau<\infty}\|(A(t)-A(s))A(\tau)^{-1}\|,$$

then, by our assumption, $\eta(\mu)\to\infty$ as $\mu\to\infty$. From this and (5.5), it follows that the inequality

$$\|(A(t)-A(s))A(\tau)^{-1}\|\leqslant C\sqrt{\eta(\mu)}(t-s)^{\alpha/2} \qquad (5.172)$$

holds for $\mu\leqslant s<t$.

Lemma 5.6.1 *There exists a function $\eta_1(\mu)>0$, satisfying $\lim_{\mu\to\infty}\eta_1(\mu)=0$, such that*

$$\|R(t,s)\|\leqslant C\sqrt{\eta(\mu)}(t-s)^{\alpha/2-1}e^{-\delta(t-s)}\exp\{\eta_1(\mu)(t-s)\}. \qquad (5.173)$$

Proof By (5.171) and (5.172), there exists a positive constant C_1 such that

$$\|R_1(t, s)\| \leqslant C_1 \sqrt{\eta(\mu)}(t-s)^{\alpha/2-1} e^{-\delta(t-s)}$$

holds for $0 \leqslant \mu \leqslant s < t$. By induction, we obtain

$$\|R_m(t, s)\| \leqslant (C_1 \sqrt{\eta(\mu)})^m \Gamma(\alpha/2)^m \Gamma(\alpha m/2)^{-1} (t-s)^{m\alpha/2-1} e^{-\delta(t-s)}$$

for $m = 1, 2, \ldots,$ from which the conclusion of the lemma is easily derived. $\square$

Lemma 5.6.2 *For any θ and any β satisfying $0 < \theta < \delta$ and $0 < \beta < \alpha/2$, respectively, there exist positive numbers $\mu_{\beta,\theta}$ and $C_{\beta,\theta}$ such that the following inequality holds for $\mu_{\beta,\theta} \leqslant \mu < s < t$:*

$$\|R(t, s) - R(\tau, s)\| \leqslant C_{\beta,\theta} \sqrt{\eta(\mu)}(t-\tau)^{\beta}(\tau-s)^{\alpha/2-\beta-1} e^{-\theta(\tau-s)}. \tag{5.174}$$

Proof As in the proof of (5.23), it is found that

$$\|R_1(t, s) - R_1(\tau, s)\| \leqslant C\sqrt{\eta(\mu)}(t-\tau)^{\beta}(\tau-s)^{\alpha/2-\beta-1} e^{-\delta(\tau-s)}. \tag{5.175}$$

Since $1 - \beta/\alpha > \frac{1}{2}$, we have

$$\|(A(t) - A(s))A(\tau)^{-1}\| \leqslant C\{\eta(\mu)\}^{1-\beta/\alpha}\{(t-s)^{\alpha}\}^{\beta/\alpha}$$

$$\leqslant C\sqrt{\eta(\mu)}(t-s)^{\beta},$$

which, combined with (5.171), gives

$$\|R_1(t, s)\| \leqslant C\sqrt{\eta(\mu)}(t-s)^{\beta-1} e^{-\delta(t-s)}. \tag{5.176}$$

From (5.173) and (5.176), it follows that

$$\left\|\int_{\tau}^{t} R_1(t, \sigma)R(\sigma, s)\,d\sigma\right\| \leqslant C\eta(\mu)(t-\tau)^{\beta}(\tau-s)^{\alpha/2-1}$$

$$\times \exp\{-(\delta - \eta_1(\mu))(t-s)\}.$$

After an elementary calculation, we obtain (5.174). $\square$

Lemma 5.6.3 *For $\mu \leqslant s < t < \infty$, we have the result*

$$\|S(t, s)\| \leqslant C\sqrt{\eta(\mu)}(t-s)^{\alpha/2-1} e^{-\delta(t-s)}. \tag{5.177}$$

Proof Put $\Gamma = \{\lambda : |\arg \lambda| = \theta\}$ and $\Gamma_{\delta} = \{\lambda : \lambda - \delta \in \Gamma\}$. We represent

$S(t, s)$ by the integral

$$S(t, s) = \frac{1}{2\pi i} \int_{\Gamma_s} \lambda e^{-\lambda(t-s)} \{(A(t)-\lambda)^{-1} - (A(s)-\lambda)^{-1}\} \, d\lambda$$

and make a change of the variable, $\lambda \to \lambda + \delta$, obtaining

$$S(t, s) = \frac{e^{-\delta(t-s)}}{2\pi i} \int_{\Gamma} (\lambda + \delta) e^{-\lambda(t-s)} \{(A(t)-\lambda-\delta)^{-1} - (A(s)-\lambda-\delta)^{-1}\} \, d\lambda.$$

From this and the inequality

$$\|(A(t)-\lambda-\delta)^{-1} - (A(s)-\lambda-\delta)^{-1}\|$$
$$\leq \|(A(t)-\lambda-\delta)^{-1}\| \|(A(t)-A(s))A(s)^{-1}\| \|A(s)(A(s)-\lambda-\delta)^{-1}\|$$
$$\leq C\sqrt{\eta(\mu)}(t-s)^{\alpha/2}/|\lambda+\delta|,$$

the result (5.177) follow immediately. $\square$

Lemma 5.6.4 *For $0 < \theta < \delta$ and $0 < \beta < \alpha/2$, the following inequality holds in $\mu_{\beta,\theta} \leq \mu < s < t$:*

$$\|(\partial/\partial t) W(t, s)\| \leq C_{\beta,\theta} \sqrt{\eta(\mu)}(t-s)^{\alpha/2-1} e^{-\theta(t-s)}.$$

Proof The lemma can easily be proved by the use of (5.29), (5.31), (5.171) and Lemmas 5.6.1–5.6.3. $\square$

Lemma 5.6.5

$$\lim_{t \to \infty} \|(\partial/\partial t) U(t, s)\| = 0$$

for all s.

Proof It follows directly from (5.171) and Lemma 5.6.4. $\square$

After these preparations, we are going to prove Theorem 5.6.1. Let us begin with the proof of (5.168). Assume $\mu_{\beta,\theta} \leq \mu \leq s < t$ and represent

$$u(t) = v(t) + w(t), \qquad v(t) = U(t, s)u(s), \qquad w(t) = \int_s^t U(t, \sigma)f(\sigma) \, d\sigma.$$

By the proof of Theorem 5.2.3, we have

$$dw(t)/dt = I + II + III + IV,$$

where

$$I = \int_s^t S(t, \sigma) f(\sigma) \, d\sigma,$$

$$II = -\int_s^t A(t) \exp\left(-(t-\sigma)A(t)\right)(f(\sigma) - f(t)) \, d\sigma,$$

$$III = \exp\left(-(t-s)A(t)\right)f(t),$$

$$IV = \int_s^t (\partial/\partial t) W(t, \sigma) f(\sigma) \, d\sigma.$$

On account of Lemmas 5.6.3 and 5.6.4, and (5.170), apart from II, these terms have the estimates

$$\|I\| \leq C\sqrt{\eta(\mu)}, \qquad \|IV\| \leq C\sqrt{\eta(\mu)}, \qquad \|III\| \leq Ce^{-\delta(t-s)}.$$

Next, if we write

$$\delta(\mu) = \sup_{\mu \leq s < t} \|f(t) - f(s)\|,$$

then, by our assumption, $\delta(\mu) \to 0$ as $\mu \to \infty$. By (5.165),

$$\|f(t) - f(s)\| \leq C\sqrt{\delta(M)}\,(t-s)^{\gamma/2}$$

holds for $\mu \leq s < t$, so that by (5.171) it is found that $\|II\| \leq C\sqrt{\delta(\mu)}$. Also, $\|dv(t)/dt\| \to 0$ as $t \to \infty$ by Lemma 5.6.5. Collecting these results, we obtain (5.168), which, together with (5.1), implies that

$$A(t)u(t) \to f(\infty) \tag{5.178}$$

strongly. Since, by our assumption,

$$\|I - A(\infty)A(t)^{-1}\| = \|(A(t) - A(\infty))A(t)^{-1}\| \to 0$$

as $t \to \infty$, it is found that a bounded inverse of $A(\infty)A(t)^{-1}$ exists if t is sufficiently large and, moreover, that $A(\infty)^{-1}$ also exists and is bounded. Since $A(\infty)u(t) = A(\infty)A(t)^{-1} \cdot A(t)u(t) \to f(\infty)$ strongly, the limit $u(\infty) = $ s-$\lim_{t \to \infty} u(t)$ exists and (5.167) holds. The statement (5.169) follows from (5.178) and (5.167). This completes the proof of Theorem 5.6.1. Also in the cases considered in Sections 5.3 and 5.4 we can prove similar theorems.

Pazy [147] showed that if $A(t)$ and $f(t)$ have asymptotic expansions as $t \to \infty$,

$$A(t) \sim A_0 + t^{-1}A_1 + t^{-2}A_2 + \ldots, \qquad f(t) \sim f_0 + t^{-1}f_1 + t^{-2}f_2 + \ldots. \tag{5.179}$$

respectively, then the solution can also be expanded as

$$u(t) \sim u_0 + t^{-1} u_1 + t^{-2} u_2 + \dots, \tag{5.180}$$

the coefficients of which are uniquely determined by those of (5.179). He applied this result to the mixed problem for parabolic equations.

5.7 Regularity of solutions

5.7.1 Complex analyticity

Komatsu [95] showed that, if Assumptions 5.1.1 and 5.1.2 are satisfied in a complex neighbourhood Δ of $[0, T]$, and if $A(t)A(0)^{-1}$ and $f(t)$ are regular functions of t in Δ, then a solution of (5.1) and (5.2) is regular in a complex neighbourhood of $(0, T)$. The same conclusion is also true for the case when the assumption in Section 5.3 is satisfied in Δ and $A(t)^{-1}$ is a regular function of t there. In the present subsection, we will prove this statement, following the method of Komatsu [95].

Let Δ be a convex complex neighbourhood of the closed interval $[0, T]$.

Assumption 5.7.1 $A(t)$ *for each* $t \in \bar{\Delta}$ *is a closed operator defined densely in* X. *There exists an angle* $\theta \in (0, \pi/2)$ *such that* $\rho(A(t))$ *contains the closed sector* $\Sigma = \{\lambda : |\arg \lambda| \geq \theta\} \cup \{0\}$ *and that*

$$\|(A(t) - \lambda)^{-1}\| \leq K/(1 + |\lambda|) \tag{5.181}$$

holds for $t \in \bar{\Delta}$ *and* $\lambda \in \Sigma$, *where* K *is independent of* t *as well as* λ.

Assumption 5.7.2 $A(t)^{-1}$ *is a regular function in* $\bar{\Delta}$ *with values in* $B(X)$.

As is seen from Cauchy's integral representation,

$$(A(t) - \lambda)^{-1} = \frac{1}{2\pi i} \int (\tau - t)^{-1} (A(\tau) - \lambda)^{-1} \, d\tau,$$

if Assumption 5.7.1 is satisfied, then so is Assumption 5.3.3 with $t \in \bar{\Delta}$ and $\rho = 1$. Thus, there exists a constant N such that the following inequality holds for $t \in \bar{\Delta}$ and $\lambda \in \Sigma$:

$$\|(\partial/\partial t)(A(t) - \lambda)^{-1}\| \leq N/|\lambda|. \tag{5.182}$$

It is easy to see that, if $A(t)$ obeys Assumption 5.7.1, $D(A(t))$ is independent of t in $\bar{\Delta}$, and $A(t)A(0)^{-1}$ is a regular function of t in $\bar{\Delta}$, then Assumption 5.7.2 is satisfied.

Put $\phi = \pi/2 - \theta$. For two complex numbers t and s we write, following Friedman [6], $t > s \pmod{\phi}$ if $t \neq s$ and $|\arg(t-s)| < \phi$. Further, $t \geqslant s$ $\pmod{\phi}$ means that $t > s \pmod{\phi}$ or $t = s$. $\exp(-tA(s))$ for each $s \in \bar{\Delta}$, as a function of t, is regular in $t > 0 \pmod{\phi}$ and strongly continuous in $t \geqslant 0$ $\pmod{\phi}$.

Theorem 5.7.1 *Under Assumptions 5.7.1 and 5.7.2, the fundamental solution $U(t, s)$ of (5.1) can be continued to a regular function of two variables t, s in $t \in \Delta$, $s \in \Delta$, $t > s \pmod{\phi}$. Moreover, $U(t, s)$ is strongly continuous in $t \in \Delta$, $s \in \Delta$, $t \geqslant s \pmod{\phi}$.*

Theorem 5.7.2 *Suppose that Assumptions 5.7.1 and 5.7.2 are satisfied. If $u_0 \in X$ and f is regular in some complex neighbourhood of $[0, T]$, then the solution of (5.1) and (5.2) is regular in a complex neighbourhood of $(0, T]$.*

Lemma 5.7.1 *Let $P(t, s)$ and $Q(t, s)$ be uniformly bounded regular functions with values in $B(X)$ which are defined in $\Xi = \{(t, s): t \in \Delta, s \in \Delta, t > s \pmod{\phi}\}$. Then the integral*

$$\int_s^t P(t, \tau) Q(\tau, s) \, d\tau \tag{5.183}$$

can be defined as a uniformly bounded regular function in Ξ.

Proof The uniform boundedness of (5.183) is evident. It suffices to prove the regularity in t for s fixed. Let $t_0 \in \Delta$, $s \in \Delta$ and $t_0 > s \pmod{\phi}$. Denote by N_ε an ε-neighbourhood of t_0. Take $\varepsilon_0 > 0$ and τ_0 in such a way that $t > \tau_0 > s \pmod{\phi}$ for $t \in N_{\varepsilon_0}$. With this τ_0, we divide the integral (5.183) into two parts:

$$\int_s^t P(t, \tau) Q(\tau, s) \, d\tau = \int_s^{\tau_0} + \int_{\tau_0}^t, \tag{5.184}$$

the first of which is obviously regular in t in N_{ε_0}. In the second term, we change the variable into $\sigma = t - \tau$ to obtain the integral $\int_0^{t-\tau_0} S(t, \sigma, s) \, d\sigma$, where $S(t, \sigma, s) = P(t, t-\sigma) Q(t-\sigma, s)$. $S(t, \sigma, s)$ is regular in $\sigma > 0$ $\pmod{\phi}$ and $t - \sigma > s \pmod{\phi}$. If $r > 0 \pmod{\phi}$ with $|r|$ sufficiently small, the integral is regular in t and, as $r \to 0$, it converges in the norm uniformly in t, so that the integral $\int_0^{t_0-\tau_0} S(t, \sigma, s) \, d\sigma$ is regular in $t \in N_{\varepsilon_0}$. Hence we only need to prove the regularity of

$$K(t, s) = \int_{t_0-\tau_0}^{t-\tau_0} S(t, \sigma, s) \, d\sigma.$$

We take $\varepsilon \leqslant \varepsilon_0$ so small that if $t \in N_\varepsilon$, then $t - \sigma > s$ (mod ϕ) for any σ on the segment connecting $t_0 - \tau_0$ and $t - \tau_0$. By noting that

$$\frac{1}{h}\left(K(t+h, s) - K(t, s)\right) = \frac{1}{h} \int_{t-\tau_0}^{t+h-\tau_0} S(t+h, \sigma, s)\,d\sigma$$
$$+ \frac{1}{h} \int_{t_0-\tau_0}^{t-\tau_0} \{S(t+h, \sigma, s) - S(t, \sigma, s)\}\,d\sigma$$

for $t \in N_\varepsilon$ and h sufficiently small, we see that $(\partial/\partial t)K(t, s)$ exists and is equal to

$$S(t, t - \tau_0, s) + \int_{t_0-\tau_0}^{t-\tau_0} (\partial/\partial t)S(t, \sigma, s)\,d\sigma. \quad \square$$

Proof of Theorems 5.7.1 and 5.7.2   From the fact that, under Assumptions 5.7.1 and 5.7.2,

$$\exp\left(-(t-s)A(t)\right), \qquad R_1(t, s) = -(\partial/\partial t + \partial/\partial s)\exp\left(-(t-s)A(t)\right)$$

satisfy the condition of Lemma 5.7.1 and from the construction of $U(t, s)$ the conclusion of Theorem 5.7.1 follows easily. The proof of Theorem 5.7.2 is also easy. $\square$

5.7.2 Estimation of successive derivatives of the solution

In the preceding subsection, it has been proved that the solution can be holomorphically continued to a complex neighbourhood. In the present subsection, we will show that, if known functions belong to Gevrey's class, so does the solution; in particular, if known functions are analytic, so is the solution.

Let $\{M_k\}_{k=0}^\infty$ be a sequence of positive numbers satisfying the following conditions: there exist positive numbers d_0, d_1 and d_2 such that

$$M_{k+1} \leqslant d_0^k M_k \qquad \text{for all} \quad k \geqslant 0, \tag{5.185}$$

$$\binom{k}{j} M_{k-j} M_j \leqslant d_1 M_k \quad \text{for} \quad 0 \leqslant j \leqslant k, \tag{5.186}$$

$$M_k \leqslant M_{k+1} \qquad \text{for all} \quad k \geqslant 0, \tag{5.187}$$

$$M_{j+k} \leqslant d_2^{j+k} M_j M_k \qquad \text{for all} \quad j, k \geqslant 0. \tag{5.188}$$

Evidently, the conditions (5.185)–(5.187) are satisfied by $M_k = k!$. Moreover, by the binomial theorem,

$$2^{j+k} = (1+1)^{j+k} = \sum_{l=0}^{j+k} \binom{j+k}{l} \geqslant \binom{j+k}{j} = \frac{(j+k)!}{j!\,k!},$$

so that (5.188) is also satisfied. Similarly, it is seen that $M_k = (k!)^s$ satisfies the conditions (5.185) to (5.188) for arbitrary $s > 1$. Let (a, b) be an open interval on the real axis and let $f \in C^\infty(a, b)$. A function f is said to belong to the *class* $\{M_k\}$ in (a, b) if, for each compact set K of (a, b), there exist positive constants C_0 and C such that

$$\sup_{t \in K} |f^{(k)}(t)| \leqslant C_0 C^k M_k$$

holds for all integers $k \geqslant 0$, The totality of such functions will be denoted by $D(a, b; \{M_k\})$. A similar definition is available for functions with values in Banach space. If $M_k = k!$, the $D(a, b; \{M_k\})$ coincides with the whole of functions analytic in (a, b). If $M_k = (k!)^s$ with $s > 1$ it is the totality of functions of *Gevrey's class*. By Stirling's formula, we have

$$D(a, b; \{(k!)^s\}) = D(a, b; \{\Gamma(sk + 1)\}).$$

$D(a, b; \{M_k\})$ is said to be *quasi-analytic* if the condition that $f \in D(a, b; \{M_k\})$ and f itself and all its derivatives are equal to 0 at a point of (a, b) implies that $f \equiv 0$. A necessary and sufficient condition for the quasi-analyticity is that $\sum_{k=0}^{\infty} (M_k)^{-1/k} = \infty$ (Mandelbrojt [16]). For more details about functions of class $\{M_k\}$, see Friedman [5] and Lions and Magenes [118, 119].

Putting $j = k - 1$ in (5.188), we have

$$kM_{k-1} \leqslant \frac{d_1}{M_1} M_k, \tag{5.189}$$

which gives $M_j \geqslant (M_1/d_1)^j j! \, d_1$, so that (again by (5.186))

$$M_{k-j} \leqslant \left(\frac{d_1}{M_1}\right)^j \frac{(k-j)!}{k!} M_k \quad \text{for} \quad \leqslant j \leqslant k. \tag{5.190}$$

Putting $j = k$ in (5.190), we finally obtain

$$k! \leqslant \left(\frac{d_1}{M_1}\right)^k \frac{M_k}{M_0}. \tag{5.191}$$

In this section, besides Assumption 5.1.1, the following assumption will be used.

Assumption 5.7.3 $A(t)^{-1}$ *is an infinitely differentiable function in* $0 \leqslant t \leqslant T$ *with values in* $B(X)$.

Assumption 5.7.4 *There exist positive numbers* K_0 *and* K *such that the inequality*

$$\|(\partial/\partial t)^n (A(t) - \lambda)^{-1}\| \leqslant K_0 K^n M_n / |\lambda| \tag{5.192}$$

holds for every $\lambda \in \Sigma$, $0 \leqslant t \leqslant T$ *and every integer* $n \geqslant 0$.

Here also, Σ denotes the sector given in Section 5.1.

By Theorem 5.3.3, the fundamental solution $U(t, s)$ of (5.1) can be constructed under these assumptions. The two main theorems in the present section are:

Theorem 5.7.3 *Suppose that Assumptions 5.1.1., 5.7.3 and 5.7.4 are satisfied. Then the fundamental solution $U(t, s)$ of (5.1) is infinitely differentiable with respect to s, t in $0 \leqslant s < t \leqslant T$ and there exist positive numbers L_0 and L such that the inequality*

$$\left\| \left(\frac{\partial}{\partial t}\right)^n \left(\frac{\partial}{\partial t} + \frac{\partial}{\partial s}\right)^m \left(\frac{\partial}{\partial s}\right)^l U(t, s) \right\| \leqslant L_0 L^{n+m+l} M_{n+m+l} (t-s)^{-n-l} \quad (5.193)$$

holds for all non-negative integers n, m, l.

Theorem 5.7.4 *Let $f \in C^\infty([0, T]; X)$ and assume that there exist positive constants F_0 and F such that*

$$\|d^n f(t)/dt^n\| \leqslant F_0 F^n M_n \quad (5.194)$$

holds in $0 \leqslant t \leqslant T$ for every integer $n \geqslant 0$. Then, with v being a solution of (5.1) which satisfies $v(0) = 0$, there exist positive numbers $\bar{F}_0$ and $\bar{F}$ such that

$$\|d^n v(t)/dt^n\| \leqslant \bar{F}_0 \bar{F}^n M_n t^{1-n} \quad (5.195)$$

holds in $0 \leqslant t \leqslant T$ for every $n \geqslant 0$. Hence, for a solution u of (5.1) which satisfies $u(0) = u_0$, there exist positive numbers H_0 and H such that the following inequality holds:

$$\|d^n u(t)/dt^n\| \leqslant H_0 H^n M_n \|u_0\| \, t^{-n} + \bar{F}_0 \bar{F}^n M_n t^{1-n}. \quad (5.196)$$

As was described in Section 5.3, the fundamental solution can be constructed in two different ways. We use the same notation as in Section 5.3 and, in addition, put

$$R_{0,n,m}(t, s) = -(\partial/\partial t)^n (\partial/\partial t + \partial/\partial s)^m \exp\left(-(t-s)A(t)\right),$$

$$R_{l,n,m}(t, s) = (\partial/\partial t)^n (\partial/\partial t + \partial/\partial s)^m R_l(t, s),$$

$$Q_{0,n,m}(t, s) = (\partial/\partial t)^n (\partial/\partial t + \partial/\partial s)^m \exp\left(-(t-s)A(s)\right),$$

$$Q_{l,n,m}(t, s) = (\partial/\partial t)^n (\partial/\partial t + \partial/\partial s)^m Q_l(t, s),$$

where n, m and l are non-negative integers. The relations

$$R_{1,n,m}(t, s) = (\partial/\partial t + \partial/\partial s) R_{0,n,m}(t, s) \quad (5.197)$$

and

$$Q_{1,n,m}(t, s) = (\partial/\partial t + \partial/\partial s) Q_{0,n,m}(t, s) \quad (5.198)$$

are obvious.

Lemma 5.7.2 *There exist positive numbers N_0 and N such that the inequalities*

$$\|R_{l,n,m}(t,s)\| \leqslant N_0 N^{n+m} M_n M_m (t-s)^{-n} \tag{5.199}$$

and

$$\|Q_{l,n,m}(t,s)\| \leqslant N_0 N^{n+m} M_n M_m (t-s)^{-n} \tag{5.200}$$

hold for $l = 0, 1$ and every non-negative integer n and m.

Proof Let Γ be a piecewise smooth contour running in Σ from $\infty e^{-i\theta}$ to $\infty e^{i\theta}$. We represent $R_{0,n,m}(t,s)$ in the form

$$R_{0,n,m}(t,s) = \mathrm{I} + \mathrm{II},$$

where

$$\mathrm{I} = -\frac{1}{2\pi i} \int_\Gamma e^{-\lambda(t-s)} \left(\frac{\partial}{\partial t}\right)^{n+m} (A(t)-\lambda)^{-1}\, d\lambda \tag{5.201}$$

and

$$\mathrm{II} = -\frac{1}{2\pi i} \int_\Gamma \sum_{k=0}^{n-1} \binom{n}{k}(-\lambda)^{n-k} e^{-\lambda(t-s)} \left(\frac{\partial}{\partial t}\right)^{m+k} (A(t)-\lambda)^{-1}\, d\lambda. \tag{5.202}$$

Deforming the contour into $\Gamma = \Gamma_1 \cup \Gamma_2 \cup \Gamma_3$ with

$$\Gamma_1 = \{re^{-i\theta} : (t-s)^{-1} \leqslant r < \infty\},$$
$$\Gamma_2 = \{(t-s)^{-1} e^{i\varphi} : \theta \leqslant \varphi \leqslant 2\pi - \theta\},$$
$$\Gamma_3 = \{re^{i\theta} : (t-s)^{-1} \leqslant r < \infty\},$$

we have

$$\|\mathrm{I}\| \leqslant c_0 K_0 (d_2 K)^{n+m} M_n M_m$$
$$\leqslant c_0 K_0 (d_2 KT)^n (d_2 K)^m M_n M_m (t-s)^{-n},$$

by (5.192), where

$$c_0 = \frac{1}{\pi} \int_{\cos\theta}^\infty e^{-\xi}\frac{d\xi}{\xi} + \frac{1}{2\pi} \int_\theta^{2\pi-\theta} e^{-\cos\varphi}\, d\varphi.$$

In (5.202), we take the boundary of Σ as Γ and represent $\lambda = re^{i\theta}$ or $re^{-i\theta}$, obtaining

$$\|\mathrm{II}\| \leqslant \frac{1}{\pi} \int_0^\infty \sum_{k=0}^{n-1} \binom{n}{k} r^{n-k} e^{-(t-s)r\cos\theta} K_0 K^{m+k} M_{m+k} r^{-1}\, dr$$
$$= \frac{K_0}{\pi} \sum_{k=0}^{n-1} \binom{n}{k} K^{m+k} M_{m+k} \frac{(n-k-1)!}{\{(t-s)\cos\theta\}^{n-k}}.$$

The inequalities (5.188) and (5.190) imply that

$$M_{m+k} \leqslant d_2^{m+k} M_m M_k \leqslant d_2^{m+k} M_m \left(\frac{d_1}{M_1}\right)^{n-k} \frac{k!}{n!} M_n,$$

so that, by putting $c_1 = \max(1, d_1^{-1} M_1 d_2 KT \cos\theta)$, we have

$$\|\text{II}\| \leqslant \frac{K_0}{\pi} \left(\frac{d_1}{M_1}\right)^n \frac{(d_2 K)^m M_n M_m}{\{(t-s)\cos\theta\}^n} \sum_{k=0}^{n-1} \{d_1^{-1} M_1 d_2 K(t-s)\cos\theta\}^k$$

$$\leqslant \frac{K_0}{\pi} \frac{n}{c_1} \left(\frac{c_1 d_1}{M_1}\right)^n \frac{(d_2 K)^m M_n M_m}{\{(t-s)\cos\theta\}^n}$$

$$\leqslant \frac{K_0}{\pi c_1} \left(\frac{c_1 d_1 e}{M_1}\right)^n \frac{(d_2 K)^m M_n M_m}{\{(t-s)\cos\theta\}^n},$$

Hence, if we put

$$N_0 = \frac{K_0}{\pi c_1} + c_0 K_0, \qquad N = \max\left(\frac{c_1 d_1 e}{M_1 \cos\theta}, d_2 K, d_2 KT\right), \qquad (5.203)$$

we obtain (5.199) with $l = 0$. (5.197) may be used to get (5.199) with $l = 1$ by replacing N_0 and N by other numbers, if necessary. The proof of (5.200) is similar. $\square$

First, we will prove that $U(t, s)$ is differentiable any number of times. For this purpose let us replace Assumption 5.7.4 by the following weaker one.

Assumption 5.7.4' *There exists a sequence of positive numbers $\{B_n\}$ such that the inequality*

$$\|(\partial/\partial t)^n (A(t) - \lambda)^{-1}\| \leqslant B_n/|\lambda| \qquad (5.204)$$

holds for every $n \geqslant 0$, $\lambda \in \Sigma$ and $0 \leqslant t \leqslant T$.

Lemma 5.7.3 *Let $F(t, s)$ and $G(t, s)$ be m-times continuously differentiable functions in $0 \leqslant s < t \leqslant T$ with values in $B(X)$ and put*

$$F_{k,l}(t, s) = (\partial/\partial t)^k (\partial/\partial t + \partial/\partial s)^l F(t, s);$$

$G_{k,l}(t, s)$ is defined similarly. Assume that $F_{0,j}(t, s)(0 \leqslant j \leqslant m)$ is uniformly bounded in $0 \leqslant s < t \leqslant T$. Then for $s < r < t$, we have

$$(\partial/\partial t)^m \int_r^t F(t, \tau) G(\tau, s)\, d\tau$$

$$= \sum_{k=0}^{m-1} \sum_{j=0}^{m-1-k} \binom{m-1-k}{j} F_{k,m-1-k-j}(t, r) \cdot G_{j,0}(r, s)$$

$$+ \int_r^t \sum_{k=0}^m \binom{m}{k} F_{0,m-k}(t, \tau) \cdot G_{k,0}(\tau, s)\, d\tau \qquad (5.205)$$

and, for $s < \rho < r < t$, we have

$$(\partial/\partial t)^m \int_\rho^r F(t, \tau) G(\tau, s) \, d\tau$$

$$= -\sum_{k=0}^{m-1} \sum_{j=1}^{m-1-k} \binom{m-1-k}{j} [F_{k,m-1-k-j}(t, \tau) \cdot G_{j,0}(\tau, s)]_{\tau=\rho}^{\tau=r}$$

$$+ \int_\rho^r \sum_{k=0}^m \binom{m}{k} F_{0,m-k}(t, \tau) G_{k,0}(\tau, s) \, d\tau. \qquad (5.206)$$

Proof With ε being a sufficiently small positive number, consider

$$(\partial/\partial t) \int_r^{t-\varepsilon} F(t, \tau) G(\tau, s) \, d\tau$$

$$= F(t, t-\varepsilon) G(t-\varepsilon, s) + \int_r^{t-\varepsilon} \left(\frac{\partial}{\partial t} + \frac{\partial}{\partial \tau} \right) F(t, \tau) \cdot G(\tau, s) \, d\tau$$

$$- \int_r^{t-\varepsilon} \frac{\partial}{\partial \tau} F(t, \tau) \cdot G(\tau, s) \, d\tau,$$

which, by the integration by parts in the last term, turns into

$$= F(t, r) G(r, s)$$

$$+ \int_r^{t-\varepsilon} \left(\frac{\partial}{\partial t} + \frac{\partial}{\partial \tau} \right) F(t, \tau) \cdot G(\tau, s) \, d\tau + \int_r^{t-\varepsilon} F(t, \tau) \frac{\partial}{\partial \tau} G(\tau, s) \, d\tau.$$

Hence, in the limit as $\varepsilon \to 0$, we get

$$(\partial/\partial t) \int_r^t F(t, \tau) G(\tau, s) \, d\tau$$

$$= F(t, r) G(r, s) + \int_r^t \left(\frac{\partial}{\partial t} + \frac{\partial}{\partial \tau} \right) F(t, \tau) \cdot G(\tau, s) \, d\tau$$

$$+ \int_r^t F(t, \tau) \frac{\partial}{\partial \tau} G(\tau, s) \, d\tau.$$

The relation (5.205) is obtained by repeating this operation. The proof of (5.206) is similar. This time it is simpler than that for (5.205) to the extent that we need not introduce a number $\varepsilon > 0$ to diminish the region of integration. □

It is easy to see that

$$R_l(t, s) = \int_s^t R_{l-1}(t, \tau) R_1(\tau, s) \, d\tau \qquad (5.207)$$

holds for $l = 2, 3, \ldots,$

Lemma 5.7.4 $R_l(t, s)$ *is an infinitely differentiable function of t, s for each $l \geq 1$, and there exists a sequence of positive numbers $\{C_{l,n,m}\}$ such that the inequality*

$$\|R_{l,n,m}(t, s)\| \leq C_{l,n,m}(t-s)^{l-n-1} \tag{5.208}$$

holds in $0 \leq s < t \leq T$ for every $l \geq 1$ and $m, n \geq 0$.

Proof   The proof for the case $l = 1$ is similar to that of (5.199). For general l, we suppose that the lemma has been proved up to $l - 1$. By (5.207) and the method we used in the proof of Lemma 5.7.3, we find that

$$R_{l,0,m}(t, s) = \int_s^t \sum_{i=0}^m \binom{m}{i} R_{l-1,0,m-i}(t, \tau) R_{1,0,i}(\tau, s)\, d\tau. \tag{5.209}$$

With $s < r < t$, we divide the integral into two parts, one from s to r and the other from r to t, and apply Lemma 5.7.3 to the latter, obtaining

$$
\begin{aligned}
R_{l,n,m}(t, s) = \sum_{i=0}^m \binom{m}{i}\bigg\{ & \sum_{k=0}^{n-1}\sum_{j=0}^{n-1-k}\binom{n-1-k}{j} \\
& \times R_{l-1,k,n-1-k-j+m-i}(t, r) R_{1,j,i}(r, s) \\
& + \int_r^t \sum_{k=0}^n \binom{n}{k} R_{l-1,0,n-k+m-i}(t, \tau) R_{1,k,i}(\tau, s)\, d\tau \\
& + \int_s^r R_{l-1,n,m-i}(t, \tau) R_{1,0,i}(\tau, s)\, d\tau \bigg\}.
\end{aligned}
\tag{5.210}
$$

By taking $r = (t + s)/2$ and by the assumption of induction, it is easy to prove (5.208). $\square$

Lemma 5.7.5   *There exists a sequence of positive numbers $\{B_{n,m}\}$ such that*

$$\|R_{l,m,n}(t, s)\| \leq B_{n,m}^{l-n}(t-s)^{l-n-1}/(l-n-1)! \tag{5.211}$$

for $n \geq 0$, $m \geq 0$ and $l \geq n+1$.

Proof   We will show that the lemma is proved if we take $B_{n,m}$ in such a way that it satisfies

$$C_{n+1,n,m} \leq B_{n,m}, \qquad 2^m \max_{0 \leq i \leq m} C_{1,0,i} \leq B_{n,m} \tag{5.212}$$

and forms an increasing sequence with respect to m. First, by the preceding lemma and (5.212), the inequality (5.211) holds for $l = n+1$. Next, we assume that (5.211) has been established for $l - 1$. We differentiate both sides of (5.209) n times with respect to t. By noting that

$l - n - 2 \geqslant 0$, and on account of (5.208), we obtain

$$R_{l,n,m}(t, s) = \int_s^t \sum_{i=0}^m \binom{m}{i} R_{l-1,n,m-i}(t, \tau) R_{1,0,i}(\tau, s) \, d\tau,$$

which, by the assumption of induction, implies

$$\|R_{l,n,m}(t, s)\| \leqslant \int_s^t \sum_{i=0}^m \binom{m}{i} B_{n,m-i}^{l-n-1} \frac{(t-\tau)^{l-n-2}}{(l-n-2)!} C_{1,0,i} \, d\tau$$

$$\leqslant 2^m B_{n,m}^{l-n-1} \max_{0 \leqslant i \leqslant m} C_{1,0,i} \frac{(t-s)^{l-n-1}}{(l-n-1)!}$$

From this, (5.211) follows immediately, since the inequalities (5.212) are satisfied. $\square$

Lemma 5.7.6 $R(t, s)$ is infinitely differentiable in $0 \leqslant s < t \leqslant T$ and there exists a sequence of positive numbers $\{C_{n,m}\}$ such that the following inequalities hold for every $n, m \geqslant 0$ and for $0 \leqslant s < t \leqslant T$:

$$\|(\partial/\partial t)^n (\partial/\partial t + \partial/\partial s)^m R(t, s)\| \leqslant C_{n,m} (t-s)^{-n}. \qquad (5.213)$$

Proof By Lemmas 5.7.4 and 5.7.5, we have

$$\text{left-hand side of (5.213)} \leqslant \sum_{l=1}^\infty \|R_{l,n,m}(t, s)\|$$

$$\leqslant \sum_{l=1}^n C_{l,n,m} (t-s)^{l-n-1}$$

$$+ \sum_{l=n+1}^\infty B_{n,m}^{l-n} (t-s)^{l-n-1}/(l-n-1)!$$

$$\leqslant \sum_{l=1}^n C_{l,n,m} (t-s)^{l-n-1} + B_{n,m} \exp (B_{n,m}(t-s)),$$

from which (5.213) follows immediately. $\square$

Lemma 5.7.7 $U(t, s)$ is infinitely differentiable in $0 \leqslant s < t \leqslant T$ and there exists a sequence of positive numbers $\{\bar{C}_{n,m}\}$ such that the following inequalities hold for every $n, m \geqslant 0$ and for $0 \leqslant s < t \leqslant T$:

$$\|(\partial/\partial t)^n (\partial/\partial t + \partial/\partial s)^m U(t, s)\| \leqslant \bar{C}_{n,m} (t-s)^{-n}. \qquad (5.214)$$

Proof As in the proof of Lemma 5.7.2, there exists a sequence of positive numbers $\{C'_{n,m}\}$ such that the inequality

$$\|R_{0,n,m}(t, s)\| \leqslant C'_{n,m}(t-s)^{-n} \qquad (5.215)$$

holds. Define

$$R_{k,l}(t, s) = (\partial/\partial t)^k (\partial/\partial t + \partial/\partial s)^l R(t, s).$$

As in the proof of (5.209), we have

$$\left(\frac{\partial}{\partial t}+\frac{\partial}{\partial s}\right)^m W(t, s) = \int_s^t \sum_{i=0}^m \binom{m}{i} R_{0,0,m-i}(t, \tau) R_{0,i}(\tau, s)\, \mathrm{d}\tau. \tag{5.216}$$

With $s < r < t$ it is found, as in the proof of (5.210), that

$$(\partial/\partial t)^n(\partial/\partial t + \partial/\partial s)^m W(t, s)$$

$$= \sum_{i=0}^m \binom{m}{i} \sum_{k=0}^{n-1} \sum_{j=0}^{n-1-k} \binom{n-1-k}{j} R_{0,k,n-1-k-j+m-i}(t, r) R_{j,i}(t, s)$$

$$+ \sum_{i=0}^m \binom{m}{i} \int_r^t \sum_{k=0}^n \binom{n}{k} R_{0,0,n-k+m-i}(t, \tau) R_{k,i}(\tau, s)\, \mathrm{d}\tau$$

$$+ \sum_{i=0}^m \binom{m}{i} \int_s^r R_{0,n,m-i}(t, \tau) R_{0,i}(\tau, s)\, \mathrm{d}\tau.$$

Again, putting $r = (t + s)/2$, and on account of Lemmas 5.7.4 and 5.7.6, we have

$$\|(\partial/\partial t)^n(\partial/\partial t + \partial/\partial s)^m W(t, s)\| \leq C''_{n,m}(t - s)^{1-n}, \tag{5.217}$$

where $\{C''_{n,m}\}$ is a certain sequence of positive numbers. From (5.215) and (5.217), the result (5.214) then follows. $\square$

Lemma 5.7.8

$$Z(t, s) = \int_s^t Q_1(t, \tau) U(\tau, s)\, \mathrm{d}\tau. \tag{5.218}$$

Proof By induction, for $l \geq 2$, we have

$$Q_l(t, s) = \int_s^t Q_1(t, \tau) Q_{l-1}(\tau, s)\, \mathrm{d}\tau.$$

By (5.35),

$$Q(t, s) = Q_1(t, s) + \int_s^t Q_1(t, \tau) Q(\tau, s)\, \mathrm{d}\tau, \tag{5.219}$$

which, being substituted into (5.70), gives (5.218). $\square$

Lemma 5.7.9 With $s = r_0 < r_1 < \cdots < r_n < r_{n+1} = t$, we have

$$\left(\frac{\partial}{\partial t}\right)^n Z(t, s) = \sum_{i=1}^n \sum_{j=0}^{i-1} \binom{i-1}{j} Q_{1,n-i,i-1-j}(t, r_i)\left(\frac{\partial}{\partial t}\right)^j U(r_i, s)$$

$$+ \sum_{i=0}^n \int_{r_i}^{r_{i+1}} \sum_{j=0}^i \binom{i}{j} Q_{1,n-i,i-j}(t, \tau)\left(\frac{\partial}{\partial \tau}\right)^j U(\tau, s)\, \mathrm{d}\tau.$$

$$\tag{5.220}$$

Proof By (5.218), we get

$$
\left(\frac{\partial}{\partial t}\right)^n Z(t, s) = \left(\frac{\partial}{\partial t}\right)^n \sum_{i=0}^{n} \int_{r_i}^{r_{i+1}} Q_1(t, \tau) U(\tau, s)\, d\tau
$$

$$
= \left(\frac{\partial}{\partial t}\right)^n \int_{r_n}^{t} Q_1(t, \tau) U(\tau, s)\, d\tau
$$

$$
+ \sum_{i=1}^{n-1} \left(\frac{\partial}{\partial t}\right)^{n-i} \left\{ \left(\frac{\partial}{\partial t}\right)^i \int_{r_i}^{r_{i+1}} Q_1(t, \tau) U(\tau, s)\, d\tau \right\}
$$

$$
+ \left(\frac{\partial}{\partial t}\right)^n \int_{s}^{r_1} Q_1(t, \tau) U(\tau, s)\, d\tau,
$$

which, on applying (5.205) to the first term and (5.206) to the second, becomes

$$
= \sum_{k=0}^{n-1} \sum_{j=0}^{n-1-k} \binom{n-1-k}{j} Q_{1,k,n-1-k-j}(t, r_n) \left(\frac{\partial}{\partial t}\right)^j U(r_n, s)
$$

$$
+ \int_{r_n}^{t} \sum_{k=0}^{n} \binom{n}{k} Q_{1,0,n-k}(t, \tau) \left(\frac{\partial}{\partial \tau}\right)^k U(\tau, s)\, d\tau + \sum_{i=1}^{n-1} \left(\frac{\partial}{\partial t}\right)^{n-i}
$$

$$
\times \left\{ - \sum_{k=0}^{i-1} \sum_{j=0}^{i-1-k} \binom{i-1-k}{j} \left[Q_{1,k,i-1-k-j}(t, \tau) \left(\frac{\partial}{\partial \tau}\right)^j U(\tau, s) \right]_{\tau=r_i}^{\tau=r_{i+1}} \right.
$$

$$
\left. + \int_{r_i}^{r_{i+1}} \sum_{k=0}^{i} \binom{i}{k} Q_{1,0,i-k}(t, \tau) \left(\frac{\partial}{\partial \tau}\right)^k U(\tau, s)\, d\tau \right\}
$$

$$
+ \int_{s}^{r_1} \left(\frac{\partial}{\partial t}\right)^n Q_1(t, \tau) U(\tau, s)\, d\tau.
$$

The result is rearranged in the following form:

$$
(\partial/\partial t)^n Z(t, s) = \mathrm{I} + \mathrm{II} + \mathrm{III}, \tag{5.221}
$$

where

$$
\mathrm{I} = \sum_{i=1}^{n} \sum_{k=0}^{i-1} \sum_{j=0}^{i-1-k} \binom{i-1-k}{j} Q_{1,n-i+k,i-1-k-j}(t, r_i) \left(\frac{\partial}{\partial t}\right)^j U(r_i, s),
$$

$$
\mathrm{II} = - \sum_{i=1}^{n-1} \sum_{k=0}^{i-1} \sum_{j=0}^{i-1-k} \binom{i-1-k}{j} Q_{1,n-i+k,i-1-k-j}(t, r_{i+1}) \left(\frac{\partial}{\partial t}\right)^j U(r_{i+1}, s),
$$

$$
\mathrm{III} = \sum_{i=0}^{n} \int_{r_i}^{r_{i+1}} \sum_{k=0}^{i} \binom{i}{k} Q_{1,n-i,i-k}(t, \tau) \left(\frac{\partial}{\partial \tau}\right)^k U(\tau, s)\, d\tau.
$$

If, in II, we replace i and k by $i-1$ and $k-1$, respectively, we have

$$\text{II} = -\sum_{i=2}^{n} \sum_{k=1}^{i-1} P_{i,k},$$

where

$$P_{i,k} = \sum_{j=0}^{i-1-k} \binom{i-1-k}{j} Q_{1,n-i+k,i-1-k-j}(t, r_i)\left(\frac{\partial}{\partial t}\right)^j U(r_i, s),$$

and, hence,

$$\text{I+II} = \sum_{i=1}^{n} \sum_{k=0}^{i-1} P_{i,k} - \sum_{i=2}^{n} \sum_{k=1}^{i-1} P_{i,k}$$

$$= P_{1,0} + \sum_{i=2}^{n} \sum_{k=0}^{i-1} P_{i,k} - \sum_{i=2}^{n} \sum_{k=1}^{i-1} P_{i,k} = P_{1,0} + \sum_{i=2}^{n} P_{i,0}$$

$$= \sum_{i=1}^{n} P_{i,0} = \sum_{i=1}^{n} \sum_{j=0}^{i-1} \binom{i-1}{j} Q_{1,n-i,i-1-j}(t, r_i)\left(\frac{\partial}{\partial t}\right)^j U(r_i, s).$$

From this and (5.221) the result (5.220) then follows. $\square$

Returning to the outset, we suppose that Assumptions 5.1.1, 5.7.3 and 5.7.4 are satisfied. By Stirling's formula there exists a positive number ω such that the following inequality holds for every $n \geq 1$:

$$\omega^{-1} n^n e^{-n} \sqrt{n} \leq n! \leq \omega n^n e^{-n} \sqrt{n}. \tag{5.222}$$

Proposition 5.7.1 *There exist positive numbers H_0 and H such that the inequality*

$$\|(\partial/\partial t)^n U(t, s)\| \leq H_0 H^n M_n (t-s)^n \tag{5.223}$$

holds for every $n \geq 0$.

Proof If $H_0 = \bar{C}_{0,0}$ (see (5.214)) is taken, (5.223) holds for $n = 0$. Put

$$C_0 = 2ed_1 N_0(\omega^3 ed_1^2 M_1^{-1} + \omega^3 d_1 N + M_0 NT), \tag{5.224}$$

where N_0 and N are the same constants as in (5.199) and (5.200). The number C_0 is determined only by θ, K_0, K, T and $\{M_k\}$. Put $J = \exp(eN_0 M_0^2 T)$ and let H be some number satisfying

$$H \geq \max(2N, 2NT), \qquad H_0 H \geq 2N_0 M_0 NJ, \qquad H \geq 2C_0 JT. \tag{5.225}$$

With these H_0 and H, it will be shown that (5.223) holds. For this purpose, we assume that (5.223) is valid for 0 to $n-1$. By Lemma 5.7.9, we represent

$$(\partial/\partial t)^n U(t, s) = \mathrm{I} + \mathrm{II} + \mathrm{III} + \mathrm{IV} + \mathrm{V}, \qquad (5.226)$$

where

$\mathrm{I} = (\partial/\partial t)^n \exp\left(-(t-s)A(s)\right),$

$\mathrm{II} =$ the first term on the right-hand side of (5.220),

$$\mathrm{III} = \sum_{i=0}^{n-1} \int_{r_i}^{r_{i+1}} \sum_{j=0}^{i} \binom{i}{j} Q_{1,n-i,i-j}(t, \tau) \left(\frac{\partial}{\partial \tau}\right)^j U(\tau, s)\, d\tau,$$

$$\mathrm{IV} = \int_{r_n}^{t} \sum_{j=0}^{n-1} \binom{n}{j} Q_{1,0,n-j}(t, \tau) \left(\frac{\partial}{\partial \tau}\right)^j U(\tau, s)\, d\tau.$$

$$\mathrm{V} = \int_{r_n}^{t} Q_1(t, \tau) \left(\frac{\partial}{\partial \tau}\right)^n U(\tau, s)\, d\tau.$$

By Lemma 5.7.2, we have the estimate

$$\|\mathrm{I}\| \leqslant N_0 M_0 N^n M_n (t-s)^{-n}. \qquad (5.227)$$

Henceforth, we write

$$r_i = s + i(t-s)/(n+1), \qquad i = 1, \ldots, n. \qquad (5.228)$$

On account of (5.200) and (5.186), we obtain

$$\|\mathrm{II}\| \leqslant \sum_{i=1}^{n} \sum_{j=0}^{i-1} \binom{i-1}{j} \frac{N_0 N^{n-1-j} M_{n-i} M_{i-1-j}}{(t-r_i)^{n-i}} \frac{H_0 H^j M_j}{(r_i - s)^j}$$

$$\leqslant d_1 N_0 H_0 N^{n-1} \sum_{i=1}^{n} \frac{M_{i-1} M_{n-i}}{(t-r_i)^{n-i}} \sum_{j=0}^{i-1} \left(\frac{H}{N}\right)^j \frac{1}{(r_i - s)^j}.$$

The last sum can be estimated by (5.225) as

$$\sum_{j=0}^{i-1} \left(\frac{H}{N}\right)^j \frac{1}{(r_i - s)^j} = \left(\frac{H}{N}\right)^{i-1} \frac{1}{(r_i - s)^{i-1}}$$

$$\times \left\{ 1 + \frac{N(r_i - s)}{H} + \cdots + \left(\frac{N(r_i - s)}{H}\right)^{i-1} \right\}$$

$$\leqslant 2 \left(\frac{H}{N}\right)^{i-1} \frac{1}{(r_i - s)^{i-1}},$$

which, combined again with (5.186), gives

$$\|\text{II}\| \leqslant 2d_1^2 N_0 H_0 H^{n-1} M_{n-1}$$
$$\times \sum_{i=1}^{n} \frac{(i-1)!\,(n-i)!}{(n-1)!} (t-r_i)^{i-n}(r_i-s)^{1-i}\left(\frac{N}{H}\right)^{n-i}. \tag{5.229}$$

By (5.222) and (5.228) the sum on the right-hand side of (5.229) does not exceed

$$\left(\frac{n+1}{n}\right)^{n-1}(t-s)^{1-n}\left(\frac{N}{H}\right)^{n-1}+\left(\frac{n+1}{n}\right)^{n-1}(t-s)^{1-n}$$

$$+\sum_{i=2}^{n-1}\omega^3\left(\frac{n+1}{n-1}\right)^{n-1}\binom{i-1}{i}^{i-1}\left(\frac{n-i}{n+1-i}\right)^{n-i}\left\{\frac{(i-1)(n-i)}{n-1}\right\}^{1/2}$$

$$\times\left(\frac{N}{H}\right)^{n-i}(t-s)^{1-n}\leqslant e(t-s)^{1-n}\left(\frac{N}{H}\right)^{n-1}+e(t-s)^{1-n}$$

$$+\omega^3 e^2(t-s)^{1-n}\sum_{i=2}^{n-1}\left\{\frac{(i-1)(n-i)}{n-1}\right\}^{1/2}\left(\frac{N}{H}\right)^{n-i},$$

which, because of the inequality $(i-1)(n-1)\leqslant(n-1)^2/4$, is further dominated by

$$\leqslant e(t-s)^{1-n}2^{1-n}+e(t-s)^{1-n}+\omega^3 e^2(t-s)^{1-n}2^{-1}\sqrt{n-1}\sum_{i=2}^{n-1}2^{i-n}$$

$$\leqslant\omega^3 e^2(t-s)^{1-n}2^{-1}\sqrt{n-1}\sum_{i=1}^{n}2^{i-n}\leqslant\omega^3 e^2\sqrt{n-1}(t-s)^{1-n}.$$

The substitution of this result into (5.299) gives

$$\|\text{II}\|\leqslant 2\omega^3 e^2 d_1^2 N_0 H_0\sqrt{n-1}H^{n-1}M_{n-1}(t-s)^{1-n},$$

and, hence, by (5.189), it is found that

$$\|\text{II}\|\leqslant 2\omega^3 e^2 d_1^3 N_0 H_0 M_1^{-1}H^{n-1}n^{-1/2}M_n(t-s)^{1-n}.$$

The terms III and IV can be estimated in similar ways, e.g., for III we have

$$\|\text{III}\|\leqslant N_0 H_0 N^n\sum_{i=0}^{n-1}\int_{r_i}^{r_{i+1}}(t-\tau)^{i-n}\sum_{j=0}^{i}\binom{i}{j}M_{n-i}M_{i-j}M_j\left(\frac{H}{N}\right)^j(\tau-s)^{-j}\,d\tau,$$

in which

the part of $\sum_{i=0}^{n-1} \int_{r_i}^{r_{i+1}} \cdots$ on the right-hand side

$$\leqslant d_1^2 M_n \sum_{i=0}^{n-1} \frac{i!\,(n-i)!}{n!} \int_{r_i}^{r_{i+1}} (t-\tau)^{i-n} \sum_{j=0}^{i} \left(\frac{H}{N}\right)^j (\tau-s)^{-j}\,d\tau$$

$$\leqslant 2d_1^2 M_n \sum_{i=0}^{n-1} \frac{i!\,(n-i)!}{n!} \left(\frac{H}{N}\right)^i \int_{r_1}^{r_{i+1}} (t-\tau)^{i-n}(\tau-s)^{-i}\,d\tau$$

$$\leqslant 2d_1^2 M_n \left\{ \sum_{i=1}^{n-1} \frac{i!\,(n-i)!}{n!} \left(\frac{H}{N}\right)^i \frac{r_{i+1}-r_i}{(t-r_{i+1})^{n-i}(r_i-s)^i} + (t-r_1)^{-n}(r_1-s) \right\}$$

$$\leqslant 2d_1^2 M_n \left\{ \frac{e\omega^3}{2\sqrt{n+1}} \sum_{i=1}^{n-1} \left(\frac{H}{N}\right)^i + \frac{e}{n+1} \right\} (t-s)^{1-n}$$

$$\leqslant 2d_1^2 M_n \frac{e\omega^3}{2\sqrt{n+1}} \sum_{i=0}^{n-1} \left(\frac{H}{N}\right)^i (t-s)^{1-n},$$

so that

$$\|III\| \leqslant 2\omega^3 e d_1^2 N_0 H_0 N(n+1)^{-1/2} H^{n-1} M_n (t-s)^{1-n},$$
$$\|IV\| \leqslant 2e d_1 N_0 H_0 M_0 N(n+1)^{-1} H^{n-1} M_n (t-s)^{2-n}$$
$$\leqslant 2e d_1 N_0 H_0 M_0 N T(n+1)^{-1} H^{n-1} M_n (t-s)^{1-n}.$$

Finally, as for V, we have

$$\|V\| \leqslant N_0 M_0^2 \int_{r_n}^{t} \|(\partial/\partial\tau)^n U(\tau,s)\|\,d\tau.$$

Collecting these results, we find that

$$\|(\partial/\partial t)^n U(t,s)\| \leqslant N_0 M_0 N^n M_n (t-s)^{-n}$$

$$+ C_0 H_0 H^{n-1} M_n (t-s)^{1-n} + N_0 M_0^2 \int_{r_n}^{t} \|(\partial/\partial\tau)^n U(\tau,s)\|\,d\tau$$

and, hence, by (5.225),

$$\|(\partial/\partial t)^n U(t,s)\|$$

$$\leqslant J^{-1} H_0 H^n M_n (t-s)^{-n} + N_0 M_0^2 \int_{r_n}^{t} \|(\partial/\partial\tau)^n U(\tau,s)\|\,d\tau. \quad (5.230)$$

Put $Y(t,s) = (t-s)^n \|(\partial/\partial t)^n U(t,s)\|$. On multiplying both sides of (5.230) by $(t-s)^n$ and noting that

$$(t-s)^n < (1+n^{-1})^n (\tau-s)^n < e(\tau-s)^n$$

for $r_n < \tau < t$, we obtain

$$Y(t, s) \leqslant J^{-1} H_0 H^n M_n + e N_0 M_0^2 \int_s^t Y(\tau, s) \, d\tau.$$

By virtue of Lemma 5.7.7, the integral on the right-hand side converges. This inequality can be regarded as an integral inequality for Y, the integration of which yields

$$Y(t, s) \leqslant J^{-1} H_0 H^n M_n \exp\left(e N_0 M_0^2 (t - s)\right) \leqslant H_0 H^n M_n.$$

Thus it has turned out that (5.223) holds also for n. $\square$

Proof of Theorem 5.7.3 Like (5.209) or (5.216), we can show that

$$(\partial/\partial t + \partial/\partial s)^m U(t, s)$$

$$= Q_{0,0,m}(t, s) + \sum_{k=0}^m \binom{m}{k} \int_s^t Q_{1,0,m-k}(t, \tau) \left(\frac{\partial}{\partial \tau} + \frac{\partial}{\partial s}\right)^k U(\tau, s) \, d\tau. \quad (5.231)$$

By the application of the proof of Lemma 5.7.1 to each term on the right-hand side of (5.231), and by induction with respect to $n + m$, we are led to (5.193) with $l = 0$. In the general case, by means of (5.193) with $l = 0$, we have

$$\|(\partial/\partial t)^n (\partial/\partial t + \partial/\partial s)^m (\partial/\partial s)^l U(t, s)\|$$

$$= \left\| \sum_{k=0}^l (-1)^k \binom{l}{k} \left(\frac{\partial}{\partial t}\right)^{n+k} \left(\frac{\partial}{\partial t} + \frac{\partial}{\partial s}\right)^{m+l-k} U(t, s) \right\|$$

$$\leqslant \sum_{k=0}^l \binom{l}{k} L_0 L^{n+m+l} M_{n+k} M_{m+l-k} (t - s)^{-n-k}$$

$$\leqslant d_1 L_0 L^{n+m+l} M_{n+m+l} \sum_{k=0}^l \binom{l}{k} \left\{ \binom{n+m+l}{n+k} \right\}^{-1} (t - s)^{-k-n},$$

in which

$$\binom{l}{k} \left\{ \binom{n+m+l}{n+k} \right\}^{-1} \leqslant \binom{l}{k} \left\{ \binom{n+l}{n+k} \right\}^{-1} = \frac{l!}{(n+l)!} \frac{(n+k)!}{k!} \leqslant 1,$$

so that the estimation continues as follows:

$$\leqslant d_1 L_0 L^{n+m+l} M_{n+m+l} \sum_{k=0}^l (t - s)^{-k-n}$$

$$\leqslant d_1 L_0 L^{n+m+l} M_{n+m+l} (t - s)^{-n-l} \sum_{k=0}^l \{\max(T, 1)\}^{l-k}$$

$$\leqslant d_1 l L_0 L^{n+m+l} M_{n+m+l} (t - s)^{-n-l} \{\max(T, 1)\}^l.$$

From this, by replacing L_0 and L by other numbers if necessary, it follows that (5.193) is also valid for general l. $\square$

Proof of Theorem 5.7.4 First, we note that

$$u(t) = U(t, 0)u_0 + v(t), \qquad v(t) = \int_0^t U(t, \sigma)f(\sigma)\,d\sigma.$$

It is not hard to compute

$$\frac{d^n v(t)}{dt_n} = \sum_{i=1}^n \sum_{j=0}^{i-1} \binom{i-1}{j}\left(\frac{\partial}{\partial t}\right)^{n-i}\left(\frac{\partial}{\partial t}+\frac{\partial}{\partial s}\right)^{i-1-j} U(t, 0)f^{(j)}(0)$$
$$+ \int_0^t \sum_{j=0}^m \binom{n}{j}\left(\frac{\partial}{\partial t}+\frac{\partial}{\partial \sigma}\right)^{n-j} U(t, \sigma) \cdot f^{(j)}(\sigma)\,d\sigma.$$

By virtue of Theorem 5.7.3, we get

$$\left\|\frac{d^n v(t)}{dt^n}\right\| \leq \sum_{i=1}^n \sum_{j=0}^{i-1} \binom{i-1}{j} L_0 L^{n-1-j} M_{n-1-j} t^{i-n} F_0 F^j M_j$$
$$+ \int_0^t \sum_{j=0}^n \binom{n}{j} L_0 L^{n-j} M_{n-j} F_0 F^j M_j \, d\sigma.$$

Since the inequalities

$$\binom{i-1}{j} M_{n-1-j} M_j \leq d_1 M_{n-1} \binom{i-1}{j}\left\{\binom{n-1}{j}\right\}^{-1} \leq d_1 M_{n-1}$$

hold for $j < i \leq n$, this turns out to be

$$\leq d_1 L_0 F_0 L^{n-1} M_{n-1} \sum_{i=1}^n \sum_{j=0}^{i-1} (F/L)^j (t-s)^{i-n} + d_1 L_0 F_0 L^n M_n \sum_{j=0}^n (F/L)^j (t-s).$$

Thus it is found that (5.195) is correct if one takes

$$\bar{F}_0 = 6d_1 L_0 F_0, \qquad \bar{F} = \max\{1, FT, 2L, 2LT\}.$$

The inequality (5.196) follows immediately from Proposition 5.7.1 and (5.195). $\square$

5.7.3 Application 1. The case in which $D(A(t))$ is independent of t

Theorem 5.7.5 *Let $A(t)$ be an operator, satisfying Assumption 5.1.1, whose domain $D(A(t)) \equiv D$ is independent of t. Assume that $A(t)A(0)^{-1}$ is infinitely differentiable in $0 \leq t \leq T$ and that there exist positive numbers B_0 and B such that*

$$\|(d/dt)^n A(t)A(0)^{-1}\| = \|A^{(n)}(t)A(0)^{-1}\| \leq B_0 B^n M_n \qquad (5.232)$$

holds for every $n \geq 0$ and $0 \leq t \leq T$. Then Assumptions 5.7.3 and 5.7.4 are satisfied.

Proof For every fixed s, $A(t)A(s)^{-1}$ is also infinitely differentiable in t. On replacing B_0 by another number if necessary, we may assume that the following result holds:

$$\|A^{(n)}(t)A(t)^{-1}\| \leq B_0 B^n M_n. \tag{5.233}$$

We will show that (5.192) and

$$\|A(t)(\partial/\partial t)^n(A(t)-\lambda)^{-1}\| \leq K_0 K^n M_n \tag{5.234}$$

hold for some constants K_0 and K. They are really correct for $n=0$ if K_0 is sufficiently large. Suppose that they have been established up to n. We differentiate both sides of

$$(\partial/\partial t)(A(t)-\lambda)^{-1} = -(A(t)-\lambda)^{-1}A'(t)(A(t)-\lambda)^{-1} \tag{5.235}$$

n times, obtaining

$$(\partial/\partial t)^{n+1}(A(t)-\lambda)^{-1}$$

$$= -\sum_{j=0}^{n}\binom{n}{j}\left(\frac{\partial}{\partial t}\right)^{n-j}(A(t)-\lambda)^{-1}\sum_{k=0}^{j}\binom{j}{k}A^{(j-k+1)}(t)\left(\frac{\partial}{\partial t}\right)^{k}(A(t)-\lambda)^{-1}.$$

By (5.233) and by the assumption of induction, we have

$$\|(\partial/\partial t)^{n+1}(A(t)-\lambda)^{-1}\|$$

$$\leq \sum_{j=0}^{n}\binom{n}{j}K_0 K^{n-j}M_{n-j}\,|\lambda|^{-1}\sum_{k=0}^{j}\binom{j}{k}B_0 B^{j-k+1}M_{j-k+1}K_0 K^k M_k.$$

The conditions (5.188) and (5.186) give

$$\binom{n}{j}M_{n-j}\binom{j}{k}M_{j-k+1}M_k \leq \binom{n}{j}M_{n-j}\binom{j}{k}d_2^{j-k+1}M_{j-k}M_1 M_k$$

$$\leq d_1 M_1 d_2^{j-k+1}\binom{n}{j}M_{n-j}M_j \leq d_1^2 M_1 d_2^{j-k+1}M_n,$$

so that

$$\|(\partial/\partial t)^{n+1}(A(t)-\lambda)^{-1}\|$$

$$\leq d_1^2 B_0 K_0^2 M_1 M_n\,|\lambda|^{-1}\sum_{j=0}^{n}K^{n-j}(d_2 B)^{j+1}\sum_{k=0}^{j}(K/d_2 B)^k.$$

If $K \geq 2d_2 B$, the right-hand side does not exceed

$$2d_1^2 B_0 K_0^2 M_1 M_n\,|\lambda|^{-1}\sum_{j=0}^{n}K^{n-j}(d_2 B)^{j+1}(K/d_2 B)^j \leq \bar{B}K_0^2 K^n n M_n\,|\lambda|^{-1}.$$

where $\bar{B} = 2d_1^2 d_2 B_0 B M_1$. Thus it is found that

$$\|(\partial/\partial t)^{n+1}(A(t)-\lambda)^{-1}\| \leq \bar{B}K_0^2 K^n n M_n |\lambda|^{-1}$$

holds. Similarly, we have

$$\|A(t)(\partial/\partial t)^{n+1}(A(t)-\lambda)^{-1}\| \leq \bar{B}K_0^2 K^n n M_n.$$

On account of (5.198), it is seen that if

$$K \geq \max (2Bd_2, \bar{B}K_0 d_1/M_1).$$

then (5.192) and (5.234) hold also for $n+1$. $\square$

5.7.4 Application 2. The case in which $A(t)$ is a regularly dissipative operator

As in Section 5.4, let X, V be Hilbert spaces and let $a(t; u, v)$ be a quadratic form and assume that (5.108) is satisfied with $k = 0$. Assume further that $a(t; u, v)$ is infinitely differentiable in $0 \leq t \leq T$ and that there exist positive numbers B_0 and B such that the inequality

$$|a^{(n)}(t; u, v)| \leq B_0 B^n M_n \|u\| \|v\| \tag{5.236}$$

holds for every $n \geq 0$ and $0 \leq t \leq T$. In this case, we have

Theorem 5.7.6 The resolvent $(A(t)-\lambda)^{-1}$ with $\lambda \in \Sigma$ is infinitely differentiable in $0 \leq t \leq T$ and there exist positive numbers K_0 and K such that the following inequalities hold for every $n \geq 0$, $0 \leq t \leq T$ and $f \in X$ or V^*:

$$|(\partial/\partial t)^n (A(t)-\lambda)^{-1} f| \leq K_0 K^n M_n |f|/|\lambda|, \tag{5.237}$$

$$\|(\partial/\partial t)^n (A(t)-\lambda)^{-1} f\| \leq K_0 K^n M_n |f|/|\lambda|^{1/2}, \tag{5.238}$$

$$|(\partial/\partial t)^n (A(t)-\lambda)^{-1} f| \leq K_0 K^n M_n \|f\|_*/|\lambda|^{1/2}, \tag{5.239}$$

$$\|(\partial/\partial t)^n (A(t)-\lambda)^{-1} f\| \leq K_0 K^n M_n \|f\|_*. \tag{5.240}$$

Proof As in Section 5.4, we consider $A(T)$ as a closed operator in V^*. Then, since its domain V is independent of t and since

$$|A^{(n)}(t)u, v)| = |a^{(n)}(t; u, v)| \leq B_0 B^n M_n \|u\| \|v\|$$

by (5.236), we have

$$\|A^{(n)}(t)u\|_* \leq B_0 B^n M_n \|u\|,$$

so that $A(t)$ as an operator in V^* satisfies the assumption of Theorem 5.7.5 and $(A(t)-\lambda)^{-1}f$ with $\lambda \in \Sigma$ and $f \in V^*$ is also infinitely differentiable in V. Keeping this fact in mind, we can prove (5.237)–(5.240) in exactly the same way as for Theorem 5.7.5. $\quad\square$

5.7.5 Application 3. The mixed problem of a parabolic equation

We again consider the example (5.94) in Section 5.3. Here, however, all of the coefficients of $A(x, t, D)$ and $\{B_j(x, t, D)\}$, as functions of t, are assumed to belong to the class $\{M_k\}$. Moreover, it is assumed that each coefficient of $B_j(x, t, D)$ is also defined in the entire $\bar{\Omega}$ and the inequalities

$$|(\partial/\partial t)^l a_\alpha(x, t)| \le B_0 B^l M_l \tag{5.241}$$

and

$$|(\partial/\partial t)^l b_{j,\beta}(x, t)| \le B_0 B^l M_l \tag{5.242}$$

hold for every $x \in \bar{\Omega}$, $t \in [0, T]$, $|\alpha| \le m$, $|\beta| \le m_j$, $j = 1, \ldots, m/2$ and $l = 1, 2, \ldots$. Then we will prove in the sequel that there exist positive numbers K_0 and K such that the inequality

$$\|(\partial/\partial t)^l (A(t)-\lambda)^{-1}\| \le K_0 K^l M_l/|\lambda|$$

holds for $0 \le t \le T$, $\theta_0 \le \arg \lambda \le 2\pi - \theta_0$ and $l = 1, 2, \ldots$. First of all, as in the proof of Lemma 5.3.5, it can be shown that $(A(t)-\lambda)^{-1}$ is infinitely differentiable in t. By putting $u(t) = (A(t)-\lambda)^{-1}f$ with $f \in L^p(\Omega)$, we have, as in (5.105) and (5.106),

$$(A(x, t, D) - \lambda)(\partial/\partial t)^l u(x, t)$$
$$= -\sum_{k=0}^{l-1} \binom{l}{k} A^{(l-k)}(x, t, D) \left(\frac{\partial}{\partial t}\right)^k u(x, t) \equiv f(x, t), \quad x \in \Omega,$$

$$B_j(x, t, D)(\partial/\partial t)^l u(x, t)$$
$$= -\sum_{k=0}^{l-1} \binom{l}{k} B_j^{(l-k)}(x, t, D) \left(\frac{\partial}{\partial t}\right)^k u(x, t) \equiv g_i(x, t),$$
$$x \in \partial\Omega, \quad j = 1, \ldots, m/2,$$

where $A^{(l-k)}$ and $B_j^{(l-k)}$ are operators obtained from A and B_j, respectively, by differentiating their coefficients $l - k$ times with respect to t. We apply (5.104) with g_j replaced by $g_j(x, t)$ to $(\partial/\partial t)^l u(x, t)$, obtaining

$$\|(d/dt)^l u(t)\|_{m,p} + |\lambda| \, \|(d/dt)^l u(t)\|_p$$
$$\le C \left\{ \|f(t)\|_p + \sum_{j=1}^{m/2} |\lambda|^{(m-m_j)/m} \|g_j(t)\|_p + \sum_{j=1}^{m/2} \|g_j(t)\|_{m-m_j, p} \right\}.$$

By (5.241) and (5.242), the right-hand side equals

$$C\left\{\sum_{k=0}^{l-1} \binom{l}{k} B_0 B^{l-k} M_{l-k} \left\|(\partial/\partial t)^k u(t)\right\|_{m,p}\right.$$
$$\left. + \sum_{j=1}^{m/2} \sum_{k=0}^{l-1} \binom{l}{k} B_0 B^{l-k} M_{l-k} |\lambda|^{(m-m_j)/m} \left\|(\partial/\partial t)^k u(t)\right\|_{m_j,p}\right\}.$$

Hence, we may use the interpolation inequality

$$\|u\|_{m_j,p} \leq c \|u\|_{m,p}^{m_j/m} \|u\|_{p}^{(m-m_j)/m}$$

to get the required relation:

$$|\lambda| \left\|(d/dt)^l u(t)\right\|_p + \left\|(d/dt)^l u(t)\right\|_{m,p}$$
$$\leq C_0 \sum_{k=0}^{l-1} \binom{l}{k} B_0 B^{l-k} M_{l-k} \{\left\|(d/dt)^k u(t)\right\|_{m,p} + |\lambda| \left\|(d/dt)^k u(t)\right\|_p\}.$$

From this it is found that, if K_0 is first chosen so large that

$$|\lambda| \left\|(d/dt)^l u(t)\right\|_p + \left\|(d/dt)^l u(t)\right\|_{m,p} \leq K_0 K^l M_l \|f\|_p \qquad (5.243)$$

holds for $l = 0$ and next K is chosen so as to satisfy

$$K \geq \max(2B, 2d_1 C_0 B_0 B),$$

then, by induction, (5.243) may be shown to be valid for all l.

A description closely related to the present section can be found in Lions and Magenes [118, 119]. In these articles, for a technical reason, they treated only cases which are not quasi-analytic. Also, for the higher differentiability of the solution when Assumptions 5.1.1 and 5.3.3 are satisfied and $A(t)^{-1}$ is several times continuously differentiable, we refer the reader to an article of Suryanarayana [165].

Supplementary notes to Chapter 5

One of the main subjects which has not been mentioned in the present chapter (though it is not concerned only with Chapter 5) is the method of singular perturbation. For this, see a lecture note of Lions [14]. The reader can find the singular perturbation of parabolic equations discussed in Tanabe [168], Tanabe and Watanabe [170], degenerate equations in Friedman and Schuss [65], Sovolevskiĭ [161], Matsuzawa [127, 128], and parabolic equations of higher order in t in Lagnes [113]. The behaviour of solutions as $t \to \infty$ is also treated in Agmon and Nirenberg [20]. In [20] and [21] the properties of solutions are studied very profoundly, independently of the existence of the solutions. Concerning the smoothness of solutions, Bardos [25] found a sufficient condition for realizing a solution

u of the equation, with $u_0 \in V$ and $f \in L^2(0, T; X)$, which is treated in Section 5.4, that satisfies $u' \in L^2(0, T; X)$ and $Au \in L^2(0, T; X)$. By assuming, as opposed to Section 5.7, that an operator-valued function $U(t, s)$ satisfying the conclusion of Theorem 5.7.1 is given, Masuda [126] constructed a family of generators $\{A(t)\}$. Other articles to be noted include those by Carroll, Mazumdar and Cooper, [48, 49, 51, 52].

Burak [42, 43, 44] investigated the existence, uniqueness and regularity of the solution of the two-point problem

$$du(t)/dt = A(t)u(t) + f(t), \qquad \alpha \leqslant t \leqslant \beta,$$

$$E_1(\alpha)u(\alpha) = u_\alpha, \qquad E_2(\beta)u(\beta) = u_\beta,$$

where $E_1(t)$ and $E_2(t)$ are projection operators to restrict $A(t)$ and satisfy $X = E_1(t)X \oplus E_2(t)X$. If $A(t)$ is an operator defined by an elliptic boundary-value problem, the existence of such projection operators can be proved by means of a penetrating result of Seeley [156, 157] on the resolvent and complex powers.

6

Non-linear equations

6.1 Semilinear wave equations

Jörgens [74, 75], Browder [37] and others have studied the method for solving the semilinear wave equation

$$\partial^2 u/\partial t^2 - \Delta u + F'(|u|^2)u = 0, \tag{6.1}$$

which is important in quantum field theory. In this section, we explain it briefly. The reader is also referred to Mizohata [17], Chapter 7 and Pogorelenko and Sobolevskiĭ [150]. As preliminaries, we consider the initial-value problem of the following semilinear equation in a Hilbert space X:

$$du(t)/dt = Au(t) + f(u(t)). \qquad t \geq 0. \tag{6.2}$$

$$u(0) = u_0. \tag{6.3}$$

Assumption 6.1.1 *A is a generator of a contraction semigroup in X.*

Assumption 6.1.2 *f is a non-linear mapping from the whole of X into X. For any $C > 0$ there exists some constant $k_C > 0$ such that*

$$\|f(u)\| \leq k_C, \qquad \|f(u) - f(v)\| \leq k_C \|u - v\|$$

hold for $\|u\| \leq C$ and $\|v\| \leq C$.

The semigroup generated by A is denoted by $T(t)$. Note that $\|T(t)\| \leq 1$. By Theorem 3.1.5, we have

$$\mathrm{Re}\,(Au, u) \leq 0 \tag{6.4}$$

for each $u \in D(A)$. k_C may be assumed to be an increasing function of C.

For a solution of (6.2) and (6.3) in the wider sense, we are going to find a solution of the integral equation

$$u(t) = T(t)u_0 + \int_0^t T(t-s)f(u(s))\,\mathrm{d}s. \tag{6.5}$$

A local solution of this equation will be shown to exist by successive iteration as follows. First, put

$$u_0(t) = T(t)u_0 \tag{6.6}$$

and define $u_{j+1}(t)$ by using $u_j(t)$ as

$$u_{j+1}(t) = u_0(t) + \int_0^t T(t-s)f(u_j(s))\,\mathrm{d}s. \tag{6.7}$$

Choose $C > \|u_0\|$. Then, since $\|u_0(t)\| < C$, we have $\|f(u_0(t))\| \leqslant k_C$, so that

$$\|u_1(t)\| \leqslant C + k_C t \leqslant 2C$$

for any t in $0 \leqslant t \leqslant C/k_C$. By induction, it can be shown that for all $j = 1, 2, \ldots$

$$\|u_j(t)\| \leqslant 2C \tag{6.8}$$

in $0 \leqslant t \leqslant C/k_{2C}$. Also, from the equation

$$u_{j+1}(t) - u_j(t) = \int_0^t T(t-s)(f(u_j(s)) - f(u_{j-1}(s)))\,\mathrm{d}s$$

together with (6.8) and Assumption 6.1.2, we can observe that the inequality

$$\|u_{j+1}(t) - u_j(t)\| \leqslant \frac{(k_{2C}t)^j}{j!} \max_{0 \leqslant s \leqslant Ck_{2C}^{-1}} \|u_1(s) - u_0(s)\|$$

holds for any t in $0 \leqslant t \leqslant C/k_{2C}$ and for each $j = 0, 1, 2, \ldots$. Hence, $\{u_j(t)\}$ is strongly convergent to a function $u(t)$ uniformly on $0 \leqslant t \leqslant C/k_{2C}$. By letting $j \to \infty$ in (6.7), we obtain (6.5). Suppose v is another solution of (6.5). If $\|u(t)\| \leqslant C'$ and $\|v(t)\| \leqslant C'$, then

$$\|u(t) - v(t)\| \leqslant k_{C'} \int_0^t \|u(s) - v(s)\|\,\mathrm{d}s,$$

from which it immediately follows that $u(t) \equiv v(t)$. Thus we have the following:

Theorem 6.1.1 *Suppose Assumptions 6.1.1 and 6.1.2 are satisfied. If $\|u_0\| < C$, the solution of (6.5) exists uniquely in $0 \leqslant t \leqslant C/k_{2C}$.*

Remark 6.1.1 Even if A depends on t, similar results to that above still hold when X is a Banach space and $u'(t) = A(t)u(t)$ has a fundamental solution.

Next, we consider the global existence of a solution of (6.5). If we assume a solution $u \in C([0, b]; X)$ of (6.5) exists in a closed interval $[0, b]$, then, by Theorem 6.1.1, a solution v of

$$v(t) = T(t-b)u(b) + \int_b^t T(t-s)f(v(s))\,ds$$

exists in some interval $[b, \hat{b})$ By letting $\hat{u}(t) = u(t)$ for $0 \leqslant t \leqslant b$ and $\hat{u}(t) = v(t)$ for $b \leqslant t < \hat{b}$, it is easy to see that $\hat{u}$ is a solution in $0 \leqslant t < \hat{b}$. Therefore u can be extended over b as a solution of (6.5). Let u be a bounded solution of (6.5) in $[0, b)$: $\|u(t)\| \leqslant C' < \infty$. Then, since $\|f(u(t))\| \leqslant k_{C'}$ for $0 \leqslant t < b$ by Assumption 6.1.2, it is readily shown that

$$\int_0^b T(b-s)f(u(s))\,ds = \underset{t \to b-0}{\text{s-lim}} \int_0^t T(t-s)f(u(s))\,ds$$

exists. Therefore, if we put

$$u(b) = T(b)u_0 + \int_0^b T(b-s)f(u(s))\,ds,$$

u is continuous in $0 \leqslant t \leqslant b$ and, moreover, satisfies (6.5). Hence, u can be extended over b as a solution. With these preliminaries we want to prove the following:

Theorem 6.1.2 *Let b be a positive number and suppose that if u is a solution of (6.5) in $0 \leqslant t \leqslant b$, it satisfies*

$$\sup_{0 \leqslant t < b} \int_0^t \text{Re}\,(f(u(s)), u(s))\,ds < \infty.$$

Then, there exists a solution of (6.5) in $0 \leqslant t < \infty$.

Proof It is enough to show that if u is a solution in $0 \leqslant t < b < \infty$, then $u(t)$ is bounded in $0 \leqslant t < b$. For each natural number n, put $I_n = (1 - n^{-1}A)^{-1}$ and $u_n(t) = I_n u(t)$. Equation (3.12) implies that $\text{s-lim}_{n \to \infty}(1 - n^{-1}A)^{-1} = I$ and

$$u_n(t) = T(t)I_n u_0 + \int_0^t T(t-s)I_n f(u(s))\,ds$$

for $0 \leqslant t < b$. By Theorem 3.2.3, we find that $u_n \in C^1([0, b); X)$, $u_n \in D(A)$

for $0 \le t < b$ and

$$u'_n(t) = Au_n(t) + I_n f(u(t)).$$

Hence, from (6.4), it follows that

$$
\begin{aligned}
(d/dt)\|u_n(t)\|^2 &= 2\mathrm{Re}\,(u'_n(t), u_n(t)) \\
&= \mathrm{Re}\,(Au_n(t) + I_n f(u(t)), u_n(t)) \\
&= \mathrm{Re}\,(I_n f(u(t)), u_n(t)).
\end{aligned}
$$

Integrating this inequality, we have

$$\|u_n(t)\|^2 \le \|I_n u_0\|^2 + 2\mathrm{Re} \int_0^t (I_n f(u(s)), u_n(s))\, ds,$$

which, on letting $n \to \infty$, gives

$$\|u(t)\|^2 \le \|u_0\|^2 + 2\mathrm{Re} \int_0^t (f(u(s)), u(s))\, ds. \tag{6.9}$$

By assumption, the right-hand side of (6.9) is bounded for $0 \le t < b$. $\quad\square$

Now, as a sufficient condition for the solution of (6.5) to be really a solution of (6.2) and (6.3), we assume that f is Fréchet differentiable. Usually, a mapping f from a Banach space X into a Banach space Y is said to be Fréchet differentiable at $u_0 \in X$ if, for $z \in X$ and $\|z\| \to 0$, there exists a $T \in B(X, Y)$ such that

$$f(u_0 + z) = f(u_0) + Tz + o(\|z\|).$$

T is called the Fréchet derivative of f at u_0. When both X and Y are real Banach spaces, the above definition serves our purposes, but when they are complex Banach spaces, Fréchet differentiable functions in the definition are limited to those corresponding to regular functions. We slightly modify the definition as follows.

Definition 6.1.1 *Let X and Y be complex Banach spaces. An operator S from X into Y is called* antilinear *if $S(u + v) = Su + Sv$ and $S(\lambda u) = \bar{\lambda} Su$ for each u, $v \in X$ and for each $\lambda \in \mathbb{C}$.*

Similarly to linear operators, we can define the boundedness, norms, strong convergence and so on for antilinear operators. Let us denote the collection of all bounded antilinear operators from X into Y by $\bar{B}(X, Y)$.

Definition 6.1.2 *Let X and Y be complex Banach spaces and f a mapping from a certain neighbourhood of an element u_0 of X into Y. If* *there exist $T \in B(X, Y)$ and $S \in \bar{B}(X, Y)$ such that*

$$f(u_0 + z) = f(u_0) + Tz + Sz + o(\|z\|)$$

for $z \in X$ and $\|z\| \to 0$, then f is said to be Fréchet differentiable *at u_0 and the pair $\{T, S\}$ is called the* Fréchet derivative *of f at u_0. If f is Fréchet differentiable at each point of a set, we say that f is Fréchet differentiable on that set and the Fréchet derivative of f at u is denoted by $\{Df(u), \bar{D}f(u)\}$.*

If $X = Y = \mathbb{C}$, we have $Df = \partial f / \partial z$ and $\bar{D}f = \partial f / \partial \bar{z}$.

Assumption 6.1.3 *f is Fréchet differentiable on X and the Fréchet derivative $\{Df(u), \bar{D}f(u)\}$ is strongly continuous in u. Further, for any $C > 0$ there exists some $k_C > 0$ such that*

$$\|Df(u)\| \leq k_C/2, \qquad \|\bar{D}f(u)\| \leq k_C/2$$

for $\|u\| \leq C$.

Next, we will show that the solution of (6.5) is also the solution of (6.2) and (6.3) if Assumption 6.1.3 is satisfied as well as Assumptions 6.1.1 and 6.1.2 and if $u_0 \in D$. First, it is easy to see that if u belongs to $C^1([0, T]; X)$, so does $f(u(\cdot))$ and

$$(\mathrm{d}/\mathrm{d}t) f(u(t)) = Df(u(t)) u'(t) + \bar{D}f(u(t)) \overline{u'(t)}. \tag{6.10}$$

Let $\|u\| \leq C$ and $\|v\| \leq C$. Since $\|tu + (1-t)v\| \leq C$ for each $t \in [0, 1]$, we have

$$\|f(u) - f(v)\| = \left\| \int_0^1 (\mathrm{d}/\mathrm{d}t) f(tu + (1-t)v) \, \mathrm{d}t \right\|$$

$$= \left\| \int_0^1 \{ Df(tu + (1-t)v)(u-v) + \bar{D}f(tu + (1-t)v)\overline{(u-v)} \} \, \mathrm{d}t \right\|$$

$$\leq k_C \|u - v\|.$$

Hence, Assumption 6.1.2 follows from Assumption 6.1.3 by replacing k_C by another constant if necessary.

Theorem 6.1.3 *Suppose Assumptions 6.1.1 and 6.1.3 are satisfied. If $u_0 \in D(A)$, then the solution u of (6.5) in $[0, b)$ is also the solution of (6.2) and (6.3) in $[0, b)$.*

Proof Let u be assumed to be strongly continuously differentiable.

Calculations similar to those in the proof of Theorem 3.2.2 give

$$u'(t) = \psi(t) + \int_0^t T(t-s)\{Df(u(s))u'(s) + \bar{D}f(u(s))u'(s)\}\,ds,$$

$$\psi(t) = T(t)\{Au_0 + f(u_0)\}. \tag{6.11}$$

Now let v be a solution of

$$v(t) = \psi(t) + \int_0^t T(t-s)\{Df(u(s))v(s) + \bar{D}f(u(s))v(s)\}\,ds \tag{6.12}$$

Since $\psi \in C([0, \infty); X)$ and since $Df(u(s))$ and $\bar{D}f(u(s))$ are strongly continuous functions with values in $B(X, Y)$ and $\bar{B}(X, Y)$, respectively, the equation (6.12) can be solved by successive iteration and it has a unique solution belonging to $C([0, b); X)$. Let $\|u_0\| < C$ and we first assume $C/k_{2C} < b$. $u_j(t)$ for each j, defined by (6.6) and (6.7), is differentiable in $0 \le t \le C/k_{2C}$ and we have

$$u_j'(t) = \psi(t) + \int_0^t T(t-s)\{Df(u_{j-1}(s))u_{j-1}'(s) + \bar{D}f(u_{j-1}(s))u_{j-1}'(s)\}\,ds. \tag{6.13}$$

From (6.12) and (6.13), it follows that

$$\begin{aligned}
u_j'(t) - v(t) &= \int_0^t T(t-s)Df(u_{j-1}(s))(u_{j-1}'(s) - v(s))\,ds \\
&\quad + \int_0^t T(t-s)\{Df(u_{j-1}(s)) - Df(u(s))\}v(s)\,ds \\
&\quad + \int_0^t T(t-s)\bar{D}f(u_{j-1}(s))(u_{j-1}'(s) - v(s))\,ds \\
&\quad + \int_0^t T(t-s)\{\bar{D}f(u_{j-1}(s)) - \bar{D}f(u(s))\}v(s)\,ds \\
&= \mathrm{I} + \mathrm{II} + \mathrm{III} + \mathrm{IV}.
\end{aligned}$$

Put

$$\omega_j(t) = \|u_j'(t) - v(t)\| \tag{6.15}$$

for each $j \ge 0$. The integrand on the right-hand side of

$$\|\mathrm{II}\| \le \int_0^T \|\{Df(u_{j-1}(s)) - Df(u(s))\}v(s)\|\,ds$$

is uniformly bounded by (6.8) and Assumption 6.1.2, and, for each s, it converges to 0 as $j \to \infty$ by Assumption 6.1.3. Therefore, s-$\lim_{j\to\infty} \mathrm{II} = 0$

uniformly on $[0, C/k_{2C}]$. The same holds for IV. Hence, there exists a sequence of positive numbers $\{\varepsilon_j\}$ satisfying $\lim_{j \to \infty} \varepsilon_j = 0$ such that

$$\omega_j(t) \le \varepsilon_j + k_{2C} \int_0^t \omega_{j-1}(s)\, ds \qquad (6.16)$$

for all $j = 1, 2, \ldots$ and $t \in [0, C/k_{2C}]$. If $0 < j \le l$, we obtain, from (6.16),

$$\omega_l(t) \le \sum_{i=0}^{j-1} \frac{(k_{2C}t)^i}{i!} \varepsilon_{l-j} + (k_{2C})^j \int_0^t \frac{(t-s)^{j-1}}{(j-1)!} \omega_{l-j}(s)\, ds. \qquad (6.17)$$

Let

$$\varepsilon_0 = \max_{0 \le t \le C/k_{2C}} \omega_0(t).$$

By setting $l = j$ in (6.17), we obtain

$$\omega_j(t) \le \sum_{i=0}^{j} (k_{2C}t)^i (i!)^{-1} \max(\varepsilon_0, \varepsilon_1, \ldots, \varepsilon_j),$$

so that $\omega_j(t)$ is uniformly bounded on $0 \le t \le C/k_{2C}$. Hence, if we let $\omega_j(t) \le K$ for each j and t, the inequality

$$\omega_l(t) \le \sum_{i=0}^{j-1} (k_{2C}t)^i (i!)^{-1} \varepsilon_{l-i} + K(k_{2C}t)^j (j!)^{-1}. \qquad (6.18)$$

then follows from (6.17). Let ε be an arbitrarily given positive number. Choose j sufficiently large that $KC^j(j!)^{-1} < \varepsilon/2$, and next choose l sufficiently large so that

$$e^C \max(\varepsilon_{l-j+1}, \varepsilon_{l-j+2}, \ldots, \varepsilon_l) < \varepsilon/2.$$

Then, (6.18) implies $\omega_l(t) < \varepsilon$ in $0 \le t \le C/k_{2C}$. Thus, we have $\omega_j(t) \to 0$ as $j \to \infty$ uniformly on $[0, C/k_{2C}]$, i.e., $u_j'(t) \to v(t)$ strongly, so that the strong derivative $u'(t)$ exists in $[0, C/k_{2C}]$ and is equal to $v(t)$. Hence, $f(u(t))$ also is differentiable, $u(t) \in D(A)$, and u satisfies (6.2) and (6.3) in $[0, C/k_{2C}]$. Furthermore the equation (6.11) holds. If $C/k_{2C} \ge b$, the above arguments will have completed the proof. If $C/k_{2C} < b$, we may repeat a similar process by taking $t = C/k_{2C}$ as an initial time. Let u be a solution of (6.2) and (6.3) in $[0, b_1)$ for some $b_1 \in (0, b)$. Since u' satisfies (6.11) in $[0, b_1)$, we have $u'(t) = v(t)$ on that interval. Therefore, if we let $u'(b_1) = v(b_1)$, then $u \in C^1([0, b_1]; X)$, $Au \in C([0, b_1]; X)$, and u is a solution of (6.2) and (6.3) in $[0, b_1]$. Repeat the first part of the proof by taking b_1 as an initial time. Then we find that there exists a certain $b_2 > b_1$ such that u is a solution of (6.2) and (6.3) in $[0, b_2)$. Thus u is shown to be a solution of (6.2) and (6.3) in $[0, b)$. $\square$

As an application, we consider the following problem:

$$d^2u(t)/dt^2 + Au(t) + M(u(t)) = 0, \tag{6.19}$$

$$u(0) = \varphi, \qquad (d/dt)u(0) = \psi, \tag{6.20}$$

where A is a positive definite self-adjoint operator in a Hilbert space X, M is a non-linear operator with $D(A^{1/2})$ as its domain, and $\varphi \in D(A)$ and $\psi \in D(A^{1/2})$ are given elements. If we let $\|u\|_W = \|A^{1/2}u\|$ for each element of $D(A^{1/2})$, the domain $D(A^{1/2})$ becomes a Hilbert space W under this norm. By assumption, there exists a positive constant c such that $\|u\|_W \geq c\|u\|$ for all $u \in W$. We make the following assumptions on M.

(I) For any positive number C, there exists some constant $k_C \geq 0$ such that $\|M(u)\| \leq k_C$ and $\|M(u) - M(v)\| \leq k_C\|u - v\|_W$ for $\|u\|_W, \|v\|_W \leq C$.

(II) If $0 < b < \infty$ and $u \in C^1([0, b); X) \cap C([0, b); W)$,

$$\text{Re} \int_0^t (M(u(s)), u'(s)) \, ds$$

is bounded from below on $0 \leq t < b$.

(III) M is Fréchet differentiable on W and the Fréchet derivative $\{DM(u), \bar{D}M(u)\}$ is strongly continuous in u. For any positive number C there exists some constant $k_C > 0$ such that

$$\|DM(u)\| \leq k_C/2, \qquad \|\bar{D}M(u)\| \leq k_C/2.$$

for $\|u\|_W \leq C$.

Both $DM(u)$ and $\bar{D}M(u)$ are mappings from W into X. By replacing k_C by another constant if necessary, we can show by the same arguments as before that (I) follows from (III).

Definition 6.1.3 *If $u \in C^2([0, b); X)$, $u' \in C([0, b); W)$, $u(t) \in D(A)$ for each $t \in [0, b)$ and $Au \in C([0, b); X)$, and if u satisfies (6.19) in $0 \leq t < b$ and (6.20) at $t = 0$, then u is called a solution of (6.19) and (6.20) in $[0, b)$.*

If we let

$$\mathfrak{X} = W \times X, \qquad \|U\|^2 = \|u_0\|_W^2 + \|u_1\|^2 \quad \text{for} \quad U = {}^t(u_0, u_1) \in \mathfrak{X},$$

$\mathfrak{X}$ is a Hilbert space. Let $\mathfrak{A}$ be an operator defined by $D(\mathfrak{A}) = D(A) \times D(A^{1/2})$ and

$$\mathfrak{A}\begin{pmatrix} u_0 \\ u_1 \end{pmatrix} = \begin{pmatrix} 0 & 1 \\ -A & 0 \end{pmatrix}\begin{pmatrix} u_0 \\ u_1 \end{pmatrix} = \begin{pmatrix} u_1 \\ -Au_0 \end{pmatrix},$$

then $\sqrt{-1}\,\mathfrak{A}$ is a self-adjoint operator in $\mathfrak{X}$. Indeed, for each ${}^t(u_0, u_1) \in D(\mathfrak{A})$

$$\mathrm{Re}\left(\mathfrak{A}\begin{pmatrix}u_0\\u_1\end{pmatrix},\begin{pmatrix}u_0\\u_1\end{pmatrix}\right) = \mathrm{Re}\,\{(u_1, Au_0) - (Au_0, u_1)\} = 0.$$

Also, with a non-zero real number λ, consider

$$u_0 = -(A + \lambda^2)^{-1}(\lambda f_0 + f_1), \qquad u_1 = -\lambda(A + \lambda^2)^{-1}(\lambda f_0 + f_1) + f_0$$

for each $F = {}^t(f_0, f_1) \in \mathfrak{X}$. Then it can be easily shown that $U = {}^t(u_0, u_1)$ is a solution of $(\mathfrak{A} - \lambda)U = F$. If we put $F(U) = {}^t(0, -M(u_0))$ and $U_0 = {}^t(\varphi, \psi)$ for $U = {}^t(u_0, u_1)$, then (6.19) and (6.20) are equivalent to

$$dU(t)/dt = \mathfrak{A}U(t) + F(U(t)). \tag{6.21}$$

It is also easy to show that if M satisfies (I) and (III), then F satisfies Assumptions 6.1.2 and 6.1.3, respectively. The Fréchet derivative of F is given by

$$DF\left(\begin{pmatrix}u_0\\u_1\end{pmatrix}\right) = \begin{pmatrix}0 & 0\\-DM(u_0) & 0\end{pmatrix}, \qquad \bar{D}F\left(\begin{pmatrix}u_0\\u_1\end{pmatrix}\right) = \begin{pmatrix}0 & 0\\-\bar{D}M(u_0), & 0\end{pmatrix}.$$

Therefore, by denoting by $\mathfrak{T}(t)$ the semigroup which $\mathfrak{A}$ generates, we obtain the following.

Theorem 6.1.4 *Suppose M satisfies* (I). *If $\|\Phi\| < C$ for each $\Phi \in \mathfrak{X}$, then the solution of*

$$U(t) = \mathfrak{T}(t)\Phi + \int_0^t \mathfrak{T}(t - s)F(U(s))\,ds \tag{6.22}$$

exists uniquely in $0 \le t \le C/k_{2C}$.

The next theorem states the global existence of the solution of (6.22).

Theorem 6.1.5 *If M satisfies* (I) *and* (II), *then for each $\Phi \in \mathfrak{X}$ the solution of* (6.22) *exists uniquely in $[0, \infty)$.*

Proof It is enough to show that the conditions of Theorem 6.1.2 are satisfied. Let $U(t) = {}^t(u_0(t), u_1(t))$ be a solution of (6.22) in $[0, b)$. By putting $\mathfrak{I}_n = (1 - n^{-1}A)^{-1}$ and $U_n(t) = {}^t(u_{0n}(t), u_{1n}(t)) = \mathfrak{I}_n U(t)$, we obtain

$$U_n(t) = \mathfrak{T}(t)\mathfrak{I}_n\Phi + \int_0^t \mathfrak{T}(t - s)\mathfrak{I}_n F(U(s))\,ds,$$

so that $U_n(t)$ is differentiable, $\mathfrak{A}U_n(t)$ is defined and continuous, and

$$U_n'(t) = \mathfrak{A}U_n(t) + \mathfrak{I}_n F(U(t))$$

holds for $0 \leq t < b$. Hence, writing $\mathfrak{I}_n F(U(t)) = {}^t(g_{0n}(t), g_{1n}(t))$, we have

$$u'_{0n}(t) = u_{1n}(t) + g_{0n}(t),$$
$$u'_{1n}(t) = -Au_{0n}(t) + g_{1n}(t), \tag{6.23}$$

from which it follows that

$$u_{0n}(t) = u_{0n}(0) + \int_0^t (u_{1n}(s) + g_{0n}(s)) \, ds \tag{6.24}$$

holds for $0 \leq t < b$. As $n \to \infty$, $u_{0n}(t) \to u_0(t)$ and $g_{0n}(t) \to 0$ in W, $u_{1n}(t) \to u_1(t)$ in X, and $u_{0n}(0) \to \varphi$ in W, so that from (6.24) we obtain

$$u_0(t) = \varphi + \int_0^t u_1(s) \, ds,$$

and, hence, $u'_0(t) = u_1(t)$. Consequently,

$$\text{Re} \int_0^t (F(U(s)), U(s)) \, ds = \text{Re} \int_0^t \left(\begin{pmatrix} 0 \\ -M(u_0(s)) \end{pmatrix}, \begin{pmatrix} u_0(s) \\ u'_0(s) \end{pmatrix} \right) ds$$
$$= -\text{Re} \int_0^t (M(u_0(s)), u'_0(s)) \, ds$$

is bounded from above for $0 \leq t < b$. $\quad\square$

From the discussions above we immediately obtain the following.

Theorem 6.1.6 *If M satisfies* (I), (II) *and* (III), *then for each $\varphi \in D(A)$ and each $\psi \in D(A^{1/2})$ the solution of* (6.19) *and* (6.20) *exists uniquely in* $[0, \infty)$.

As an application of the theorem, we consider the following mixed problem:

$$\partial^2 u / \partial t^2 - \Delta u + F'(|u|^2)u = 0, \qquad x \in \Omega, \qquad 0 \leq t < \infty, \tag{6.25}$$

$$u(x, t) = 0, \qquad\qquad\qquad x \in \partial\Omega, \qquad 0 \leq t < \infty. \tag{6.26}$$

$$u(x, 0) = \varphi(x), \qquad (\partial/\partial t)u(x, 0) = \psi(x), \qquad x \in \Omega, \tag{6.27}$$

where Ω is a bounded region in $\mathbb{R}^3$ and F is a real-valued function belonging to $C^2([0, \infty))$ which satisfies the conditions

(i) $F(0) = 0$, $F(r) \geq 0$ for $r > 0$,

(ii) $|F'(r)| \leq c(r + 1)$ and $|F''(r)| \leq c$ for $r \geq 0$.

For each $u, v \in \mathring{H}_1(\Omega)$. let

$$a(u, v) = \int_\Omega \sum_{i=1}^3 \frac{\partial u}{\partial x_i} \frac{\overline{\partial v}}{\partial x_i} \, dx. \tag{6.28}$$

If we denote by A the operator defined by (2.6) in terms of $a(u, v)$, A is a positive definite self-adjoint operator in $L^2(\Omega)$ by Theorem 2.2.3. From

Lemma 1.2.1, we have $\mathring{H}_1(\Omega) \subset L^6(\Omega)$. If we represent

$$(Mu)(x) = F'(|u(x)|^2)u(x),$$

M is a mapping from the whole of $\mathring{H}_1(\Omega)$ into $L^2(\Omega)$. Thus, we can express (6.25) to (6.27) in the abstract form of (6.19) and (6.20) as equations in $L^2(\Omega)$. Observe that M is Fréchet differentiable, its Fréchet derivative being given by

$$(DM(u)z)(x) = \{F'(|u(x)|^2) + F''(|u(x)|^2)\,|u(x)|^2\}z(x),$$

$$(\bar{D}M(u)z)(x) = F''(|u(x)|^2)u(x)^2\overline{z(x)}$$

and that the following results hold:

$$\|DM(u)\| \le c_0\left\{\int_\Omega |F'(|u|^2) + F''(|u|^2)\,|u|^2|^3\,\mathrm{d}x\right\}^{1/3}$$
$$\le cc_0(2\,\|u\|_6^2 + |\Omega|^{1/3}),$$

$$\|\bar{D}M(u)\| \le c_0\left\{\int_\Omega |F''(|u|^2)|^3\,|u|^6\,\mathrm{d}x\right\}^{1/3} \le cc_0\,\|u\|_6^2,$$

where c_0 is a constant such that $\|u\|_6 \le c_0\,\|u\|_W$. Then we can see that M satisfies (I) and (III). As for (II), we first note that $F(r) \le c(r^2/2 + r)$ implies that

$$\int_\Omega F(|u(x)|^2)\,\mathrm{d}x \le \tfrac{1}{2}c\int_\Omega |u(x)|^4\,\mathrm{d}x + c\int_\Omega |u(x)|^2\,\mathrm{d}x$$
$$\le (c/2)\,\|u\|_2\,\|u\|_6^3 + c\,\|u\|_2^2$$

For $u \in C^1([0, b); L^2(\Omega)) \cap C([0, b); W)$ and for $0 < t < b$, we have

$$2\,\mathrm{Re}\int_0^t (M(u(s)), u'(s))\,\mathrm{d}s = 2\,\mathrm{Re}\int_0^t\int_\Omega F'(|u|^2)u(\overline{\partial u/\partial s})\,\mathrm{d}x\,\mathrm{d}s$$

$$= \int_0^t\int_\Omega (\partial/\partial s)F(|u(x, s)|^2)\,\mathrm{d}x\,\mathrm{d}s$$

$$= \int_0^t (\partial/\partial s)\int_\Omega F(|u(x, s)|^2)\,\mathrm{d}x\,\mathrm{d}s$$

$$= \int_\Omega F(|u(x, t)|^2)\,\mathrm{d}x - \int_\Omega F(|u(x, 0)|^2)\,\mathrm{d}x$$

$$\le -\int_\Omega F(|u(x, 0)|^2)\,\mathrm{d}x,$$

which indicates that M satisfies (II). Thus, Theorem 6.1.6 has been shown to be applicable to (6.25)–(6.27).

As an example of F in the above, we can choose $F(r) = \mu^2 r + \eta^2 r^2/2$. Equation (6.25) then becomes

$$\partial^2 u/\partial t^2 - \Delta u + \mu^2 u + \eta^2 |u|^2 u = 0$$

which is a *meson equation*. For $\mu \neq 0$, we denote by A the operator defined by a quadratic form

$$a_1(u, v) = \int_{\Omega} \left(\sum_{i=1}^{3} \frac{\partial u}{\partial x_i} \frac{\overline{\partial v}}{\partial x_i} + \mu^2 u \bar{v} \right) dx$$

on $\mathring{H}_1(\Omega) \times \mathring{H}_1(\Omega)$. Even though Ω is not bounded, A becomes positive definite and Theorem 6.1.6 can be applied with $M(u) = \eta^2 |u|^2 u$. When $\mu = 0$, we do not have $0 \in \rho(A)$ unless Ω is bounded, where A is an operator determined by (6.28). In this case, if a norm defined by

$$\|u\|_W^2 = \int_{\Omega} \left(\sum_{i=1}^{3} \left| \frac{\partial u}{\partial x_i} \right|^2 + |u|^2 \right) dx$$

is introduced in $W = \mathring{H}_1(\Omega) = D(A^{1/2})$, similar calculations to those above give

$$
\begin{aligned}
|\text{Re} \, (\mathfrak{A} U, U)| &= |\text{Re} \{ a_1(u_1, u_0) - (Au_0, u_1) \}| \\
&= |\text{Re} \{ (u_1, Au_0) + (u_1, u_0) - (Au_0, u_1) \}| \\
&= |\text{Re} \, (u_1, u_0)| \leq \|u_0\| \, \|u_1\|_W
\end{aligned}
$$

for $U = {}^t(u_0, u_1) \in D(\mathfrak{A}) = D(A) \times W$, so that $\mathfrak{A} - \frac{1}{2}$ generates a contraction semigroup. In such a case, by modifying the previous arguments slightly, we also obtain results similar to Theorem 6.16. Since (6.25) is invariant under the replacement of t by $-t$, similar conclusions hold when solving it for the past.

Even under the weakest assumption on non-linear terms, the existence of weak solution can be shown, though the uniqueness is not certain. For example, the reader is referred to Strauss [164]. For the case when the region varies with time, *see* Bardos and Cooper [27], Medeiros [129], Inoue [73], etc. On the behaviour of solutions for $t \to \infty$, *see* Strauss [163], Glassy [67], etc.; for linear cases, *see* Bobisud and Calvert [29], Duffin [58], Goldstein [69, 70], etc. Recently a paper by Heinz and von Wahl [72] was published.

6.2 Monotone operators

Hereafter, in this chapter, we consider only real Banach spaces. The theory of monotone operators initiated by Minty [130–136] has become one of the more important branches in the theory of non-linear equations. Let X be a real Hilbert space and A a generally non-linear operator in X. The operator A is called a monotone operator if

$$(Au - Av, u - v) \geq 0 \tag{6.29}$$

for all $u, v \in D(A)$. Sometimes A is also called a monotone operator if

$$(Au - Av, u - v) \leq 0 \tag{6.30}$$

holds instead of (6.29). When A is linear, it is monotone if and only if it is an accretive operator; A is monotone in the sense of (6.30) if and only if it is a dissipative operator. When $X = \mathbb{R}$ monotone operators reduce to increasing functions. The theory of non-linear semigroups initiated by Komura [95] is closely related to Minty's theory and has made us recognize that the study of multi-valued mappings is indispensable in the theory of monotone operators. A mapping A from X into Y is called a multi-valued mapping if to each element u of a subset $D(A)$ of X there corresponds a subset Au of Y. Let a pair of $u \in X$ and $v \in Y$ be denoted by $[u, v]$.

$$G(A) = \{[u, v] : v \in Au, u \in D(A)\}$$

is called the *graph* of a multi-valued mapping A. Thus there is one-to-one correspondence between a multi-valued mapping A from X into Y and the subset $G(A)$ of $X \times Y$, so that from here on we will not distinguish between them. Both of them will be denoted simply by A. When A is a subset of $X \times Y$, the mapping corresponding to A has a *domain*

$$D(A) = \{u \in X : \text{there exists a } v \in Y \text{ such that } [u, v] \in A\}$$

and the set of all values of A at $u \in D(A)$ is given by $Au = \{v \in Y : [u, v] \in A\}$. The *range* of A is $R(A) = \bigcup_{u \in D(A)} Au$. Also, the *inverse mapping* of A is given by

$$A^{-1} = \{[v, u] : v \in Au, u \in D(A)\}.$$

For $u \notin D(A)$, we define Au to be the empty set.

Definition 6.2.1 *Let A be a generally multi-valued mapping from a Hilbert space X into itself. If*

$$(u' - v', u - v) \geq 0 \tag{6.31}$$

for all $u, v \in D(A)$, $u' \in Au$ and $v' \in Av$, then A is called a monotone operator.

Remark 6.2.1 When X is a complex Hilbert space, arguments similar to those for the real Hilbert space are possible if the monotone operator is defined by Re $(u'-v', u-v) \geq 0$. The same remark applies to monotone accretive operators in a Banach space to be defined later.

Let φ be an increasing function defined on $(-\infty, \infty)$, so that there exist at most a countable number of discontinuities. Define a mapping A from $\mathbb{R}$ into itself by

$$Au = \begin{cases} \varphi(u) & \text{if } \varphi \text{ is continuous at } u, \\ [\varphi(u-0), \varphi(u+0)] & \text{if } \varphi \text{ is discontinuous at } u. \end{cases}$$

Then A is monotone. It is clear that $R(1+A)=\mathbb{R}$, $(1+A)^{-1}$ is single-valued and that $|(1+A)^{-1}u-(1+A)^{-1}v| \leq |u-v|$ for each $u, v \in \mathbb{R}$.

Definition 6.2.2 *A single-valued mapping T from a Banach space X into itself is called a* contraction operator *if $\|Tu-Tv\| \leq \|u-v\|$ for each $u, v \in D(T)$.*

Proposition 6.2.1 *If A is a monotone mapping from a Hilbert space X into itself, then $(1+A)^{-1}$ is a contraction operator.*

Proof Let $u' \in (1+A)^{-1}u$ and $v' \in (1+A)^{-1}v$. Since $u-u' \in Au'$ and $v-v' \in Av'$, we have $((u-u')-(v-v'), u'-v') \geq 0$. From this,

$$\|u'-v'\|^2 \leq (u-v, u'-v') \leq \|u-v\| \|u'-v'\|$$

follows immediately. $\square$

Monotone operators can be naturally generalized to Banach spaces as follows.

Definition 6.2.3 *Let A be a generally multi-valued mapping from a real Banach space X into its conjugate space X^*. A is called* monotone *if*

$$(f-g, u-v) \geq 0 \tag{6.32}$$

for all $u, v \in D(A)$, $f \in Au$ and $g \in Av$. Here $(\cdot, \cdot)$ on the left-hand side means the value of the element $f-g$ in X^* at $u-v$. From now on in this chapter the value of $f \in X^*$ at $u \in X$ is denoted by (f, u).

Definition 6.2.4 *If $A \subset X \times X^*$ and $B \subset X \times X^*$ are both monotone and if $A \subset B$, it is said that B is a* monotone extension *of A. If an operator A has no monotone extension other than itself, then it is called* maximal monotone. *Let K be a subset of X. If $A \subset K \times X^*$ is monotone and the only monotone extension of A contained in $K \times X^*$ is A itself, then A is called* maximal monotone in $K \times X^*$.

From Zorn's lemma, it is clear that any monotone operator $A \subset K \times X^*$ has a maximal monotone extension in $K \times X^*$.

Lemma 6.2.1   *Let $A \subset K \times X^*$ be monotone. The mapping A is maximal monotone if and only if $[u, f] \subset K \times X^*$ satisfying*

$$(g - f, v - u) \geq 0 \quad \text{for every} \quad [v, g] \in A \tag{6.33}$$

belongs to A.

Proof   If A is maximal monotone in $K \times X^*$ and the condition (6.33) holds for $[u, f]$, then the mapping $\tilde{A}$ obtained by adding $[u, f]$ to A is a monotone extension of A in $K \times X^*$. Hence, $[u, f] \in \tilde{A} = A$. Next, if $\tilde{A} \subset K \times X^*$ is monotone and $A \subsetneqq \tilde{A}$, then there exists some $[u, f]$ which does not belong to A but to $\tilde{A}$. It is clear that $[u, f]$ satisfies (6.33). $\square$

Definition 6.2.5   *Let L be a monotone linear operator from X into X^*. If the only monotone linear extension of L is L itself, then it is called* maximal monotone linear.

Proposition 6.2.2   *Let L be a maximal monotone linear operator from X into X^* and its domain $D(L)$ be dense, then L is maximal monotone.*

Proof   Clearly, a linear mapping L is monotone if and only if $(Lu, u) \geq 0$ for every $u \in D(L)$. Let $u \in X$, $f \in X^*$ and

$$(f - Lv, u - v) \geq 0 \tag{6.34}$$

for all $v \in D(L)$. By Lemma 6.2.1, it is enough to show that $u \in D(L)$ and $Lu = f$. Suppose $u \notin D(L)$. Let $\tilde{L}(v + \lambda u) = Lv + \lambda f$ for every $v \in D(L)$ and $\lambda \in \mathbb{R}$. Then $\tilde{L}$ is linear and, if $\lambda \neq 0$, we obtain

$$(\tilde{L}(v + \lambda u), v + \lambda u) = (Lv + \lambda f, v + \lambda u)$$

$$= \lambda^2 (f - L(-\lambda^{-1} v), u - (-\lambda^{-1} v)) \geq 0$$

from (6.34), so that $\tilde{L}$ is monotone. Since $\tilde{L}$ is a proper extension of L, it is a contradiction and, hence, $u \in D(L)$. Let v be an arbitrary element of $D(L)$. Since $(1 - \theta)u + \theta v \in D(L)$ for each $\theta \in (0, 1)$, by substituting it into (6.34) we obtain

$$(f - (1 - \theta)Lu - \theta Lv, \theta(u - v)) \geq 0.$$

Divide the above equation by θ and let $\theta \to 0$. Then we have $(f - Lu, u - v) \geqslant 0$. Since $D(L)$ is dense, $f = Lu$. $\square$

Another generalization of monotone operators to Banach spaces can be formulated as follows.

Definition 6.2.6 *Let A be a generally multi-valued mapping from a Banach space X into itself. Let F denote the duality mapping on X. If for each $u, v \in D(A)$, $u' \in Au$ and $v' \in Av$, there exists an $f \in F(u - v)$ such that $(f, u' - v') \geqslant 0$, then A is called an* accretive operator.

For linear operators, it is obvious that accretive operators in the sense of this definition coincide with those given by Definition 2.1.2. By Lemma 2.1.1, A is an accretive operator if and only if

$$\|(u + \lambda u') - (v + \lambda v')\| \geqslant \|u - v\|$$

for all $u, v \in D(A)$, $u' \in Au$, $v' \in Av$ and all $\lambda > 0$. Therefore, if A is an accretive operator, $(1 + A)^{-1}$ is a contraction operator.

Crandall and Liggett [54] showed the following. Let X be a general real Banach space and $A \subset X \times X$, and suppose that $A + \omega$ is an accretive operator for some real number ω and that $R(1 + \lambda A) \supset \overline{D(A)}$ for all sufficiently small $\lambda > 0$. Then, for every $u \in \overline{D(A)}$ and $t > 0$, the strong limit

$$S(t)u = \text{s-} \lim_{n \to \infty} (1 + tn^{-1}A)^{-n}u$$

exists. $S(t)$ maps $\overline{D(A)}$ into $\overline{D(A)}$, $S(t + s) = S(t)S(s)$ for each t, $s \geqslant 0$, $S(t)u \to u$ as $t \to 0$ for each $u \in D(A)$, and

$$\|S(t)u - S(t)v\| \leqslant e^{\omega t}\|u - v\|$$

holds for each $t > 0$ and $u, v \in \overline{D(A)}$. If X is not reflexive, there occur cases in which $S(t)u$ is not differentiable for any $t > 0$ even if $u \in D(A)$. It has been shown, however, that if $S(t)u$ is differentiable at almost all t, then $u(t) = S(t)u$ is a solution of $-du(t)/dt \in Au(t)$ and $u(0) = u$. On this subject the reader should also be referred to Miyadera [137]. This provides a decisive result in the theory of non-linear semigroups; applications of this result in various forms can be found in Aizawa [23], Crandall [53], Konishi [98–110], Kurts [112], etc. In this connection, besides those, *see also* Crandall and Liggett [55], Crandall and Pazy [56], Kato [84, 86], Oharu [141], Oharu and Takahashi [142], etc.; for equations in a Hilbert space, *see* Brezis [2], Komura [97], etc. In this chapter, we give some further discussion of monotone operators.

6.3 Various kinds of continuities and pseudo-monotone operators

A neighbourhood in a weak topology will hereafter simply be called a *weak neighbourhood*. In this chapter, strong convergence is denoted by $\rightarrow$ and weak convergence by $\rightharpoonup$. Let $\mathfrak{U}$ be a directed set. If to each $\alpha \in \mathfrak{U}$ there corresponds a $u_\alpha \in X$, then $\{u_\alpha : \alpha \in \mathfrak{U}\}$ is called a *directed family of points* in X. When, for any weak neighbourhood V of u, there exists an $\alpha_0 \in \mathfrak{U}$ such that $u_\alpha \in V$ for all $\alpha \geq \alpha_0$, $\{u_\alpha\}$ is said to be *weakly convergent* to u. Let f be a real-valued functional defined on X and $\{u_\alpha : \alpha \in \mathfrak{U}\}$ be a directed family of points in X. Then, $\lim \sup f(u_\alpha) = a$ is defined by:

(i) For any positive constant ε there exists an $\alpha_0 \in \mathfrak{U}$ such that $f(u_\alpha) < a + \varepsilon$ for all $\alpha \geq \alpha_0$;

(ii) For any positive constant ε and for any $\beta \in \mathfrak{U}$ there exists an $\alpha \geq \beta$ such that $f(u_\alpha) > a - \varepsilon$.

Similarly $\lim \inf f(u_\alpha)$ and $\lim f(u_\alpha)$ are defined.

Definition 6.3.1 *Let X and Y be Banach spaces and T a mapping from X into Y. The domain $D(T)$ of T is assumed to be convex. T is called* hemicontinuous *if $T((1-\lambda)u + \lambda v)$ for any $u, v \in D(T)$ is continuous in $0 \leq \lambda \leq 1$ in the weak topology of Y.*

Linear operators are obviously hemicontinuous.

Definition 6.3.2 *Let X and Y be Banach spaces and T a single-valued mapping from X into Y. T is called* demicontinuous *if $u_n \in D(T)$ and $u_n \rightarrow u \in D(T)$ imply that $Tu_n \rightharpoonup Tu$.*

Definition 6.3.3 *Let T be a mapping from a Banach space X into its conjugate space X^*. T is said to be* pseudo-monotone *if the following condition is satisfied. If $\{u_i\}$ is a directed family of points, contained in $D(T)$, which converges weakly to an element u of $D(T)$ and if $\lim \sup (Tu_i, u_i - u) \leq 0$, then $\lim \inf (Tu_i, u_i - v) \geq (Tu, u - v)$ for all $v \in D(T)$.*

Remark 6.3.1 In the above definition, $\lim (Tu_i, u_i - u) = 0$ is seen by taking $v = u$.

Proposition 6.3.1 *Hemicontinuous monotone mappings from a Banach space X into X^* are pseudo-monotone.*

Proof Assume that a mapping T from X into X^* is hemicontinuous and monotone. Let $\{u_i\}$ be a directed family of points as given in Definition 6.3.3. Since T is monotone, we have $(Tu_i, u_i - u) \geq (Tu, u_i - u)$, so that

$$\liminf (Tu_i, u_i - u) \geq \lim (Tu, u_i - u) = 0.$$

Hence,

$$\lim (Tu_i, u_i - u) = 0. \tag{6.35}$$

Let v be an arbitrary element of $D(T)$. Again, by the monotonicity of T, we have

$$\liminf (Tu_i, u_i - v) \geq \lim (Tv, u_i - v) = (Tv, u - v). \tag{6.36}$$

Let $w = (1 - \theta)u + \theta v$ with $0 < \theta < 1$. Since $w \in D(T)$, the substitution of w in place of v in (6.36) gives

$$\liminf (Tu_i, u_i - u + \theta(u - v)) \geq (Tw, \theta(u - v)).$$

By (6.35), we have

$$\liminf (Tu_i, u - v) \geq (Tw, u - v), \tag{6.37}$$

which implies, as $\theta \to 0$, that

$$\liminf (Tu_i\, u - v) \geq (Tu, u - v). \tag{6.38}$$

From (6.35) and (6.38), there follows the result

$$\liminf (Tu_i, u_i - v) = \liminf \{(Tu_i, u_i - u) + (Tu_i, u - v)\}$$

$$\geq (Tu, u - v).$$

The discussions in this section were mainly taken from Brezis [30] and Browder [40]. For relations between hemicontinuity, demicontinuity and monotonicity, *see* Kato [85] or Carroll [3].

6.4 Duality mappings

Definition 6.4.1 *A function j is called a* gauge *function if it is a real-valued continuous function, defined on $0 \leq r < \infty$, which is strictly monotone increasing and satisfies $j(0) = 0$ and $\lim_{r \to \infty} j(r) = \infty$.*

Theorem 6.4.1 *Let X be a Banach space and j a gauge function. Then, for any element u of X, there exists a function $f \in X^*$ satisfying*

$$(f, u) = \|f\| \|u\|, \qquad \|f\| = j(\|u\|).$$

Proof By Theorem 1.1.2, there exists an $f_0 \in X^*$ such that $(f_0, u) = \|u\|$ and $\|f_0\| = 1$. The desired function can be constructed by $f = j(\|u\|)f_0$. $\square$

If we denote by Fu the set of all f satisfying the result of Theorem 6.4.1, then $F \subset X \times X^*$.

Definition 6.4.2 *The F thus defined is called a* duality mapping *from X into* X^* *with a gauge function j.*

The duality mapping in Definition 1.1.1 is a special case for $j(r) \equiv r$.

Theorem 6.4.2 *Duality mappings are monotone.*

Proof Let $u, v \in X$, $f \in Fu$ and $g \in Fv$. We have

$$(f - g, u - v) = (f, u) - (f, v) - (g, u) + (g, v)$$
$$\geq \|f\| \|u\| - \|f\| \|v\| - \|g\| \|u\| + \|g\| \|v\|$$
$$= (\|f\| - \|g\|)(\|u\| - \|v\|)$$
$$= (j(\|u\|) - j(\|v\|))(\|u\| - \|v\|). \tag{6.39}$$

Since j is increasing, the right-hand side of (6.39) is non-negative. □

Definition 6.4.3 *Let X be a Banach space. If* $\|(1 - \lambda)u + \lambda v\| < \|u\| = \|v\|$ *for all* $u, v \in X$ *satisfying* $\|u\| = \|v\|$ *and for all* $\lambda \in (0, 1)$, *then X is said to be* strictly convex. *If for any* $\varepsilon > 0$ *there exists a* $\delta > 0$ *such that* $\|(u + v)/2\| \leq 1 - \delta$ *for* $\|u\| \leq 1$, $\|v\| \leq 1$ *and* $\|u - v\| \geq \varepsilon$, *then X is called* uniformly convex.

Remark 6.4.1 Asplund [24] has shown that when X is reflexive, by replacing the norm of X by a suitable equivalent norm, both X and X^* become strictly convex.

Actually, rather than saying that X is strictly convex or uniformly convex, we should say the ball in X is strictly convex or uniformly convex. Obviously, uniform convexity implies strict convexity. It is known that if a space is uniformly convex the duality mapping is strongly continuous uniformly on a bounded set (Kato [84]) and that a uniformly convex space is reflexive. Hilbert spaces are clearly uniformly convex. It is also known that $L^p(\Omega)$ for $1 < p < \infty$ is uniformly convex.

Theorem 6.4.3 *If* X^* *is strictly convex, then the duality mapping F is single-valued and hemicontinuous in the* w^* *topology. That is,* $F((1 - t)u + tv)$ *for* $u, v \in X$ *is continuous in* $0 \leq t \leq 1$ *in the* w^* *topology.*

Proof If $f, g \in Fu$, then $(f, g) = (g, u) = j(\|u\|) \|u\|$ and $\|f\| = \|g\| = j(\|u\|)$.

For $0 < \lambda < 1$, we have

$$((1-\lambda)f + \lambda g, u) = j(\|u\|)\, \|u\| \tag{6.40}$$

and

$$\|(1-\lambda)f + \lambda g\| \leq j(\|u\|). \tag{6.41}$$

Since $\|(1-\lambda)f + \lambda g\| \geq j(\|u\|)$ by (6.39), the result

$$\|(1-\lambda)f + \lambda g\| = j(\|u\|) = \|f\| = \|g\|$$

follows from (6.40) and, hence, $f = g$. Next, put $f_\lambda = F((1-\lambda)u + \lambda v)$ for $u, v \in X$ and $0 \leq \lambda \leq 1$. As $\lambda \to \lambda_0$, the right-hand side of

$$(f_\lambda, (1-\lambda)u + \lambda v) = j(\|(1-\lambda)u + \lambda v\|)\, \|(1-\lambda)u + \lambda v\| \tag{6.42}$$

converges to $j(\|(1-\lambda_0)u + \lambda_0 v\|)\, \|(1-\lambda_0)u + \lambda_0 v\|$. Also, $\{f_\lambda\}$ is bounded since $\|f_\lambda\| = j(\|(1-\lambda)u + \lambda v\|)$. Therefore, by virtue of Theorem 1.1.5, there exists an $f \in X^*$, and for any w^* weak neighbourhood V of f and for any $\varepsilon > 0$ there exists some λ such that $|\lambda - \lambda_0| < \varepsilon$ and $f_\lambda \in V$. Hence, from (6.42), it follows that

$$(f, (1-\lambda_0)u + \lambda_0 v) = j(\|(1-\lambda_0)u + \lambda_0 v\|)\, \|(1-\lambda_0)u + \lambda_0 v\|,$$

so that $\|f\| \geq j(\|(1-\lambda_0)u + \lambda_0 v\|) = \|f_{\lambda_0}\|$. On the other hand, $\|f\| \leq \|f_{\lambda_0}\|$ is clear from the definition of f. Therefore, we have $\|f\| = \|f_{\lambda_0}\|$. Consequently, $f = f_{\lambda_0}$ and f_λ converges to f_{λ_0} as $\lambda \to \lambda_0$ in the w^* topology. $\square$

Corollary *In addition to the assumptions of the theorem, suppose X is reflexive. Then the duality mapping is monotone and hemicontinuous, and, hence, it is pseudo-monotone.*

Theorem 6.4.4 *Let X be reflexive and both X and X^* be strictly convex. Then the duality mapping F is an injection from X onto X^* and, if F is a duality mapping with a gauge function j, F^{-1} is a duality mapping from X^* into X with the gauge function j^{-1}, where j^{-1} is the inverse of j.*

Proof Let $u, v \in X$ and $f = Fu = Fv$. Since $\|u\| = \|v\|$,

$$(f, (1-\lambda)u + \lambda v) = (1-\lambda)(f, u) + \lambda(f, v)$$

$$= (1-\lambda)\|f\|\,\|u\| + \lambda\|f\|\,\|v\| = \|f\|\,\|u\|$$

for $0 < \lambda < 1$, so that $\|u\| \leq \|(1-\lambda)u + \lambda v\| \leq \|u\|$, which implies that $\|(1-\lambda)u + \lambda v\| = \|u\| = \|v\|$, and, hence, $u = v$. Next, let f be an arbitrary element of X^*. Since j^{-1} is a gauge function and X is reflexive, by Theorem 6.4.1 there exists a $u \in X$ satisfying $(f, u) = \|f\| \cdot \|u\|$ and $\|u\| = j^{-1}(\|f\|)$, and, accordingly, $\|f\| = j(\|u\|)$. Therefore, $R(F) = X^*$. $\square$

Proposition 6.4.1   *Let X be a reflexive Banach space and both X and X^* be strictly convex. Further, let $M \subset X \times X^*$ be monotone and F a duality mapping from X into X^*. If $R(M+F) = X^*$, then M is maximal monotone.*

Proof   Assume $R(M+F) = X^*$. Let $[u, f]$ be an element of $X \times X^*$ and suppose that

$$(g-f, v-u) \geqslant 0 \tag{6.43}$$

holds for all $[v, g] \in M$. By assumption, there exists some $[v, g] \in M$ satisfying $f + Fu = g + Fv$. The inequality (6.43) implies that

$$0 \leqslant (Fv - Fu, v - u) = (f - g, v - u) \leqslant 0,$$

so that $(Fv - Fu, v - u) = 0$, which yields

$$\|Fv\| \|v\| + \|Fu\| \|u\| = (Fv, u) + (Fu, v). \tag{6.44}$$

Hence, by (6.39), we obtain

$$(j(\|v\|) - j(\|u\|))(\|v\| - \|u\|) \leqslant (Fv - Fu, v - u) = 0,$$

so that $\|v\| = \|u\|$. Since

$$\|Fv\| \|u\| + \|Fu\| \|v\| = (Fv, u) + (Fu, v)$$

by (6.44), we should have $\|Fv\| \|u\| = (Fv, u)$ and $\|Fu\| \|v\| = (Fu, v)$. On the other hand, since $\|Fv\| = \|Fu\| = j(\|u\|)$, we have $Fv = Fu$, so that $v = u$ by the preceding theorem. Therefore, $[u, f] = [v, g] \in M$. $\square$

When a space like $\mathring{W}_p^1(\Omega)$ $(1 < p < \infty)$ is considered, it is usual to take $j(r) = r^{p-1}$ as a gauge function. In some cases, we must choose an appropriate gauge function depending on the problem in question; *see*, for example, Brezis [32].

6.5   Existence of solutions of monotone operator equations

Two fundamental existence theorems on the solutions of equations containing monotone operators will be described, following Browder [38, 39].

Definition 6.5.1   *A mapping which maps a bounded set into a bounded set is called a* bounded mapping.

Theorem 6.5.1 *Let X be a reflexive Banach space, K be a convex closed subset of X, $M \subset X \times X^*$ be monotone, and T be a pseudo-monotone bounded mapping from $D(T) = K$ into X^*. Assume that there exists $[u_0, f_0] \in M$ and $u_0 \in K$ such that*

$$\lim_{u \in K, \|u\| \to \infty} (Tu + f_0, u - u_0) = \infty. \tag{6.45}$$

Then, there exists an element u of K such that the inequality

$$(g + Tu, v - u) \geqslant 0 \tag{6.46}$$

for all $[v, g] \in M$ and $v \in K$.

For the proof of this theorem we need the following two lemmas.

Lemma 6.5.1 *Let X be a finite-dimensional Banach space, K be a bounded convex closed subset of X, $M \subset K \times X^*$ be monotone, and T be a pseudo-monotone bounded mapping from K into X^*. Then, there exists a $u \in K$ such that the inequality (6.45) holds for all $[v, g] \in M$.*

Proof Suppose such a u does not exist. Then, for any $u \in K$, we can find some $[v, g] \in M$ such that $(g + Tu, v - u) < 0$. For each $[v, g] \in M$, we put

$$N(v, g) = \{u \in K : (g + Tu, v - u) < 0\}.$$

From the assumption, $K = \bigcup_{[v, g] \in M} N(v, g)$. We claim that each $N(v, g)$ is an open subset of K. To show this, let us choose $\{u_j\}$ in such a way that $u_j \in K \setminus N(v, g)$ and $u_j \to u$. Since $\{u_j\}$ is bounded, $\{Tu_j\}$ is also bounded, so that $(Tu_j, u_j - u) \to 0$. Since T is pseudo-monotone, we obtain $\liminf (Tu_j, u_j - v) \geqslant (Tu, u - v)$. Hence,

$$(g + Tu, v - u) = (g, v - u) - (Tu, u - v)$$

$$\geqslant \lim (g, v - u_j) - \liminf (Tu_j, u_j - v)$$

$$= \limsup (g + Tu_j, v - u_j) \geqslant 0,$$

so that $u \in K \setminus N(v, g)$. Consequently, $\{N(v, g) : [v, g] \in M\}$ is an open covering of the compact space K. Therefore, there exist $[v_i, g_i] \in M$ ($i = 1, \dots, n$) such that $K = \bigcup_{i=1}^n N(v_i, g_i)$. Let $\{\phi_i\}_{i=1}^n$ be a resolution of the identity belonging to $\{N(v_i, g_i)\}$. For each $u \in K$, put

$$p(u) = \sum_{i=1}^n \phi_i(u) v_i, \qquad q(u) = \sum_{i=1}^n \phi_i(u) g_i.$$

Since K is convex, p is a continuous mapping from K into K. Therefore, there exists a fixed point $\bar{u} \in K$ of p. If we let $\psi(u) = (q(u) + Tu, p(u) - u)$, then

$$\psi(u) = \left(\sum_{i=1}^{n} \phi_i(u)(g_i + Tu), \sum_{j=1}^{n} \phi_j(u)(v_j - u) \right)$$

$$= \sum_{i=1}^{n} \phi_i(u)^2(g_i + Tu, v_i - u) + \sum_{i<j} \phi_i(u)\varphi_j(u)$$

$$\times \{(g_i + Tu, v_i - u) + (g_j + Tu, v_j - u) - (g_i - g_j, v_i - v_j)\} < 0$$

holds for every $u \in K$. On the other hand, $\psi(\bar{u}) = 0$, which is a contradiction. $\square$

Lemma 6.5.2 *Let X be a reflexive Banach space, K be a bounded convex closed subset of X, $M \subset X \times X^*$ be monotone, and T be a pseudo-monotone bounded mapping from K into X^*. Then, there exists a $u \in K$ such that the equality (6.46) holds for all $[v, g] \in M$ and $v \in K$.*

Proof We will prove this lemma by a *generalized Galerkin method.* $M|_K = \{[v, g] : v \in K, [v, g] \in M\}$ is a monotone subset of $K \times X^*$. Let us denote the maximal monotone extension of $M|_K$ in $K \times X^*$ by $\tilde{M}$. It is enough to show that (6.46) holds for all $[v, g] \in \tilde{M}$. Denote the collection of all finite-dimensional subspaces of X by $\mathcal{Y}$. For each $Y \in \mathcal{Y}$, let ι_Y be an embedding of Y into X. Clearly, $\iota_Y \in B(Y, X)$ and $\iota_{Y^*} \in B(X^*, Y^*)$. If we put

$$M_Y = \{[v, \iota_{Y^*}g] : v \in Y, [v, g] \in \tilde{M}\}, \qquad T_Y = \iota_{Y^*} T \iota_Y,$$

then $M_Y \subset (Y \cap K) \times Y^*$ is monotone and T_Y is a pseudo-monotone bounded mapping from $Y \cap K$ into Y^*. Therefore, by Lemma 6.5.1, there exists a $u_Y \in Y \cap K$ such that

$$(g + Tu_Y, v - u_Y) = (\iota_{Y^*}g + T_Y u_Y, v - u_Y) \geqslant 0 \qquad (6.47)$$

holds for all $[v, g] \in \tilde{M}$ and $v \in Y$. $\{u_Y\}$ is bounded since it is contained in K. By assumption, $\{Tu_Y\}$ is also bounded. By virtue of Theorem 1.1.6, there exists a $u \in K$ and an $f \in X^*$ such that, for any $Y \in \mathcal{Y}$, for any weak neighbourhood V of u, and for any weak neighbourhood W of f, one can find a $Y_1 \in \mathcal{Y}$ for which $Y_1 \supset Y$, $u_{Y_1} \in V$ and $Tu_{Y_1} \in W$. By choosing such a Y_1 for each $U = (Y, V, W)$, we put

$$u_U = u_{Y_1}. \qquad (6.48)$$

Denote the collection of all $U = (Y, V, W)$ by $\mathfrak{U}$. For two elements $U = (Y, V, W)$ and $U' = (Y', V', W')$ of $\mathfrak{U}$, define $U \leqslant U'$ if and only if $Y \subset Y'$, $V \supset V'$ and $W \supset W'$. Then, $\mathfrak{U}$ is a directed set and $\{u_U\}$ is a directed family of points. Obviously, $u_U \rightharpoonup u$ and $Tu_U \rightharpoonup f$. Let $[v, g]$ be an arbitrary element of $\tilde{M}$. For any positive number ε,

$$V_0 = \{w \in X: |(g, w - u)| < \varepsilon/2\},$$

$$W_0 = \{h \in X^*: |(h - f, u - v)| < \varepsilon/2\}$$

are weak neighbourhoods of u and f, respectively. Let Y_0 be a subspace generated by v and put $U_0 = (Y_0, V_0, W_0)$. When $U = (Y, V, W) \geqslant U_0$, by expressing $u_U = u_{Y_1}$ according to (6.48), we obtain

$$(g + Tu_U, v - u_U) = (g + Tu_{Y_1}, v - u_{Y_1}) \geqslant 0, \tag{6.49}$$

since $Y_1 \supset Y \supset Y_0 \ni v$. Because of $u_U \in V \subset V_0$ and $Tu_U \in W \subset W_0$, by (6.49) we have

$$(Tu_U, u_U - u) = (Tu_U, u_U - v) + (Tu_U, v - u)$$

$$\leqslant (g, v - u_U) + (Tu_U, v - u) < (g + f, v - u) + \varepsilon,$$

so that

$$\limsup (Tu_U, u_U - u) \leqslant (g + f, v - u).$$

Since $[v, g] \in \tilde{M}$ is arbitrary, it follows that

$$\limsup (Tu_U, u_U - u) \leqslant \inf_{[v,g] \in \tilde{M}} (g + f, v - u).$$

If $\inf_{[v,g] \in \tilde{M}} (g + f, v - u) > 0$, we get $[u, -f] \in \tilde{M}$ since $\tilde{M}$ is maximal monotone in $K \times X^*$. This is a contradiction. Therefore, $\limsup (Tu_U, u_U - u) \leqslant 0$. Since T is pseudo-monotone, the inequality

$$\liminf (Tu_U, u_U - w) \geqslant (Tu, u - w) \tag{6.50}$$

holds for all $w \in K$. Again we choose $[v, g] \in \tilde{M}$ arbitrarily. If $U = (Y, V, W)$ and $Y \ni v$, since $(g + Tu_U, v - u_U) \geqslant 0$ by (6.47), we obtain

$$(g, v - u) = \lim (g, v - u_U) \geqslant \liminf (Tu_U, u_U - v) \geqslant (Tu, u - v),$$

from (6.50), which leads to (6.46). $\quad \square$

Proof of Theorem 6.5.1 As before, let us denote the maximal monotone extension of $M|_K$ in $K \times X^*$ by $\tilde{M}$. For each positive integer n, $K_n = \{u \in K: \|u\| \leqslant n\}$ is a bounded convex closed set, and the restriction T_n of T to K_n is a pseudo-monotone bounded mapping from K_n into X^*. Therefore, by Lemma 6.5.2, there exists a $u_n \in K_n$ such that

$$(g + Tu_n, v - u_n) = (g + T_n u_n, v - u_n) \geqslant 0 \tag{6.51}$$

holds for all $[v, g] \in \tilde{M}$ and $v \in K_n$. Since $[u_0, f_0] \in M|_K$, the inequality $n \geqslant \|u_0\|$ implies that $(f_0 + Tu_n, u_0 - u_n) \geqslant 0$ by (6.51). Hence, $\{u_n\}$ is bounded by (6.45) and, $\{Tu_n\}$ being also bounded, there exists a subsequence $\{u_{n_j}\}$ such that $u_{n_j} \rightharpoonup u$ and $Tu_{n_j} \rightharpoonup f$. Let $[v, g]$ be an arbitrary element of $\tilde{M}$. If $\|v\| \le n_j$, we have

$$(g + Tu_{n_j}, v - u_{n_j}) \ge 0 \qquad (6.52)$$

by (6.51), so that

$$\limsup (Tu_{n_j}, u_{n_j} - u) = \limsup (Tu_{n_j}, u_{n_j} - v) + \lim (Tu_{n_j}, v - u)$$
$$\le \lim \{(g, v - u_{n_j}) + (Tu_{n_j}, v - u)\} = (g + f, v - u).$$

Consequently, we find

$$\limsup (Tu_{n_j}, u_{n_j} - u) \le \inf_{[v,g] \in \tilde{M}} (g + f, v - u),$$

the right-hand side of which is not positive, as was shown in the proof of Lemma 6.5.2. Hence, $\limsup (Tu_n, u_n - u) \le 0$. By the pseudo-monotonicity of T and (6.51), we have

$$(Tu, u - v) \le \liminf (Tu_{n_j}, u_{n_j} - v) \le (g, v - u).$$

Thus, we have established (6.46). $\square$

Corollary 1   *Let X be a reflexive Banach space, $M \subset X \times X^*$ be maximal monotone and T be a pseudo-monotone bounded mapping from $D(T) = X$ into X^*. If there exists a $[u_0, f_0] \in M$ such that*

$$\lim_{\|u\| \to \infty} (Tu + f_0, u - u_0) = \infty,$$

then $(M + T)u \ni 0$ has a solution $u \in D(M)$.

Proof   Let $K = X$ and apply the above theorem. There exists a $u \in X$ such that (6.46) holds for all $[v, g] \in M$. Hence, by Lemma 6.2.1, we have $[u, -Tu] \in M$. $\square$

Corollary 2   *Let X be a reflexive Banach space, $M \subset X \times X^*$ be maximal monotone and T be a pseudo-monotone bounded mapping from $D(T) = X$ into X^*. If there exists a $[u_0, f_0] \in M$ such that*

$$\lim_{\|u\| \to \infty} (Tu + f_0, u - u_0)/\|u\| = \infty,$$

then $R(M + T) = X^*$, that is, for every $f \in X^*$

$$(M + T)u \ni f \qquad (6.53)$$

has a solution $u \in D(M)$.

Proof   Put $T_1 u = Tu - f$, where f is an arbitrary element of X^*. Then T_1

is a pseudo-monotone bounded mapping from $D(T_1) = X$ into X^*. Since

$$(T_1 u + f_0, u - u_0) = (Tu + f_0 - f, u - u_0)$$

$$= \|u\| \left\{ \frac{(Tu + F_0, u - u_0)}{\|u\|} - \frac{(f, u - u_0)}{\|u\|} \right\} \to \infty$$

as $\|u\| \to \infty$, the assumption of Corollary 1 is satisfied with T_1 replacing T. Hence, there exists a $u \in D(M)$ such that $(M + T_1)u \ni 0$, i.e., $(M + T)u \ni f$. $\square$

Corollary 3 *Let X be a reflexive Banach space, and let L be a maximal monotone linear operator from X into X^* with a dense domain $D(L)$. If T is a pseudo-monotone bounded mapping from $D(T) = X$ into X^* and if there exists a $u_0 \in D(L)$ such that*

$$\lim_{\|u\| \to \infty} (Tu + Lu_0, u - u_0)/\|u\| = \infty,$$

then $R(L + T) = X^*$; that is, for all $f \in X^*$,

$$(L + T)u = f \tag{6.54}$$

has a solution $u \in D(L)$.

Proof Since L is maximal monotone by Proposition 6.2.2, this corollary is a direct consequence of Corollary 2. $\square$

Corollary 4 (G. J. Minty) *Let X be a Hilbert space and $A \subset X \times X$ be maximal monotone. Then, $R(1 + A) = X$ and, hence, by Proposition 6.2.1, $(1 + A)^{-1}$ is a contraction operator defined on the whole of X.*

Proof The assumption of Corollary 2 is satisfied if we put $M = A$ and $T = I$ and let $[u_0, f_0]$ be an arbitrary element of A. Therefore, $R(1 + A) = X$. $\square$

Remark 6.5.1 If T is hemicontinuous and monotone in Corollary 2, then T is pseudo-monotone by Proposition 6.3.1. However, if in addition T is *strictly monotone*, i.e., if $(Tu - Tv, u - v) = 0$ holds only for $u = v$, then the solution of (6.53) is unique. Similar remarks apply to (6.54).

Definition 6.5.2 *Let T be a mapping from a Banach space X into its conjugate space X^*. T is called coercive if there exists a $u_0 \in D(T)$ such that*

$$\lim_{u \in D(T), \|u\| \to \infty} \frac{(Tu, u - u_0)}{\|u\|} = \infty. \tag{6.55}$$

Theorem 6.5.2 *Let X be a reflexive Banach space and K a convex closed subset of X. If T is a hemicontinuous monotone operator from $D(T) = K$ into X^* and, moreover, if it is coercive, then for any $f \in X^*$ there exists some $u \in K$ such that*

$$(f - Tu, v - u) \leq 0 \tag{6.56}$$

holds for all $v \in K$. If T is strictly monotone, the solution is unique.

Proof *Step 1.* We claim that

$$G = \{[u, Tu + w] : u \in K, (w, u - v) \geq 0 \text{ for all } v \in K\}$$

is maximal monotone if K has an interior point v_0.

It is easy to see that G is monotone. By virtue of Lemma 6.2.1, we need only show that $[v_1, f_1] \in G$ if $[v_1, f_1] \in X \times X^*$ satisfies

$$(f_1 - Tu - w, v_1 - u) \geq 0 \tag{6.57}$$

for all $[u, Tu + w] \in G$. If $v_1 \notin K$, we can find $v_2 \in \partial K$ and $s > 1$ satisfying $v_1 - v_0 = s(v_2 - v_0)$. By Theorem 1.1.9, there exists a non-zero $h_0 \in X^*$ such that

$$(h_0, v_2) \geq \sup_{v \in K} (h_0, v). \tag{6.58}$$

Let v be an arbitrary element of K. From (6.58), it follows that

$$(h_0, v_2 - v_0) \geq (h_0, v - v_0),$$

but since v_0 is an interior point of K, we have

$$(h_0, v_2 - v_0) > 0. \tag{6.59}$$

Observe that $v_2 \in K$ and, by (6.58), $(h_0, v_2 - v) \geq 0$ for all $v \in K$. Therefore, $[v_2, Tv_2 + \lambda h_0] \in G$ for any positive number λ. From (6.57), we obtain $(f_1 - Tv_2 - h_0, v_1 - v_2) \geq 0$, so that $(f_1 - Tv_2, v_1 - v_2) \geq \lambda(h_0, v_1 - v_2)$, but, since $v_1 - v_2 = (s - 1)(v_2 - v_0)$, we have $(f_1 - Tv_2, v_2 - v_0) \geq \lambda(h_0, v_2 - v_0)$. Since $\lambda > 0$ is arbitrary, this implies that $(h_0, v_2 - v_0) \leq 0$, which contradicts (6.59). Hence,

$$v_1 \in K. \tag{6.60}$$

Put $h_1 = f_1 - Tv_1$. Let v be an arbitrary element of K. Since $[v, Tv] \in G$, we have

$$(Tv_1 + h_1 - Tv, v_1 - v) \geq 0 \tag{6.61}$$

by (6.57). Because of (6.60), $(1 - \theta)v_1 + \theta v \in K$ for all $0 < \theta < 1$, so that, by replacing v by $(1 - \theta)v_1 + \theta v$ in (6.61), we find

$$(Tv_1 + h_1 - T((1 - \theta)v_1 + \theta v)), v_1 - v) \geq 0,$$

which gives $(h_1, v_1 - v) \geqslant 0$ by letting $\theta \to 0$, since T is hemicontinuous. Since $v \in K$ is arbitrary, we finally obtain $[v_1, f_1] = [v_1, Tv_1 + h_1] \in G$.

Remark 6.5.2 If $X = \mathbb{R}$, K is a finite closed interval $[a, b]$ and T is a monotone increasing continuous function φ, then G becomes as follows:

$$Gu = \begin{cases} (-\infty, \varphi(a)] & \text{if } u = a, \\ \varphi(u) & \text{if } a < u < b, \\ [\varphi(b), \infty) & \text{if } u = b. \end{cases}$$

Step 2. We claim that if X is finite-dimensional and K has an interior point, then the conclusion of the theorem holds.

We may assume X is a Hilbert space. Since G in Step 1 is maximal monotone, we have $R(1 + nG) = X$ for all positive integers n by Corollary 4 of the preceding theorem. Therefore, for any $f \in X = X^*$ and any n, there exist $u_n \in K$ and $h_n \in X^*$ such that

$$(h_n, u_n - v) \geqslant 0 \quad \text{for all} \quad v \in K, \tag{6.62}$$

that is, $[u_n, Tu_n + h_n] \in G$ and $u_n + n(Tu_n + h_n) = nf$. Hence,

$$Tu_n = f - n^{-1}u_n - h_n. \tag{6.63}$$

If the condition (6.55) is assumed to be valid for $u_0 \in K$, from (6.62) and (6.63) it follows that

$$(Tu_n, u_n - u_0)/\|u_n\| = \frac{(f, u_n - u_0)}{\|u_n\|} - \frac{\|u_n\|}{n} + \frac{(u_n, u_0)}{n\|u_n\|} - \frac{(h_n, u_n - u_0)}{\|u_n\|}$$

$$\leqslant \frac{(f, u_n - u_0)}{\|u_n\|} + \frac{\|u_0\|}{n}.$$

The right-hand side of this inequality is bounded as $n \to \infty$. Hence $\{u_n\}$ is bounded, so that, by substituting $\{u_n\}$ by its subsequence, we may suppose $u_n \to u \in K$. Since $[v, Tv] \in G$ for all $v \in K$, it follows from (6.63) that

$$(Tv - f + n^{-1}u_n, v - u_n) \geqslant (h_n, u_n - v) \geqslant 0.$$

By letting $n \to \infty$, we obtain

$$(Tv - f, v - u) \geqslant 0. \tag{6.64}$$

We can replace v in (6.64) by $(1 - \theta)u + \theta v$, since the latter belongs to K for $0 < \theta < 1$. By dividing the resultant inequality and letting $\theta \to 0$, we obtain (6.56).

Step 3. We claim that the conclusion of the theorem holds if X is finite-dimensional.

Let Y be a subspace generated by K. Since K is convex, there exists an interior point of K in Y. Let ι be an embedding from Y into X. $\iota^*T\iota$ is a hemicontinuous coercive monotone operator from $D(\iota^*T\iota) = K$ into Y. Hence, by Step 2, for any $f \in X$ there exists some $u \in K$ such that

$$(f - Tu, v - u) = (\iota^*f - \iota^*T\iota u, v - u) \leq 0$$

for all $v \in K$.

Step 4. Let $\mathcal{Y}$ be the collection of all finite-dimensional subspaces of X containing u_0. For each $Y \in \mathcal{Y}$ let ι_Y be an embedding from Y into K. Then $T_Y = \iota_Y^*T\iota_Y$ is a hemicontinuous coercive monotone mapping from $D(T_Y) = K \cap Y$ into Y^*. Therefore, by Step 3, there exists a $u_Y \in K \cap Y$ such that $(\iota_Y^*f - T_Y u_Y, v - u_Y) \leq 0$ for all $v \in K \cap Y$, i.e.,

$$(f - Tu_Y, v - u_Y) \leq 0. \tag{6.65}$$

Since $u_0 \in K \cap Y$, it follows from (6.65) that

$$(Tu_Y, u_Y - u_0)/\|u_Y\| \leq (f, u_Y - u_0)/\|u_Y\|,$$

the right-hand side of which is bounded. Therefore $\{u_Y\}$ is bounded. By virtue of Theorem 1.1.6, there exists a $u \in K$ such that, for any weak neighbourhood V of u and for any $Y \in \mathcal{Y}$, one can find a $Y_1 \supset Y$ satisfying $u_{Y_1} \in V$. Let v be an arbitrary element of K. Since $Y \ni v$ implies that $v \in K \cap Y_1$, by (6.65) we obtain

$$(f - Tv, v - u) = (f - Tv, v - u_{Y_1}) + (f - Tv, u_{Y_1} - u)$$

$$\leq (f - Tu_{Y_1}, v - u_{Y_1}) + (f - Tv, u_{Y_1} - u)$$

$$\leq (f - Tv, u_{Y_1} - u)$$

so that $(f - Tv, v - u) \leq 0$. Thus, similarly to the last part of Step 2, we obtain (6.56). The rest of the proof is easy. $\square$

Corollary *Let X be a reflexive Banach space and T a hemicontinuous coercive monotone operator from $D(T) = X$ into X^*. Then $R(T) = X^*$.*

Proof Apply the theorem with $K = X$. $\square$

Example 1 Let Ω be a bounded region in $\mathbb{R}^n$ with a smooth boundary $\partial\Omega$ and let $2 \leq p < \infty$. By means of

$$(Au, v) = \int_\Omega (1 + |\text{grad } u|^2)^{(p-2)/2} \sum_{i=1}^n \frac{\partial u}{\partial x_i} \frac{\partial v}{\partial x_i} \, dx$$

for each $u, v \in \mathring{W}_p^1(\Omega)$, we define an operator

$$Au = -\sum_{i=1}^n \frac{\partial}{\partial x_i} \left\{ (1 + |\text{grad } u|^2)^{(p-2)/2} \frac{\partial u}{\partial x_i} \right\} \tag{6.66}$$

from $X = \mathring{W}_p^1(\Omega)$ into $X^* = W_{p'}^{-1}(\Omega)$. It is easy to see that A is a hemicontinuous monotone operator. By observing that the norm of $\mathring{W}_p^1(\Omega)$ is equivalent to $\{\int |\text{grad } u|^p \, dx\}^{1/p}$ since Ω is bounded, we also find that A satisfies (6.55) for each u_0. Let ψ be a function in $W_p^1(\Omega)$ satisfying $\psi \leq 0$ on $\partial \Omega$ and put

$$K = \{u \in W_p^1(\Omega) : u \geq \psi \text{ on } \Omega\}.$$

Then K is a closed convex set in $\mathring{W}_p^1(\Omega)$. By Theorem 6.5.2, for every $f \in W_{p'}^{-1}(\Omega)$ there exists a $u \in K$ such that

$$(f - Au, v - u) \leq 0 \quad \text{for all} \quad v \in K. \tag{6.67}$$

Let $v = u + w \in K$ if $w \in C_0^\infty(\Omega)$ and $w \geq 0$, and let $v = u \pm w \in K$ if $u > \psi$ on an open set $\Omega_1 \subset \Omega$, where $w \in C_0^\infty(\Omega_1)$ and its absolute value is sufficiently small. By a formal calculation, it is found that u is a solution in the wider sense of

$$\begin{cases} Au \geq f & \text{in} \quad \Omega, \\ Au = f & \text{if} \quad u > \psi, \\ u = 0 & \text{on} \quad \partial \Omega. \end{cases} \tag{6.68}$$

Conversely, a solution of (6.68) can be shown to satisfy (6.67).

As the following example shows, Theorem 6.5.2 is related to the problem of minimizing a certain functional, that is, to the *variational problem*.

Example 2   Let Ω be the same region as in Example 1 and put

$$K = \{u \in H_1(\Omega) : u \geq 0 \quad \text{on} \quad \partial \Omega\}.$$

Then K is a closed convex subset of $H_1(\Omega)$. Let A be an operator from $H_1(\Omega)$ into $H_1(\Omega)^*$ defined as in Section 2.2 by the quadratic form

$$a(u, v) = \int_\Omega \left(\sum_{i=1}^n \frac{\partial u}{\partial x_i} \frac{\partial v}{\partial x_i} + uv \right) dx$$

on $H_1(\Omega) \times H_1(\Omega)$, i.e., $(Au, v) = a(u, v)$. Clearly, A is strictly monotone. Let $f \in H_1(\Omega)^*$ and consider the problem of minimizing

$$J(u) = \frac{1}{2} \int_\Omega \left\{ \sum_{i=1}^n \left(\frac{\partial u}{\partial x_i} \right)^2 + u^2 \right\} dx - (f, u)$$

on K. Let $u \in K$ be its solution:

$$J(u) = \min_{v \in K} J(v). \tag{6.69}$$

For an arbitrary element v of K, we have

$$0 \geq J(u) - J(v) = a(u+v, u-v)/2 - (f, u-v).$$

Since $(1-\theta)u + \theta v \in K$ for $v \in K$ and $0 < \theta < 1$, we substitute it into the above inequality instead of v. By dividing the resultant inequality by θ and letting $\theta \to 0$, we obtain

$$(Au - f, u - v) \leq 0 \quad \text{for all} \quad v \in K, \tag{6.70}$$

which is an inequality of the form (6.56). Conversely, if u satisfies (6.70), it is easy to see that u is a solution of (6.69). If we take X for $H_1(\Omega)$ and T for A, then the assumption of Theorem 6.5.2 is satisfied. Assuming u is smooth, we integrate (6.70) by parts to obtain

$$\int_{\partial\Omega} \frac{\partial u}{\partial \nu}(u-v)\,\mathrm{d}S + (-\Delta u + u - f, u - v) \leq 0, \tag{6.71}$$

where $\partial/\partial\nu$ is a derivative in the direction of the outward normal. $v = u + w \in K$ for any element w of $C_0^\infty(\Omega)$. The substitution of this into (6.71) yields $-\Delta u + u = f$ in Ω. Therefore, combining with (6.71), we obtain

$$\int_{\partial\Omega} \frac{\partial u}{\partial \nu}(u-v)\,\mathrm{d}S \leq 0$$

for every $v \in K$. Now, putting $v = u + w$, where $w \in C^1(\bar{\Omega})$ and $w \geq 0$ on $\partial\Omega$, we obtain $\partial u/\partial\nu \geq 0$ on $\partial\Omega$. Further, the choice $v = u/2$ yields $u \cdot (\partial u/\partial\nu) \geq 0$ on $\partial\Omega$. Collecting these results, we find that the solution u of (6.69) or (6.70) is a solution in the wider sense of the following boundary-value problem:

$$\begin{cases} -\Delta u + u = f & \text{in} \quad \Omega, \\ u \geq 0, \quad \partial u/\partial\nu \geq 0 \quad \text{and} \quad u(\partial u/\partial\nu) = 0 \quad \text{on} \quad \partial\Omega. \end{cases} \tag{6.72}$$

Conversely, if u is a solution of (6.72), it is not hard to see that u also satisfies (6.69) or (6.70).

For the smoothness of the solution when f belongs to $L^q(\Omega)$ in the above two examples, *see* Brezis [34] and Brezis and Stampacchia [35].

Lemma 6.5.3 *Let X be a reflexive Banach space and L a closed monotone linear operator from X into X^*. If L^* is monotone, then L is maximal monotone.*

Proof On account of Remark 6.4.1, we may suppose that X and X^* are both strictly convex. Let F be a duality mapping with a gauge function $j(r) \equiv r$. By Proposition 6.4.1, it is enough to show that $R(L+F) = X^*$.

$D(L)$ becomes a reflexive Banach space Y with the norm $\|u\|_Y = \|u\| + \|Lu\|$. Let f be an arbitrary element of X^* and assume $\varepsilon > 0$. For each $u, v \in Y$, set

$$(B_\varepsilon u, v) = \varepsilon (Lv, F^{-1}Lu) + (Lu, v) + (Fu, v) - (f, v).$$

Then B_ε is a hemicontinuous monotone mapping from $D(B_\varepsilon) = Y$ into Y^*. Also, B_ε is coercive since

$$\frac{(B_\varepsilon u, u)}{\|u\|_Y} = \frac{1}{\|u\|_Y} \{\varepsilon \|Lu\|^2 + (Lu, u) + \|u\|^2 - (f, u)\}$$

$$\geq \frac{1}{\|u\|_Y} \{\varepsilon \|Lu\|^2 + \|u\|^2 - \|f\| \|u\|\} \to \infty$$

as $\|u\|_Y \to \infty$. Therefore, by the corollary of Theorem 6.5.2, we have $R(B_\varepsilon) = Y^*$ and, hence, there exists a $u_\varepsilon \in Y$ satisfying $B_\varepsilon u_\varepsilon = 0$. Since

$$\varepsilon (Lv, F^{-1}Lu_\varepsilon) + (Lu_\varepsilon, v) + (Fu_\varepsilon, v) = (f, v) \tag{6.73}$$

for every $v \in D(L)$, we obtain $\|u_\varepsilon\| \leq \|f\|$ by putting $u = u_\varepsilon$. Also, since

$$\varepsilon L^* F^{-1} Lu_\varepsilon + Lu_\varepsilon + Fu_\varepsilon = f$$

by (6.73), we have

$$\varepsilon (L^* F^{-1} Lu_\varepsilon, F^{-1} Lu_\varepsilon) + (Lu_\varepsilon, F^{-1} Lu_\varepsilon) + (Fu_\varepsilon, F^{-1} Lu_\varepsilon) = (f, F^{-1} Lu_\varepsilon).$$

The first term on the left-hand side is non-negative from the assumption, so that $\|Lu_\varepsilon\| \leq \|u_\varepsilon\| + \|f\| \leq 2 \|f\|$. Hence, $\{u_\varepsilon\}$ and $\{Lu_\varepsilon\}$ are both bounded. Therefore, on account of Corollary 4 of Theorem 1.1.8, we may assume $u_\varepsilon \rightharpoonup u$ and $Lu_\varepsilon \rightharpoonup Lu$. Set $v = u_\varepsilon - u$ in (6.73). Then we have

$$(Fu_\varepsilon, u_\varepsilon - u) = (f, u_\varepsilon - u) - \varepsilon (L(u_\varepsilon - u), F^{-1} Lu_\varepsilon) - (Lu_\varepsilon, u_\varepsilon - u)$$

$$\leq (f, u_\varepsilon - u) - \varepsilon (L(u_\varepsilon - u), F^{-1} Lu_\varepsilon) - (Lu, u_\varepsilon - u),$$

so that $\limsup (Fu_\varepsilon, u_\varepsilon - u) \leq 0$. Since F is pseudo-monotone by the corollary of Theorem 6.4.3,

$$\liminf (Fu_\varepsilon, u_\varepsilon - v) \geq (Fu, u - v)$$

for every $v \in X$. Suppose a certain subset of the directed family of points $\{Fu_\varepsilon\}$ is weakly convergent to w. Then, the following result holds:

$$0 \geq \limsup (Fu_\varepsilon, u_\varepsilon - u)$$

$$\geq \limsup \{(Fu_\varepsilon - Fv, u_\varepsilon - v) + (Fv, u_\varepsilon - v) + (Fu_\varepsilon, v - u)\}$$

$$\geq \lim \{(Fv, u_\varepsilon - v) + (Fu_\varepsilon, v - u)\} = (w - Fv, v - u)$$

for every $v \in X$. Similarly to the last part of Step 2 in the proof of Theorem 6.5.2, we obtain $(w - Fu, v - u) \leq 0$ for every $v \in X$, so that

$w = Fu$. Hence, $\{Fu_\varepsilon\}$ converges weakly to Fu. Thus, by letting $\varepsilon \to 0$ in (6.73), we obtain

$$(Lu, v) + (Fu, v) = (f, v),$$

namely, $(L + F)u = f$. $\square$

Remark 6.5.3 The above lemma is a part of the following theorem due to Brezis [31]. Let L be a monotone linear operator. Then the following three conditions are equivalent:

(1) L is maximal monotone.
(2) L is a closed operator, $D(L)$ is dense and L^* is monotone.
(3) L is a closed operator, $D(L)$ is dense and L^* is maximal monotone.

Corollary *Let H be a Hilbert space and V a reflexive Banach space. Suppose that V is a dense subspace of H and that V has a stronger topology than H. Therefore, $V \subset H \subset V^*$, as in Section 2.2. Let $T > 0$ and $X = L^p(0, T; V)$ with $2 \leqslant p < \infty$. Then, the operator L defined by*

$$\begin{cases} D(L) = \{u \in X: u' \in X^*, u(0) = 0\}, \\ Lu = u' \quad \text{for each} \quad u \in D(L) \end{cases}$$

is maximal monotone linear.

Proof Since $X^* = L^{p'}(0, T; V^*)$ with $p' = p/(p-1)$, it follows from arguments paralleling Lemma 5.5.1 that $u \in D(L)$ implies that $u \in C([0, T]; H)$. Therefore, we find that $u(0)$ is well defined as an element of H. As is easily shown, L is a closed linear operator from X into X^* and $D(L)$ is dense in X. Further, L is monotone, since

$$(Lu, u) = (u', u) = \|u(T)\|_H^2 / 2 \geqslant 0.$$

It is also readily seen that the adjoint operator of L is given by

$$\begin{cases} D(L^*) = \{u \in X: u' \in X^*, u(T) = 0\}, \\ L^*u = -u' \quad \text{for each} \quad u \in D(L^*). \end{cases}$$

Hence, L^* is also monotone. Therefore, from the lemma, it is concluded that L is maximal monotone linear. $\square$

Let Ω be a region of $\mathbb{R}^n$, $T > 0$ and $2 \leqslant p < \infty$. If we regard the operator A given by (6.66) as an operator from $X = L^p(0, T; \mathring{W}_p^1(\Omega))$ into $X^* = L^{p'}(0, T; W_{p'}^{-1}(\Omega))$, A is a hemicontinuous, coercive and monotone operator from $D(A) = X$ into X^*. Also, there exists a constant C such that $\|Au\| \leqslant C\|u\|^{p-1}$ for all $u \in X$. In the corollary of Lemma 6.5.3, we can choose $H = L^2(\Omega)$ and $V = \mathring{W}_p^1(\Omega)$ so that the operators L and A

carrying X to X^* satisfy the assumption of Corollary 3 of Theorem 6.5.1 with $u_0 = 0$. Hence, for every $f \in X^*$ there exists a solution of $(L + A)u = f$, which is a solution in the wider sense of the problem

$$\frac{\partial u}{\partial t} = \sum_{i=1}^{n} \frac{\partial}{\partial x_i} \left\{ (1 + |\text{grad } u|^2)^{(p-2)/2} \frac{\partial u}{\partial x_i} \right\} + f, \qquad \Omega \times (0, T),$$

$$u(x, t) = 0, \qquad x \in \partial\Omega, \qquad 0 < t < T, \qquad u(x, 0) = 0, \qquad x \in \Omega.$$

The subdifferential of a convex function is an important example of non-linear monotone mappings. A real-valued functional φ defined on a Banach space X is called a *properly convex function* if it satisfies the following conditions.

$$-\infty < \varphi(u) \leqslant \infty \quad \text{for all} \quad u \in X, \tag{6.74}$$

$$\begin{cases} \varphi \text{ is a convex function, i.e., } \varphi((1-\lambda)u + \lambda v) \\ \leqslant (1-\lambda)\varphi(u) + \lambda\varphi(v) \text{ for all } u, v \in X \text{ and} \\ 0 < \lambda < 1. \end{cases} \tag{6.75}$$

φ is defined on the whole of X, but $D(\varphi) = \{u \in X : \varphi(u) < \infty\}$ in particular is called the *effective domain* of φ. Let φ be properly convex and lower semicontinuous, i.e., the set $\{u \in X : \varphi(u) \leqslant c\}$ be closed for any real number c. When u is an element of $D(\varphi)$ and there exists an $f \in X^*$ such that

$$\varphi(v) - \varphi(u) \geqslant (f, v - u) \tag{6.76}$$

for all $v \in X$, we write $u \in D(\partial\varphi)$ and $f \in \partial\varphi(u)$. $\partial\varphi$ is called the *subdifferential* of φ. $\partial\varphi$ is, in general, a multi-valued mapping, but it is monotone, as is easily shown. Since $\{(u, \lambda) : \varphi(u) \leqslant \lambda\}$ is a convex closed subset of $X \times \mathbb{R}$, it follows from Theorem 1.1.8 that there exist $f \in X^*$ and $c \in \mathbb{R}$ such that $\varphi(u) \geqslant f(u) + c$ for every $u \in X$. That is, there exist linear functions which bound φ from below, and, furthermore, it is found that φ is identical with the upper limit of such linear functions. For more details on convex functions, *see* Moreau [140].

In Example 2, following Theorem 6.5.2, we define

$$\varphi(u) = \begin{cases} J(u), & u \in K, \\ \infty, & u \notin K, \end{cases} \tag{6.77}$$

which is properly convex, lower semicontinuous in $H_1(\Omega)$ and satisfies $\lim_{\|u\| \to \infty} \varphi(u) = \infty$. By Corollary 2 of Theorem 1.1.8, a lower semicontinuous convex function is also lower semicontinuous in the weak topology. From this and Theorem 1.1.7 it follows that there exists an element $u \in K$, minimizing φ defined by (6.77), which is a solution of (6.69) or (6.70).

The problem of solving an equation of the form

$$Au + \partial\varphi(u) \ni f, \qquad du/dt + Au + \partial\varphi(u) \ni f, \qquad u(0) = u_0$$

with A being pseudo-monotone is called a *unilateral* problem and has wide applications to a variety of non-linear boundary and mixed problems. When K is a closed convex subset of a Banach space,

$$\varphi_K(u) = \begin{cases} 0, & u \in K, \\ \infty, & u \notin K, \end{cases}$$

is a properly convex, lower semicontinuous function, which we call the *indicatrix* of K. By the use of the indicatrix, (6.56) can be expressed as $Tu + \partial\varphi_K(u) \ni f$, so that it is also a unilateral problem. Brezis [33] studied the initial-value problem of the equation

$$du/dt + \partial\varphi(u) \ni f$$

in a Hilbert space and showed that this equation has properties similar to a linear parabolic equation. Watanabe [176] extended Brezis' results to the case when ψ depends on t as well as u and f also depends on u.

6.6 Semilinear equations

We prove the following theorem due to Kato [83] on the initial-value problem of a semilinear equation

$$du(t)/dt + A(t)u(t) + f(t, u(t)) = 0 \qquad 0 < t \leqslant T, \tag{6.78}$$

$$u(0) = u_0. \tag{6.79}$$

Theorem 6.6.1 *Let X be a Hilbert space and f a demicontinuous bounded mapping from $[0, T] \times X$ into X. Assume $f(t, \cdot)$ is monotone for each $t \in [0, T]$:*

$$(f(t, u) - f(t, v), u - v) > 0.$$

Assume further that $-A(t)$ is a generator of a contraction semigroup and that $(A(t) + 1)^{-1}$ is strongly continuously differentiable. For linear equations corresponding to (6.78) and (6.79), we assume the following. There exists a fundamental solution $U(t, s)$ of the linear equation

$$du(t)/dt + A(t)u(t) = 0,$$

and if $u_0 \in X$ and $h \in C([0, T]; X)$, then the solution of

$$du(t)/dt + A(t)u(t) = h(t), \qquad 0 \leqslant t \leqslant T, \tag{6.80}$$

$$u(0) = u_0 \tag{6.81}$$

is given by

$$u(t) = U(t, 0)u_0 + \int_0^t U(t, s)h(s)\, ds. \qquad (6.82)$$

If $u_0 \in D(A(0))$ *and* $h \in C^1([0, T]; X)$, *then* (6.82) *is a solution of* (6.80) *and* (6.81).

Under these conditions, there exists a solution $u \in C([0, T]; X)$ of the integral equation

$$u(t) = U(t, 0)u_0 - \int_0^t U(t, s)f(s, u(s))\, ds, \qquad 0 \le t \le T, \qquad (6.83)$$

corresponding to (6.78) and (6.79), and it is unique. Let u_1 and u_2 be the solutions with initial values u_{01} and u_{02}, respectively. Then the estimate

$$\|u_1(t) - u_2(t)\| \le \|u_{01} - u_{02}\| \qquad (6.84)$$

holds on $0 \le t \le T$. Hence, the mapping which carries the initial value u_0 to the solution u is a continuous mapping from X into $C([0, T]; X)$.

In Kato [83], the space X is assumed to be separable for simplicity. We don't assume this here. By way of preparation, we consider the initial-value problem:

$$du(t)/dt + f(t, u(t)) = 0, \qquad 0 \le t \le T, \qquad (6.85)$$

$$u(0) = u_0 \qquad (6.86)$$

i.e., the case in which $A(t) \equiv 0$. The corresponding integral equation is

$$u(t) = u_0 - \int_0^t f(s, u(s))\, ds, \qquad 0 \le t \le T. \qquad (6.87)$$

By letting $v(t) = u(t) - u_0$ and $g(t, v) = f(t, v + u_0)$, (6.87) becomes

$$v(t) + \int_0^t g(s, v(s))\, ds = 0, \qquad 0 \le t \le T. \qquad (6.88)$$

Proposition 6.6.1   *The solution* $v \in C([0, T]; X)$ *of* (6.88) *exists uniquely.*

Proof   Let us define an operator L as follows:

$$\begin{cases} D(L) = \{u \in L^2(0, T; X) : u' \in L^2(0, T; X),\ u(0) = 0\}, \\ Lu = u' \quad \text{for each} \quad u \in D(L). \end{cases}$$

By the corollary of Lemma 6.5.3, L is a maximal monotone linear operator in $L^2(0, T; X)$. L^{-1} exists and is bounded; it can be expressed as

$$(L^{-1}u)(t) = \int_0^t u(t)\, dt.$$

Let $(Gv)(t) = g(t, v(t))$ for each $v \in C([0, T]; X)$. $D(L) \subset D(G)$ since $Gv \in L^\infty(0, T; X) \subset L^2(0, T; X)$ and $D(L) \subset C([0, T]; X)$. Equation (6.88) is equivalent to

$$Lv + Gv = 0. \tag{6.89}$$

Lemma 6.6.1 *G is monotone as an operator from $L^2(0, T; X)$ into itself, and is a demicontinuous bounded mapping as an operator from $C([0, T]; X)$ into $L^2([0, T]; X)$.*

Since the proof is easy, it is omitted.

From now on, the inner product and norm of X and $L^2(0, T; X)$ are denoted by $(\cdot, \cdot)$ and $\|\cdot\|$, respectively. Let the collection of all finite-dimensional subspaces of X be denoted by $\mathcal{Y}$ and when $Y \in \mathcal{Y}$ the orthogonal projection on Y will be denoted by P_Y. For $u \in L^2(0, T; X)$ let us define $(P_Y u)(t) = P_Y u(t)$; thus P_Y also denotes the orthogonal projection in $L^2(0, T; X)$. According to the theory of ordinary differential equations, the solution of

$$dv_Y(t)/dt + P_Y g(t; v_Y(t)) = 0, \qquad v_Y(0) = 0 \tag{6.90}$$

exists in a certain neighbourhood of $t = 0$. Since

$$\frac{1}{2}\frac{d}{dt}\|v_Y(t)\|^2 = (v_Y'(t), v_Y(t))$$

$$= -(P_Y g(t, v_Y(t)), v_Y(t)) = -(g(t, v_Y(t)), v_Y(t))$$

$$= -(g(t, v_Y(t)) - g(t, 0), v_Y(t)) - (g(t, 0), v_Y(t))$$

$$\leqslant \|g(t, 0)\|\, \|v_Y(t)\|,$$

we have

$$\|v_Y(t)\| \leqslant \int_0^t \|g(s, 0)\|\, ds \leqslant C. \tag{6.91}$$

Hence, the solution of (6.90) exists in $[0, T]$ and has the bound (6.91) there. Equation (6.90) can be expressed as $Lv_Y + P_Y Gv_Y = 0$. By Lemma 6.6.1, $\{Gv_Y\}$ is bounded on $L^2(0, T; X)$:

$$\|Gv_Y\| \leqslant C_1. \tag{6.92}$$

Therefore, one can find an element z of $L^2(0, T; X)$, satisfying $\|z\| \leq C_1$, such that for any weak neighbourhood of z and for any $Y \in \mathcal{Y}$ there exists an element Y_1 of $\mathcal{Y}$ which contains Y and for which $Gv_{Y_1} \in V$. Next, we will show that $-Lv_{Y_1} \in V$ is realizable if Y_i is chosen appropriately. First, when V is given in terms of $\varepsilon > 0$ and step functions $w_1, \ldots, w_n$ by

$$V = \{u \in L^2(0, T; X): |(u - z, w_i)| < \varepsilon, i = 1, \ldots, n\},$$

we choose $Y_0 \in \mathcal{Y}$ which contains both Y and $\{w_i(t); 0 \leq t \leq T, i = 1, \ldots, n\}$. For V and Y_0 thus defined, there exists an element Y_1 of $\mathcal{Y}$, containing Y_0, such that $Gv_{Y_1} \in V$. Since

$$|(-Lv_{Y_1} - z, w_i)| = |(P_{Y_1}Gv_{Y_1} - z, w_i)|$$

$$= |(Gv_{Y_1} - z, w_i)| < \varepsilon,$$

we obtain $-Lv_{Y_1} \in V$. When $w_1, \ldots, w_n$ are general elements of $L^2(0, T; X)$, we choose step functions $w'_1, \ldots, w'_n$ in such a way that

$$\|w_i - w'_i\| < \varepsilon/(2C_1 + 1), \qquad i = 1, \ldots, n.$$

Then, by the above argument, we can find a $Y_1 \supset Y$ so that

$$|(Gv_{Y_1} - z, w'_i)| < \varepsilon/(2C_1 + 1),$$

$$|(-Lv_{Y_1} - z, w'_i)| < \varepsilon/(2C_1 + 1), \qquad i = 1, \ldots, n.$$

From (6.92) and $\|z\| \leq C_1$, it follows that

$$|(Gv_{Y_1} - z, w_i)| \leq |(Gv_{Y_1} - z, w'_i)| + |(Gv_{Y_1} - z, w_i - w'_i)|$$

$$< \varepsilon/(2C_1 + 1) + 2C_1\varepsilon/(2C_1 + 1) = \varepsilon.$$

Since $\|Lv_{Y_1}\| = \|P_{Y_1}Gv_{Y_1}\| \leq \|Gv_{Y_1}\| \leq C_1$, we similarly obtain

$$|(-Lv_{Y_1} - z, w_i)| < \varepsilon,$$

and, hence, $Gv_{Y_1} \in V$ and $-Lv_{Y_1} \in V$.

Let $v = -L^{-1}z$ and

$$V = \{u: |(u - z, w)| < \varepsilon, |(u - z, (L^{-1})^*(L + G)w)| < \varepsilon\}$$

with $w \in D(L)$ and $\varepsilon > 0$. Then V is a neighbourhood of z. Therefore, there exists a $Y \in \mathcal{Y}$ such that

$$Gv_Y \in V, \quad -Lv_Y \in V. \tag{6.93}$$

Since

$$((L + G)v_Y, v_Y) = (Lv_Y, v_Y) + (P_Y Gv_Y, v_Y)$$

$$= (Lv_Y + P_Y Gv_Y, v_Y) = 0,$$

we obtain

$$0 \leqslant ((L+G)w - (L+G)v_Y, w - v_Y)$$
$$= ((L+G)w, w) - ((L+G)w, v_Y) - ((L+G)v_Y, w). \tag{6.94}$$

From (6.93), it follows that

$$|((L+G)w, v_Y) - ((L+G)w, v)| = |((L^{-1})^*(L+G)w, Lv_Y + z)| < \varepsilon$$
$$|((L+G)v_Y, w)| \leqslant |(Lv_Y + z, w)| + |(Gv_Y - z, w)| < 2\varepsilon.$$

Therefore, (6.94) implies that

$$0 \leqslant ((L+G)w, w) - ((L+G)w, v) + 3\varepsilon.$$

By letting $\varepsilon \to 0$, we find that

$$((L+G)w, w - v) \geqslant 0 \tag{6.95}$$

holds for all $w \in D(L)$. Since $v + n^{-1}w \in D(L)$ for each positive integer n, we can replace w by $v + n^{-1}w$ in (6.95) to obtain

$$((L+G)(v + n^{-1}w), w) \geqslant 0.$$

In the limit as $n \to \infty$, this, on account of Lemma 6.6.1, gives

$$((L+G)v, w) \geqslant 0, \tag{6.96}$$

which implies (6.89), since $D(L)$ is dense in $L^2(0, T; X)$. To prove the uniqueness of the solution, suppose v_1 and v_2 satisfy (6.88). Then

$$\tfrac{1}{2}\|v_1(T) - v_2(T)\|^2 = (L(v_1 - v_2), v_1 - v_2)$$
$$= -(Gv_1 - Gv_2, v_1 - v_2) \leqslant 0,$$

so we must have $v_1(T) = v_2(T)$. On replacing $[0, T]$ by $[0, t]$, a similar argument establishes $v_1(t) = v_2(t)$ for any t in $[0, T]$. This completes the proof of Proposition 6.6.1. $\square$

Let us return to the proof of Theorem 6.6.1. Set

$$v(t) = u(t) - U(t, 0)u_0, \qquad g_1(t, v) = f(t, v + U(t, 0)u_0).$$

Equation (6.83) is transformed into

$$v(t) + \int_0^t U(t, s)g_1(s, v(s)) \, ds = 0. \tag{6.97}$$

Let $(G_1 v)(t) = g_1(t, v(t))$ for $v \in C([0, T]; X)$. Further, let

$$(Uu)(t) = \int_0^t U(t, s)u(s) \, ds$$

for $u \in L^2(0, T; X)$. Then (6.97) can be written in the form

$$v + UG_1 v = 0. \tag{6.98}$$

It is an easy task to prove the following lemma.

Lemma 6.6.2 *The statements on G in Lemma 6.6.1 hold also for G_1.*

Lemma 6.6.3 *U is a bounded linear mapping from $L^2(0, T; X)$ into $C([0, T]; X)$ and its range $R(U)$ is dense in $L^2(0, T; X)$.*

Proof The first part of the lemma is obvious. Any element u of $L^2(0, T; X)$ can be approximated by an element v of $C^1([0, T]; X)$ satisfying $v(0) = 0$. By operating on v by $(1 + n^{-1}A(t))^{-1}$, we therefore have an element w, approximating u in the strong topology of $L^2(0, T; X)$, such that $w(0) = 0$, $w \in C^1([0, T]; X)$, $w(t) \in D(A(t))$ for all $t \in [0, T]$ and $Aw \in C([0, T]; X)$. Let $w'(t) + A(t)w(t) = h(t)$; $w = Uh$ since $h \in C([0, T]; X)$. Hence, $R(U)$ is dense in $L^2(0, T; X)$. □

Lemma 6.6.4 *$G_1 U$ is a demicontinuous mapping from $L^2(0, T; X)$ into itself.*

Proof Obvious from Lemmas 6.6.2 and 6.6.3. □

Let $A_n(t) = A(t)(1 + n^{-1}A(t))^{-1}$, then $A_n(t)$ is bounded monotone linear and $A_n(t)u \to A(t)u$ for $u \in D(A(t))$. We put $(A_n u)(t) = A_n(t)u(t)$ for each $u \in L^2(0, T; X)$. Let L be the operator defined in the proof of Proposition 6.6.1. Then, by virtue of Proposition 6.6.1, the solution v_n of

$$(L + A_n + G_1)v_n = 0 \tag{6.99}$$

exists. As in the proof of Proposition 6.6.1, we have

$$\|v_n(t)\| \le \int_0^t \|f(s, U(s, 0)u_0)\| \, ds \le C_2. \tag{6.100}$$

By replacing $\{v_n\}$ by its subsequence, if necessary, we may assume that $v_n \to v$ in $L^2(0, T; X)$. It is also easy to show that

$$\|v(t)\| \le C_2 \tag{6.101}$$

for almost all t. Let w be an arbitrary element of $C^1([0, T]; X)$. Since $LUw + AUw = w$ and $A_n Uw \to AUw$ in $L^2(0, T; X)$, we obtain

$$0 \le ((L + A_n + G_1)Uw - (L + A_n + G_1)v_n, Uw - v_n)$$

$$= ((L + A_n + G_1)Uw, Uw - v_n)$$

$$\to ((L + A + G_1)Uw, Uw - v) = ((w + G_1 Uw, Uw - v)$$

as $n \to \infty$, so that

$$(w + G_1 Uw, Uw - v) \geq 0 \qquad (6.102)$$

holds for every $w \in C^1([0, T]; X)$. When $w \in L^2(0, T; X)$, by taking $w_n \in C^1([0, T]; X)$ such that $w_n \to w$ in $L^2(0, T; X)$, we obtain $G_1 Uw_n \to G_1 Uw$ in $L^2(0, T; X)$ by Lemma 6.6.1, and, hence, (6.102) holds. For each $\lambda > 0$, put

$$v_\lambda(t) = \lambda \int_0^t e^{-\lambda(t-s)} U(t, s) v(s) \, ds,$$

$$u_\lambda = \lambda(v - v_\lambda).$$

A direct calculation yields $Uu_\lambda = v_\lambda$. Also, $\|v_\lambda(t)\| \leq C_2$ by (6.100). Therefore, $\{G_1 v_\lambda\}$ is bounded in $L^2(0, T; X)$:

$$\|G_1 v_\lambda\| \leq C_3.$$

We put $w = u_\lambda$ in (6.102), obtaining

$$(\lambda(v - v_\lambda) + G_1 v_\lambda, v_\lambda - v) \geq 0,$$

so that

$$\|u_\lambda\| = \lambda \|v - v_\lambda\| \leq \|G_1 v_\lambda\| \leq C_3. \qquad (6.103)$$

Consequently, $v_\lambda \to v$ as $\lambda \to \infty$. Since $\{u_\lambda\}$ is bounded in $L^2(0, T; X)$ by (6.103), we can make $u_\lambda \to u$ by choosing an appropriate subsequence. Hence, $v = Uu$. If w is replaced by $u + n^{-1}w$ in (6.102), we have

$$(u + n^{-1}w + G_1 U(u + n^{-1}w), Uw) \geq 0,$$

which, by Lemma 6.6.4, leads to $(u + G_1 Uu, Uw) \geq 0$ for all $w \in L^2(0, T; X)$ in the limit as $n \to \infty$. Since $R(U)$ is dense in $L^2(0, T; X)$ by Lemma 6.6.3, we obtain $u + G_1 Uu = 0$ and, hence, (6.98) by operating U on both sides. Thus the existence part of the proof is completed.

Lemma 6.6.5 *For $v \in L^2(0, T; X)$ and $u_0 \in X$, let*

$$u = U(\cdot, 0)u_0 + Uv.$$

Then the following result holds

$$2(v, u) \geq \|u(T)\|^2 - \|u_0\|^2.$$

Proof When $v \in C^1([0, T]; X)$ and $u_0 \in D(A(0))$, we have $u' + Au = v$, so that

$$2(v, u) = 2(u' + Au, u) \geq 2(u', u) = \|u(T)\|^2 - \|u_0\|^2.$$

For general cases, the inequality is proved by approximating v and u_0 by sequences in $C^1([0, T]; X)$ and $D(A(0))$, respectively. This completes the proof of Lemma 6.6.5. □

Define

$$u_i(t) = U(t, 0)u_{0i} - \int_0^t U(t, s)f(s, u_i(s))\, ds, \qquad i = 1, 2.$$

By letting $u = u_1 - u_2$, $v(t) = -f(t, u_1(t)) + f(t, u_2(t))$ and $u_0 = u_{01} - u_{02}$, we have $u = U(\cdot, 0)u_0 + Uv$. Therefore, by Lemma 6.6.5,

$$\|u(T)\|^2 \leq 2(v, u) + \|u_0\|^2, \tag{6.104}$$

but, since

$$(v, u) = -\int_0^T (f(t, u_1(t)) - f(t, u_2(t)), u_1(t) - u_2(t))\, dt \leq 0,$$

we obtain

$$\|u_1(T) - u_2(T)\|^2 \leq \|u_{01} - u_{02}\|^2$$

from (6.104). We replace the interval $[0, T]$ by $[0, t]$ and repeat similar processes to arrive at (6.84). The proof of Theorem 6.6.1 is complete. □

Remark 6.6.1 As will be stated in Lemma 7.2.1 of the next chapter, there exists an inverse $S = U^{-1}$ of U. S is maximal monotone.

Next, we apply Theorem 6.6.1 to find a solution in the wider sense of (6.78) and (6.79) under somewhat different assumptions. Let X, V and a quadratic form $a(t; u, v)$ defined on $V \times V$ be all the same as in Section 5.5. Furthermore, we assume that (5.108) is satisfied for $k = 0$, i.e., there exists a positive constant δ such that

$$a(t; u, u) \geq \delta \|u\|^2$$

for all $t \in [0, T]$ and $u \in V$. As before, we denote by $A(t)$ the operator determined by $a(t; u, v)$ and, in this section, interpret it as an element of $B(V, V^*)$.

Theorem 6.6.2 Let f be a demicontinuous bounded mapping from $[0, T] \times X$ into V^*. Assume that $f(t, \cdot)$ for each t is monotone as a mapping from V into V^*. Let u_0 be an arbitrary element of X. Then there exists a solution $u \in L^2(0, T; V)$, satisfying $u' \in L^2(0, T; V^*)$, of

$$du(t)/dt + A(t)u(t) + f(t, u(t)) = 0, \qquad 0 < t \leq T, \tag{6.105}$$

$$u(0) = u_0 \tag{6.106}$$

and it is unique. The mapping which puts the initial value u_0 in correspondence to the solution u is a continuous mapping from X into $C([0, T]; X) \cap L^2(0, T; V)$.

By virtue of Theorem 5.5.1, there exists a solution $z \in L^2(0, T; V)$, satisfying $z' \in L^2(0, T; V^*)$, of

$$du(t)/dt + A(t)u(t) = 0, \qquad u(0) = u_0.$$

Put $v(t) = u(t) - z(t)$ and $g(t, v) = f(t, v + z(t))$. Then,

$$dv(t)/dt + A(t)v(t) + g(t, v(t)) = 0, \qquad v(0) = 0.$$

Since $z \in C([0, T]; X)$, g satisfies assumptions similar to those for f. Hence, as far as the existence of the solution is concerned, we may put $u_0 = 0$. By the corollary of Lemma 6.5.3, the operator defined by

$$\begin{cases} D(L) = \{u \in L^2(0, T; V) : u' \in L^2(0, T; V^*), u(0) = 0\}, \\ Lu = u' \quad \text{for all} \quad u \in D(L), \end{cases}$$

is a maximal monotone linear operator from $L^2(0, T; V)$ into $L^2(0, T; V^*)$. Let us write $(Au)(t) = A(t)u(t)$ for each $u \in L^2(0, T; V)$ and $(Gu)(t) = f(t, u(t))$ for each $u \in C([0, T]; X)$. Then both A and G are monotone operators from $L^2(0, T; V)$ into $L^2(0, T; V^*)$ and $D(G) \supset D(L)$. Since we assumed $u_0 = 0$, the equations (6.105) and (6.106) are equivalent to

$$(L + G + A)u = 0. \tag{6.107}$$

Note that $L + G$ is monotone; if it is shown to be maximal monotone, the assumption of Corollary 2 to Theorem 6.5.1 is satisfied with $M = L + G$, $T = A$ and $u_0 = 0$. Then we will have $R(L + G + A) = X^*$, which implies the existence of the solution. Let us denote by Λ the operator determined by a quadratic form $((\cdot, \cdot)) : (\Lambda u, v) = ((u, v))$. Λ is a duality mapping from V into V^*. Therefore, by expressing $(\Lambda u)(t) = u(t)$ for all $u \in L^2(0, T; V)$, this Λ becomes a duality mapping from $L^2(0, T; V)$ into $L^2(0, T; V^*)$. Hence, to see the maximal monotonicity of $L + G$, on account of Proposition 6.4.1, it is enough to verify that $R(L + G + \Lambda) = L^2(0, T; V^*)$.

By Theorems 2.2.2 and 2.2.3, if Λ is viewed as an operator in X, it is positive definite and self-adjoint; the domain of $\Lambda^{1/2}$ coincides with V. On the other hand, it is readily seen that Λ viewed as an operator in V^* is also positive definite and self-adjoint; the domain of $\Lambda^{1/2}$ coincides with X. Hence, $I_n = (1 + n^{-1}\Lambda^{1/2})^{-1}$ is a contraction operator in both X and V, and converges strongly to I as $n \to \infty$. It is also easy to see that $(I_n f, g) = (f, I_n g)$ holds for $f \in V^*$ and $g \in X$. Therefore, if we define $f_n(t, u) = I_n f(t, I_n u)$, it satisfies the assumptions for f in Theorem 6.6.1.

Let us denote the semigroup generated by Λ by $T(t)$. The following lemma is a special case of Lemma 5.5.2 and Proposition 5.5.1.

Lemma 6.6.6   *The function w defined by*

$$w(t) = \int_0^t T(t-s)\varphi(s)\,\mathrm{d}s, \tag{6.108}$$

where $\varphi \in L^2(0, T; V^*)$, is a solution of

$$(L + \Lambda)w = \varphi$$

and, moreover, satisfies

$$\tfrac{1}{2}|w(t)|^2 + \int_0^t \|w(s)\|^2\,\mathrm{d}s = \int_0^t (\varphi(s), w(s))\,\mathrm{d}s \tag{6.109}$$

and

$$|w(t)|^2 + \int_0^t \|w(s)\|^2\,\mathrm{d}s \leqslant \int_0^t \|\varphi(s)\|_*^2\,\mathrm{d}s. \tag{6.110}$$

Proof of Theorem 6.6.2   Let h be an arbitrary element of $L^2(0, T; V^*)$, $h_n \in C([0, Y]; X)$ and $h_n \to h$ in $L^2(0, T; V^*)$. Theorem 6.6.1 can be applied to the initial-value problem

$$\mathrm{d}u(t)/\mathrm{d}t + \Lambda u(t) + f_n(t, u(t)) - h_n(t) = 0, \qquad u(0) = 0$$

and ensures the existence of a solution $u_n \in C([0, T]; X)$ of the integral equation

$$u_n(t) + \int_0^t T(t-s)(f_n(s, u_n(s)) - h_n(s))\,\mathrm{d}s = 0. \tag{6.111}$$

By Lemma 6.6.6, we have

$$\begin{aligned}
\tfrac{1}{2}|u_n(t)|^2 + \int_0^t \|u_n(s)\|^2\,\mathrm{d}s &= \int_0^t (h_n(s) - f_n(s, u_n(s)), u_n(s))\,\mathrm{d}s \\
&= \int_0^t (h_n(s), u_n(s))\,\mathrm{d}s - \int_0^t (f_n(s, 0), u_n(s))\,\mathrm{d}s \\
&\quad - \int_0^t (f_n(s, u_n(s)) - f_n(s, 0), u_n(s))\,\mathrm{d}s \\
&\leqslant \int_0^t \|h_n(s)\|_*^2\,\mathrm{d}s + \int_0^t \|f(s, 0)\|_*^2\,\mathrm{d}s \\
&\quad + \tfrac{1}{2}\int_0^t \|u_n(s)\|^2\,\mathrm{d}s,
\end{aligned}$$

so that

$$|u_n(t)|^2 + \int_0^t \|u_n(s)\|^2 \, ds \leq 2 \int_0^t \|h_n(s)\|_*^2 \, ds + 2 \int_0^t \|f(s, 0)\|_*^2 \, ds$$

and $\{u_n\}$ is bounded in $C([0, T]; X) \cap L^2(0, T; V)$. Therefore, $\{f_n(\cdot, u_n)\}$ is bounded in $L^2(0, T; V^*)$. By replacing them by their subsequences, we may assume $u_n \rightharpoonup u$ in $L^2(0, T; V)$ and $f_n(\cdot, u_n) \rightharpoonup g$ in $L^2(0, T; V^*)$. By letting $n \to \infty$ in (6.111), we have

$$u(t) + \int_0^t T(t-s)(g(s) - h(s)) \, ds = 0, \tag{6.112}$$

so that $u \in C([0, T]; X)$. Let φ be an arbitrary element of $L^2(0, T; V^*)$ and w a function defined by (6.108). Since

$$u_n(t) - w(t) = \int_0^t T(t-s)(h_n(s) - f_n(s, u_n(s)) - \varphi(s)) \, ds,$$

it follows from (6.109) that

$$0 \leq \tfrac{1}{2}|u_n(T) - w(T)|^2 + \int_0^T \|u_n(t) - w(t)\|^2 \, dt$$

$$= \int_0^T (h_n(t) - f_n(t, u_n(t)) - \varphi(t), u_n(t) - w(t)) \, dt$$

$$= -\int_0^T (f(t, I_n u_n(t)) - f(t, w(t)), I_n u_n(t) - w(t)) \, dt$$

$$- \int_0^T (I_n h_n(t) - h_n(t), u_n(t) - w(t)) \, dt$$

$$- \int_0^T (f(t, I_n u_n(t)) - h_n(t), w(t) - I_n w(t)) \, dt$$

$$- \int_0^T (f(t, w(t)) - h_n(t), I_n u_n(t) - w(t)) \, dt - \int_0^T (\varphi(t), u_n(t) - w(t)) \, dt.$$

The first term on the right-hand side is non-positive. Hence, by letting $n \to \infty$, we have

$$\int_0^T (f(t, w(t)) - h(t) + \varphi(t), u(t) - w(t)) \, dt \leq 0. \tag{6.113}$$

We replace φ by $h - g - n^{-1}\varphi$ in (6.113) to obtain

$$\int_0^T (f(t, u(t) - n^{-1}w(t)) - g(t) + n^{-1}\varphi(t), w(t)) \, dt \leq 0,$$

which, in the limit as $n \to \infty$, yields

$$\int_0^T (f(t, u(t)) - g(t), w(t)) \, dt \leq 0 \qquad (6.114)$$

valid for all h, where w is a function defined by (6.108). Let w be an arbitrary function belonging to $C^1([0, T]; V)$ which satisfies $w(0) = 0$ and put $w' + \Lambda w = \varphi$. Since $\varphi \in C([0, T]; V^*)$, w is given by (6.108). Hence, the set of all functions w expressed in terms of $\varphi \in L^2(0, T; V^*)$ in the form of (6.108) is dense in $L^2(0, T; V)$. Consequently, from (6.114), it follows that $f(t, u(t)) = g(t)$. Inserting this into (6.112), we obtain

$$u(t) + \int_0^t T(t-s)(f(s, u(s)) - h(s)) \, ds = 0.$$

On account of Lemma 6.6.6, we see that u is a solution of the equation $(L + G + \Lambda)u = h$. Suppose u_1 and u_2 are solutions of (6.105) with initial conditions u_{01} and u_{02}, respectively. Then it is easy to see that

$$|u_1(t) - u_2(t)| \leq |u_{01} - u_{02}|.$$

This completes the proof of Theorem 6.6.2. $\square$

We give an example which satisfies the assumption of Theorem 6.6.2. Let Ω be a bounded region in $\mathbb{R}^n$ with $n \geq 3$, $X = L^2(\Omega)$, and let V be a closed subspace of $H_1(\Omega)$ containing $\mathring{H}_1(\Omega)$. By Lemma 1.2.1, we have

$$V \subset L^p(\Omega) \subset L^2(\Omega) \subset L^q(\Omega) \subset V^*,$$

where $p = 2n/(n-2)$ and $q = p/(p-1) = 2n/(n+2)$. Assume that $f(\lambda)$ is a continuous and increasing function defined on $-\infty < \lambda < \infty$ such that $|f(\lambda)| = 0(|\lambda|^{(n+2)/n})$ as $|\lambda| \to \infty$. If we put $f(t, u)(x) = f(u)(x) = f(u(x))$ for each $u \in X = L^2(\Omega)$, then $f(t, u) = f(u) \in L^q(\Omega)$. Clearly, f is bounded monotone as an operator from $L^p(\Omega)$ into $L^q(\Omega)$. To show that f is a demicontinuous mapping from X into $L^q(\Omega)$, let $u_n \to u$ in X. Since $\{u_n\}$ is bounded in X, so is $\{f(u_n)\}$ in $L^q(\Omega)$. Hence, there exists a subsequence $\{u_{n_j}\}$ such that $u_{n_j}(x) \to u(x)$ almost everywhere in Ω and $f(u_{n_j}) \to g$ in $L^q(\Omega)$. Since $f(\lambda)$ is a continuous function of a real variable λ, $f(u_{n_j}(x)) \to f(u(x))$ almost everywhere in Ω. Hence, from the next lemma, we obtain $f(u_n) \to f(u)$.

Lemma 6.6.7 Let $1 \leq p \leq \infty$. If $f_n \in L^p(\Omega)$, $f_n(x) \to f(x)$ almost everywhere in Ω and $f_n \to g$ in $L^p(\Omega)$, then $f = g$.

Proof If $f_n \to g$ in $L^p(\Omega)$, there exists a subsequence of $\{f_n\}$ which converges to g almost everywhere, so that $f = g$. When $f_n \to g$, from

Corollary 3 of Lemma 1.1.8, one can find an appropriate convex combination $g_k = \sum_{n \geq k} \lambda_n^k f_n$, which is strongly convergent to g in $L^p(\Omega)$. Since $g_k(x) \to f(x)$ at all x for which $f_n(x) \to f(x)$, we obtain $f = g$. $\square$

We have briefly explained the theory of monotone operators. For unilateral problems, there are many publications besides those already referred to; for example, see Lions [116, 117], Lions and Stampacchia [121], Browder [41] and Stampacchia [160]. Kenmochi [92, 93] derived new results on monotone operator equations. Ouchi [146] proved the analyticity of solutions of parabolic non-linear equations. For non-linear equations the reader should also be referred to the excellent books written by Carroll [3], Ladas and Lakshmikantham [10] and Lions [13].

7

Optimal control

In this chapter we will explain the optimal control of equations of evolution. This area has been so extensively studied by Balakrishnan, Lions, Friedman, Fattorini and others that it is impossible to explain it exhaustively; we will only give a brief description. For more details the reader is referred to Lions [12] and the references cited there. We consider only real Banach spaces in this chapter and use notational conventions different from other chapters.

7.1 Formulation of the problem

We assume that there exists a fundamental solution $S(t, s)$ of an equation of evolution in a real Banach space X:

$$dx(t)/dt = A(t)x(t), \qquad 0 \leqslant t \leqslant T. \tag{7.1}$$

Let Y be another real Banach space, and let $B(t)$ belong to $B(Y, X)$ for each $t \in [0, T]$ and be continuous in t in the norm of $B(Y, X)$. For $u \in L^\infty(0, T; Y)$ let us call

$$x(t; u) = S(t, 0)x_0 + \int_0^t S(t, s)B(s)u(s) \, ds \tag{7.2}$$

a *solution* of the initial-value problem

$$dx(t)/dt = A(t)x(t) + B(t)u(t), \qquad 0 < t \leqslant T, \tag{7.3}$$

$$x(0) = u_0. \tag{7.4}$$

Choose a subset U of Y and call it a *control set*. $u \in L^\infty(0, T; Y)$ is called an *admissible control* if $u(t) \in U$ almost everywhere, and the solution (7.2) for each admissible control u is called a *trajectory* corresponding to u. Suppose that a real-valued function $J(u)$ called a *cost functional* is defined for each admissible control u. We will consider the admissible control

230

which minimizes $J(u)$, that is, the existence and uniqueness of *optimal control*. Note that (7.2) is a strongly continuous function of t.

7.2   Distributed observation

Assume that Y is reflexive and U is a bounded convex closed subset of Y. Let Z be a real Hilbert space, and let $C(t)$ belong to $B(X, Z)$ for each t and be continuous in t in the norm of $B(X, Z)$. Let y be an element of $L^2(0, T; Z)$ and consider a case where the cost functional is defined by

$$J(u) = \int_0^T \|C(t)x(t; u) - y(t)\|^2 \, dt. \tag{7.5}$$

Theorem 7.2.1   *Under the above assumption, there exists an optimal control for the cost functional* (7.5).

Proof   Let $\{u_n\}$ be a sequence of admissible controls satisfying $\lim_{n \to \infty} J(u_n) = \inf J(u)$. Since $\{u_n\}$ is bounded in $L^2(0, T; Y)$, by replacing it by a subsequence we may assume that $u_n \to u_0$ weakly in $L^2(0, T; Y)$. It is easy to show that

$$\text{w-}\lim_{n \to \infty} \int_0^t S(t, s)B(s)u_n(s) \, ds = \int_0^t S(t, s)B(s)u_0(s) \, ds$$

for each $t \in [0, T]$, so that $x(t; u_n) \to x(t; u_0)$ weakly for all $t \in [0, T]$. We can see that u_0 is admissible as follows. Let s be a Lebesgue point of u_0 (*see* Definition 1.3.3) and put

$$w_{\varepsilon,n} = \frac{1}{\varepsilon} \int_s^{s+\varepsilon} u_n(t) \, dt$$

for each $\varepsilon > 0$ and n. Let $f \in Y^*$ and $c \in (-\infty, \infty)$ be such that $f(u) \le c$ for all $u \in U$. Then, $f(w_{\varepsilon,n}) \le c$. Since

$$w_{\varepsilon,n} \to w_\varepsilon = \frac{1}{\varepsilon} \int_s^{s+\varepsilon} u_0(t) \, dt$$

weakly as $n \to \infty$, we have $f(w_\varepsilon) \le c$. By letting $\varepsilon \to 0$, we obtain $w_\varepsilon \to u_0(s)$ and have $f(u_0(s)) \le c$, so that $u_0(s) \in U$ by Corollary 1 of Theorem 1.1.8. Since $\{Cx(\cdot; u_n) - y\}$ is weakly convergent to $Cx(\cdot; u_0) - y$ in $L^2(0, T; Z)$, we have

$$\inf J(u) \le J(u_0) \le \liminf_{n \to \infty} J(u_n) = \inf J(u).$$

From this we see that u_0 is an optimal control. $\square$

Remark 7.2.1 The fact that $u_0(t) \in U$ almost everywhere can also be concluded from the following more general statement (Lemma 8 of Kato [82], p. 152). Assume that X is a reflexive Banach space and $1 < p < \infty$. Let $u_n(t)$ be a uniformly bounded function, defined on $0 \le t \le T$, with values in X and $u_n \to u$ weakly in $L^p(0, T; X)$. Furthermore, for each $t \in [0, T]$, let $V(t)$ be the set of all limits of weakly convergent subsequences of $\{u_n(t)\}$ and $\hat{V}(t)$ be the convex hull of $V(t)$. Then, $u(t) \in \hat{V}(t)$ almost everywhere in $[0, T]$.

Lemma 7.2.1 If $u \in L^1(0, T; X)$ and

$$\int_0^t S(t, s)u(s)\, ds = 0$$

in $0 \le t \le T$, then $u(t) = 0$ for almost all $t \in [0, T]$.

Proof Let r be an arbitrary rational number in $(0, T]$. For $0 < t < r$, we have

$$\int_0^t S(r, s)u(s)\, ds = S(r, t)\int_0^t S(t, s)u(s)\, ds = 0.$$

Therefore, there exists a null set N_r contained in $[0, r]$ such that $S(r, t)u(t) = 0$ for all $t \in [0, r] - N_r$. $N = \bigcup_r N_r$ is a null set. Let $0 < t < T$ and $t \notin N$. Choose a rational number r satisfying $t < r < T$. Then $S(r, t)u(t) = 0$ since $t \in [0, r] - N_r$. By letting $r \to t$, we obtained $u(t) = 0$. $\square$

Theorem 7.2.2 If both $B(t)$ and $C(t)$ are one-to-one mappings for each t, the optimal control for the cost functional (7.5) is unique.

Proof Let $\bar{u}$ be an optimal control and put $\bar{x}(t) = x(t; \bar{u})$. Let t_0 be a Lebesgue point of $\bar{u}$, $v \in U$ and $t_0 < t_0 + \varepsilon < T$. Further, put

$$u(t) = \begin{cases} v & \text{if } t_0 < t < t_0 + \varepsilon \\ \bar{u}(t) & \text{otherwise} \end{cases} \tag{7.6}$$

Then u is an admissible control. Put $x(t) = x(t; u)$. Since

$$x(t) - \bar{x}(t) = \begin{cases} 0, & 0 \le t \le t_0, \\ \displaystyle\int_{t_0}^t S(t, s)B(s)(v - \bar{u}(s))\, ds, & t_0 < t < t_0 + \varepsilon, \\ \displaystyle\int_{t_0}^{t_0 + \varepsilon} S(t, s)B(s)(v - \bar{u}(s))\, ds, & t_0 + \varepsilon \le t \le T, \end{cases} \tag{7.7}$$

there exists a constant C such that

$$\|x(t) - \bar{x}(t)\| \le C\varepsilon \tag{7.8}$$

holds for $0 \le t \le T$. If we express

$$0 \le J(u) - J(\bar{u}) = 2 \int_0^T (C(t)(x(t) - \bar{x}(t)), C(t)\bar{x}(t) - y(t)) \, dt$$

$$+ \int_0^T \|C(t)(x(t) - \bar{x}(t))\|^2 \, dt = I + II, \tag{7.9}$$

we obtain

$$\lim_{\varepsilon \to 0} \varepsilon^{-1} II = 0 \tag{7.10}$$

from (7.8). On account of (7.7), the first term can be represented as

$$I = 2 \int_{t_0}^T (C(t)(x(t) - \bar{x}(t)), C(t)\bar{x}(t) - y(t)) \, dt$$

$$= 2 \int_{t_0}^{t_0+\varepsilon} + 2 \int_{t_0+\varepsilon}^T = I_1 + I_2.$$

It is easy to see that

$$\lim_{\varepsilon \to 0} \varepsilon^{-1} I_1 = 0. \tag{7.11}$$

Let $t > t_0$ and $\varepsilon \to 0$, then, from the fact that

$$\frac{1}{\varepsilon}(x(t) - \bar{x}(t)) = \frac{1}{\varepsilon} \int_{t_0}^{t_0+\varepsilon} S(t, s) B(s)(v - \bar{u}(s)) \, ds$$

$$\to S(t, t_0) B(t_0)(v - \bar{u}(t_0))$$

strongly and from (7.8), we obtain

$$\frac{1}{2\varepsilon} I_2 = \int_{t_0+\varepsilon}^T \left(C(t) \frac{1}{\varepsilon}(x(t) - \bar{x}(t)), C(t)\bar{x}(t) - y(t) \right) dt$$

$$\to \int_{t_0}^T (C(t)S(t, t_0)B(t_0)(v - \bar{u}(t_0)), C(t)\bar{x}(t) - y(t)) \, dt. \tag{7.12}$$

By (7.9)–(7.12), the inequality

$$\int_\varepsilon^T (C(t)S(t, s)B(s)(v - \bar{u}(s)), C(t)\bar{x}(t) - y(t)) \, dt \ge 0 \tag{7.13}$$

holds for all $v \in U$ and for all Lebesgue points s of $\bar{u}$. Therefore, if we denote two optimal controls by u_1 and u_2 and their corresponding trajectories by x_1 and x_2, then the inequalities

$$\int_s^T (C(t)S(t,s)B(s)(u_2(s)-u_1(s)), C(t)x_1(t)-y(t)) \, dt \geq 0 \qquad (7.14)$$

and

$$\int_s^T (C(t)S(t,s)B(s)(u_1(s)-u_2(s)), C(t)x_2(t)-y(t) \, dt \geq 0 \qquad (7.15)$$

hold for every Lebesgue point s common to both u_1 and u_2. Add both sides of (7.14) and (7.15), and integrate the resultant inequality from 0 to T with respect to s. Then, by noting that

$$x_2(t)-x_1(t) = \int_0^t S(t,s)B(s)(u_2(s)-u_1(s)) \, ds, \qquad (7.16)$$

we have

$$\int_0^T \|C(t)(x_2(t)-x_1(t))\|^2 \, dt \leq 0.$$

From this and the fact that $C(t)$ is one-to-one, we obtain $x_2(t)-x_1(t) \equiv 0$. Hence, by (7.16) and Lemma 7.2.1, $B(t)(u_2(t)-u_1(t))=0$ almost everywhere. Since $B(t)$ is one-to-one, $u_1(t)=u_2(t)$ holds for almost all t. $\square$

From (7.13), we have

$$\left(v-\bar{u}(s), B^*(s)\int_s^T S^*(t,s)C^*(t)(y(t)-C(t)\bar{x}(t)) \, dt\right) \leq 0.$$

Hence, if we put

$$w(s) = \int_s^T S^*(t,s)C^*(t)(y(t)-C(t)\bar{x}(t)) \, dt,$$

then

$$\max_{v \in U} (v, B^*(s)w(s)) = (\bar{u}(s), B^*(s)w(s)) \qquad (7.17)$$

holds for each Lebesgue point of $\bar{u}$ and, hence, for almost all s. This is called the *maximum principle*. $w(s)$ is a solution in some sense of the equations

$$dw(s)/ds = -A^*(s)w(s) + C^*(s)(y(s)-C(s)\bar{x}(s)),$$

$$w(T) = 0.$$

7.3 Observation of the final state

Let X be a real Hilbert space and U a bounded convex closed subset of Y. Let y be an element of X and suppose there exists no admissible control which satisfies $x(T; u) = y$. We assume a cost functional given by

$$J(u) = \|x(T; u) - y\|. \tag{7.18}$$

Theorem 7.3.1 *Under the above assumption, there exists an optimal control for the cost functional* (7.18).

The proof is similar to that of Theorem 7.2.1.

Theorem 7.3.2 *Let $\bar{u}$ be an optimal control for the cost functional* (7.18). *If we define $w(t) = S^*(T, t)(y - x(T; \bar{u}))$, the maximum principle holds in the following sense*:

$$\max_{v \in U} (v, B^*(t)w(t)) = (\bar{u}(t), B^*(t)w(t)) \tag{7.19}$$

almost everywhere in $0 \leqslant t \leqslant T$.

Proof Let t_0 be a Lebesgue point of $\bar{u}$, $v \in U$ and $t_0 < t_0 + \varepsilon < T$. Let u be an admissible control defined by (7.6). If we put $\bar{x}(t) = x(t; \bar{u})$ and $x(t) = x(t; u)$, then (7.7) yields

$$x(T) - \bar{x}(T) = \int_{t_0}^{t_0 + \varepsilon} S(T, s)B(s)(v - \bar{u}(s)) \, ds. \tag{7.20}$$

If we further define

$$0 \leqslant J(u)^2 - J(\bar{u})^2 = 2(x(T) - \bar{x}(T), \bar{x}(T) - y) + \|x(T) - \bar{x}(T)\|^2$$
$$= I + II,$$

we obtain $\varepsilon^{-1}II \to 0$ by (7.8). Also, from (7.20), it follows that

$$\frac{I}{2\varepsilon} = \left(\frac{1}{\varepsilon} \int_{t_0}^{t_0 + \varepsilon} S(T, s)B(s)(v - \bar{u}(s)) \, ds, \bar{x}(T) - y \right)$$

$$\to (S(T, t_0)B(t_0)(v - \bar{u}(t_0)), \bar{x}(T) - y),$$

so that

$$(v - \bar{u})(t_0), B^*(t_0)S^*(T, t_0)(\bar{x}(T) - y)) \geqslant 0,$$

which implies that (7.19) holds at each Lebesgue point of $\bar{u}$. $\square$

$w(t)$ is a solution in some sense of the following initial-value problem:

$$dw(t)/dt = -A^*(t)w(t), \qquad 0 \le t \le T,$$
$$w(T) = y - x(T; \bar{u}).$$

Definition 7.3.1　*Suppose that the equations*

$$dx(s)/ds = -A^*(s)x(s), \qquad 0 \le s < T$$
$$x(T) = x_0 \tag{7.21}$$

admit a solution $x(s) = S(T, s)^* x_0$ *which vanishes on a set of positive measure. If such a solution is identically equal to zero, i.e., it occurs only when* $x_0 = 0$, *the equation* (7.21) *is said to have a* weak backward uniqueness property.

Example 1　If (7.1) is a parabolic equation and satisfies the assumptions in Section 5.7.1 or the assumptions in Section 5.7.2 with $\{M_k\} = \{k!\}$, then its solution is analytic, so that it has a weak backward uniqueness property.

Example 2　Let V be a closed subspace of $H_m(\Omega)$ containing $C_0^\infty(\Omega)$, $a(t; u, v)$ for each t be a symmetric quadratic form, defined on $V \times V$, which satisfies Gårding's inequality, $A_0(t)$ be an operator in $L^2(\Omega)$ determined by $a(t; u, v)$ and $A_1(t)$ be a differential operator of degree less than m. Assume that the coefficients of $A_0(t)$ and $A_1(t)$ are sufficiently smooth. Under these conditions, (7.21) has a backward uniqueness property. *See* Mizohata [139], Lions and Malgrange [120] and Bardos and Tartar [28].

Theorem 7.3.3　(Bang-Bang Principle)　*Assume that* (7.21) *has a weak backward uniqueness property. If* $B^*(t)$ *is a one-to-one mapping for each* t, *then the optimal control* $\bar{u}$ *satisfies* $\bar{u}(t) \in \partial U$ *for almost all* t.

Proof　On account of (7.19), it is enough to show that $B^*(t)w(t) \ne 0$ for almost all t. If $B^*(t)w(t) = 0$ on a set e of positive measure, then $w(t) = 0$ for each $t \in e$. By assumption, we have $w(t) = 0$ on $0 \le t \le T$. In particular, we are led to $y - x(T; u) = w(T) = 0$, which is a contradiction.　□

Definition 7.3.2　*Let* U *be a convex set. If* $u, v \in U$ *and if* $(u + v)/2 \in \partial U$ *implies that* $u = v$, *then* U *is called* strictly convex.

Corollary 1　*Under the assumption of the above theorem, the optimal control is unique if* U *is strictly convex.*

Proof Let u_1 and u_2 be two optimal controls, and x_1 and x_2 be their corresponding trajectories. Clearly, the control $(u_1 + u_2)/2$ is admissible and its trajectory is $(x_1 + x_2)/2$. Since

$$\inf J(u) \leq \|\tfrac{1}{2}(x_1(T) + x_2(T)) - y\|$$
$$\leq \tfrac{1}{2}(\|x_1(T) - y\| + \|x_2(T) - y\|) = \inf J(u),$$

$(u_1 + u_2)/2$ is an optimal control. Hence, Theorem 7.3.3 implies that $(u_1(t) + u_2(t))/2 \in \partial U$ almost everywhere. Since U is strictly convex, we obtain $u_1(t) = u_2(t)$. $\square$

Corollary 2 *Under the assumption of Theorem 7.3.3, if U is a unit ball, then we have*

$$\bar{u}(t) = \frac{B^*(t)w(t)}{\|B^*(t)w(t)\|} \tag{7.22}$$

almost everywhere.

Proof Obviously, any ball in a Hilbert space is strictly convex. From Corollary 1 and (7.19), the result (7.22) follows immediately. $\square$

7.4 Time optimal control

Let X be a reflexive real Banach space, and $S(t, s)$ $(0 \leq s \leq t < \infty)$ be a fundamental solution of an equation of evolution

$$\mathrm{d}x(t)/\mathrm{d}t = A(t)x(t), \qquad 0 \leq t < \infty. \tag{7.23}$$

We further assume that

$$\text{s-}\lim_{t \to s+0} S^*(t, s) = I \tag{7.24}$$

for each $s \geq 0$. Similarly to the proof of (5.42), we can show that $S^*(t, s)$, if it exists, is a fundamental solution of the adjoint equation $\mathrm{d}x^*(s)/\mathrm{d}s = -A^*(s)x^*(s)$, and, hence, the condition is automatically satisfied. Let U be a bounded convex closed set containing the origin of X as an interior point, and x_0 and x_1 be two different elements of X. Suppose that an admissible control u is a strongly measurable function satisfying $u(t) \in U$ for almost all t. Then,

$$x(t; u) = S(t, 0)x_0 + \int_0^t S(t, s)u(s)\,\mathrm{d}s$$

is the trajectory under u. We assume there exists an admissible control u satisfying

$$x(\tau; u) = x_1 \tag{7.25}$$

for some $\tau > 0$. In this case, we consider the existence and uniqueness of the admissible control which attains x_1 in the shortest time. The lower limit τ_0 of τ for which there exists an admissible control satisfying (7.25) is called the *optimal time* and we ask for the existence and uniqueness of the admissible control satisfying $x(\tau_0; u) = x_1$, that is, of the *time optimal control* u with respect to $\{x_0, x_1\}$. This section is mainly adapted from Fattorini [59, 61].

Theorem 7.4.1 *Under the above assumption, there exists a time optimal control.*

Proof Let $\tau_n \to \tau_0 + 0$, u_n be an admissible control and suppose that $x(\tau_n; u_n) = x_1$. For each n choose a T such that $\tau_n < T$ and put $u_n(t) = 0$ for all $t > \tau_n$. Then, $\{u_n\}$ is bounded in $L^2(0, T; X)$. Hence, if we let $u_n \to \bar{u}$ weakly in $L^2(0, T; X)$ by replacing $\{u_n\}$ by its subsequence, it is easy to show, by virtue of Theorem 2.1, that $\bar{u}$ is an admissible control and that

$$\int_0^t S(t, s)u_n(s)\,\mathrm{d}s \to \int_0^t S(t, s)\bar{u}(s)\,\mathrm{d}s$$

weakly for each $0 \leq t \leq \tau_0$. Therefore, $x(t; u_n) \to x(t; \bar{u})$ weakly on $0 \leq t \leq \tau_0$. The first and third terms on the right-hand side of

$$x_1 = S(\tau_n, 0)x_0 + \int_0^{\tau_0} S(\tau_n, s)u_n(s)\,\mathrm{d}s + \int_{\tau_0}^{\tau_n} S(\tau_n, s)u_n(s)\,\mathrm{d}s \tag{7.26}$$

converge strongly to $S(\tau_0, 0)x_0$ and 0, respectively. We put

$$y_n = \int_0^{\tau_0} S(\tau_0, s)u_n(s)\,\mathrm{d}s,$$

obtaining

$$y_n \to y_0 = \int_0^{\tau_0} S(\tau_0, s)\bar{u}(s)\,\mathrm{d}s$$

weakly as $n \to \infty$. Since $S^*(\tau_n, \tau_0)f \to f$ strongly for each $f \in X^*$ by (7.24), we obtain

$$f\left(\int_0^{\tau_0} S(\tau_n, s)u_n(s)\,\mathrm{d}s\right) = f(S(\tau_n, \tau_0)y_n)$$

$$= (S^*(\tau_n, \tau_0)f)(y_n) \to f(y_0).$$

Therefore, by letting $n \to \infty$ in (7.26), we get $x_1 = x(\tau_0; \bar{u})$. Thus, $\bar{u}$ is a time optimal control. $\square$

7.4.1 The case in which $A(t) \equiv A$ is independent of t

Assuming $A(t)$ is an operator A which is independent of t, we denote the semigroup generated by A by $S(t)$. The trajectory under an admissible control u is given by

$$x(t; u) = S(t)x_0 + \int_0^t S(t-s)u(s)\,ds.$$

For each $t > 0$, define

$$K_t = \left\{ y \in X : y = \int_0^t S(t-s)u(s)\,ds,\ u \in L^\infty(0, t; X) \right\}.$$

Let e be a measurable set in $[0, \infty)$ and set

$$K_t(e) = \left\{ y \in X : y = \int_0^t S(t-s)u(s)\,ds,\ u \in L^\infty(0, t; X) \right.$$

$$\left. \text{and the support of } u \subseteq e \cap [0, t] \right\}.$$

Lemma 7.4.1 K_t *is independent of t.*

Proof Let $t < t'$ and suppose $y = \int_0^t S(t-s)u(s)\,ds \in K_t$. Put $\bar{u}(s) = 0$ for $0 < s < t' - t$ and $\bar{u}(s) = u(s - t' + t)$ for $t' - t < s < t'$. Then we have

$$y = \int_0^{t'} S(t'-s)\bar{u}(s)\,ds \in K_{t'}.$$

Conversely, suppose $y = \int_0^{t'} S(t'-s)u(s)\,ds \in K_{t'}$. By putting

$$v(s) = u(s + t' - t) + \frac{1}{t} S(s) \int_0^{t'-t} S(t'-t-r)u(r)\,dr,$$

we obtain $y = \int_0^t S(t-s)v(s)\,ds \in K_t$. $\square$

Since we have shown that K_t is independent of t, we simply denote it by K. The following is an important lemma proved by Fattorini [59].

Lemma 7.4.2 $K_t(e) = K$ *for almost all* $t \in e$.

Proof Since almost all points of e are points of density in e, we may assume that t is a point of density in e. Since $\lim_{r \to t-0} |[r, t] \cap e|/(t-r) = 1$, we can choose $t_1 < t$ close enough to t so that

$$\frac{|[r, t] \cap e|}{t - r} \geq \frac{2}{3} \quad \text{for} \quad t_1 \leq r < t. \tag{7.27}$$

Let t_2 be the middle point of t_1 and t, and t_{n+1} be the middle point of t_n and t, successively. If $|[t_n, t_{n+1}] \cap e| < (t_{n+1} - t_n)/3$ for some n, then, since $t_{n+1} - t_n = t - t_{n+1} = (t - t_n)/2$, we have $|[t_n, t_{n+1}] \cap e| < (t - t_n)/6$ and $|[t_{n+1}, t] \cap e| \leq t - t_{n+1} = (t - t_n)/2$. Therefore, it may be concluded that $|[t_n, t] \cap e| < 2(t - t_n)/3$, which contradicts (7.27). Hence,

$$|[t_n, t_{n+1}] \cap e| \geq (t_{n+1} - t_n)/3 \tag{7.28}$$

for all n. Clearly, $t_1 < t_2 < \cdots \to t$, and

$$\frac{t_{n+1} - t_n}{t_{n+2} - t_{n+1}} = 2. \tag{7.29}$$

Suppose $y = \int_0^t S(t - r)u(r)\, dr \in K_t$. We define $w(s)$ by

$$w(s) = \frac{1}{|[t_{n+1}, t_{n+2}] \cap e|} \int_{t_n}^{t_{n+1}} S(s - r)u(r)\, dr$$

if $s \in [t_{n+1}, t_{n+2}] \cap e$ for some n, and put $w(s) = 0$ for other s belonging to $[0, t]$. Then, we have

$$y = \sum_{n=1}^{\infty} \int_{t_n}^{t_{n+1}} S(t - r)u(r)\, dr = \sum_{n=1}^{\infty} \int_{t_{n+1}}^{t_{n+2}} S(t - s)w(s)\, ds$$

$$= \int_0^t S(t - s)w(s)\, ds.$$

The support of w is contained in $[0, t] \cap e$, and, from (7.28) and (7.29), it follows that

$$\|w(s)\| \leq \frac{t_{n+1} - t_n}{|[t_{n+1}, t_{n+2}] \cap e|} \operatorname*{ess\,sup}_{0 \leq r \leq t} \|S(s - r)u(r)\|$$

$$\leq 6 \operatorname*{ess\,sup}_{0 \leq r \leq t} \|S(s - r)u(r)\|$$

for $s \in [t_{n+1}, t_{n+2}] \cap e$. Thus $w \in L^\infty(0, t; X)$, and, hence, $y \in K_t(e)$. $\square$

Lemma 7.4.3 *Let $\bar{u}$ be a time optimal control with respect to $\{x_0, x_1\}$ and τ_0 be its optimal time. For each $\tau_1 \in (0, \tau_0)$, u is a time optimal control with respect to $\{x_0, x(\tau_1; \bar{u})\}$ and τ_1 is its optimal time.*

Proof Suppose there exists an admissible control w and $x(\tau_2; w) = x(\tau_1; \bar{u})$ at time τ_2 prior to τ_1. If we put

$$v(t) = \begin{cases} w(t), & 0 \leqslant t \leqslant \tau_2, \\ \bar{u}(t + \tau_1 - \tau_2), & \tau_2 < t \leqslant \tau_0 - \tau_1 + \tau_2, \end{cases}$$

v is an admissible control. We obtain

$$x(\tau_0 - \tau_1 + \tau_2; v) = S(\tau_0 - \tau_1 + \tau_2)x_0 + \int_0^{\tau_0 - \tau_1 + \tau_2} S(\tau_0 - \tau_1 + \tau_2 - s)v(s)\,\mathrm{d}s$$

$$= S(\tau_0 - \tau_1 + \tau_2)x_0 + \int_0^{\tau_2} S(\tau_0 - \tau_1 + \tau_2 - s)w(s)\,\mathrm{d}s$$

$$+ \int_{\tau_2}^{\tau_0 - \tau_1 + \tau_2} S(\tau_0 - \tau_1 + \tau_2 - s)\bar{u}(s + \tau_1 - \tau_2)\,\mathrm{d}s$$

$$= S(\tau_0 - \tau_1)x(\tau_2; w) + \int_{\tau_1}^{\tau_0} S(\tau_0 - s)\bar{u}(s)\,\mathrm{d}s$$

$$= S(\tau_0 - \tau_1)x(\tau_1; \bar{u}) + \int_{\tau_1}^{\tau_0} S(\tau_0 - s)\bar{u}(s)\,\mathrm{d}s$$

$$= x(\tau_0; \bar{u}) = x_1,$$

which contradicts the fact that τ_0 is the optimal time with respect to $\{x_0, x_1\}$. $\square$

Lemma 7.4.4 *Let τ_0 be the optimal time with respect to $\{x_0, x_1\}$. If u is an admissible control, dist $(u(t), \partial U) \geqslant \varepsilon > 0$ for almost all t, and if $x(\tau; u) = x_1$, then $\tau_0 < \tau$.*

Proof Let $0 < \sigma < \tau$. If we set

$$v(t) = u(\tau - \sigma + t) + \frac{1}{\sigma} S(t)\left(S(\tau - \sigma)x_0 - x_0 + \int_0^{\tau - \sigma} S(\tau - \sigma - r)u(r)\,\mathrm{d}r \right)$$

for each $t \in [0, \sigma]$, then $x(\sigma; v) = x_1$. If σ is sufficiently close to τ, then $v(t) \in U$ almost everywhere. Hence, τ is not the optimal time. $\square$

Theorem 7.4.2 (Bang-Bang Principle) *If $\bar{u}$ is a time optimal control with respect to $\{x_0, x_1\}$ and τ_0 is its optimal time, then $\bar{u}(t) \in \partial U$ for almost all $t \in [0, \tau_0]$.*

Proof Suppose $e \subset [0, \tau_0]$, $|e| > 0$, $\varepsilon > 0$ and dist $(\bar{u}(t), \partial U) \geqslant \varepsilon$ for all $t \in e$. By virtue of Lemma 7.4.2, there exists an $s \in e$ such that

$K_s(e) = K = K_s$. Therefore, we can find a function $w \in L^\infty(0, \tau_0; X)$ such that the support of w is contained in $[0, s] \cap e$ and that

$$\int_0^s S(s - \sigma)w(\sigma) \, d\sigma = \int_0^s S(s - \sigma)\bar{u}(\sigma) \, d\sigma. \tag{7.30}$$

Let $0 < \delta < 1$ and put

$$v(t) = (1 - \delta)\bar{u}(t) + \delta w(t)$$

for each $t \in (0, \tau_0)$. We claim that, for sufficiently small δ,

$$\text{dist}\,(v(t), \partial U) \geq \varepsilon_1 > 0 \tag{7.31}$$

for almost all $t \in [0, \tau_0]$. Let $C = \text{ess sup}\, \|w(t) - \bar{u}(t)\|$. If $t \in e$, we have

$$\|v(t) - z\| = \|\bar{u}(t) - z + \delta(w(t) - \bar{u}(t))\|$$
$$\geq \|\bar{u}(t) - z\| - \delta \|w(t) - \bar{u}(t)\| \geq \varepsilon - C\delta$$

for each $z \in \partial U$, so that $\text{dist}\,(v(t), \partial U) \geq \varepsilon - C\delta$. Next, choose $\rho > 0$ such that U contains a ball centred at the origin with ρ as its radius. Let $t \notin e$ and z be an arbitrary point satisfying $\|z - v(t)\| < \delta\rho$. If we put $y = (z - v(t))/\delta$, then $y \in U$ since $\|y\| < \rho$. Therefore, $z = \delta y + v(t) = \delta y + (1 - \delta)\bar{u}(t) \in U$. Hence, $\text{dist}\,(v(t), \delta U) \geq \delta\rho$. If we choose δ in such a way that $C\delta < \varepsilon$, the inequality (7.31) holds with $\varepsilon_1 = \min\,(\varepsilon - C\delta, \delta\rho)$. We have

$$x(s; v) = S(s)x_0 + \int_0^s S(s - \sigma)v(\sigma) \, d\sigma$$

$$= S(s)x_0 + (1 - \delta) \int_0^s S(s - \sigma)\bar{u}(\sigma) \, d\sigma + \delta \int_0^s S(s - \sigma)w(\sigma) \, d\sigma,$$

which, by (7.30), becomes

$$= S(s)x_0 + \int_0^s S(s - \sigma)\bar{u}(\sigma) \, d\sigma = x(s; \bar{u}).$$

From this and Lemma 7.4.3, it follows that $\bar{u}$ is the time optimal control with respect to $\{x_0; x(s; v)\}$ and its optimal time is s. On the other hand, by (7.31) and Lemma 7.4.4, there exists an admissible control which makes it possible to reach $x(s; v)$ at time earlier than s, which is a contradiction. $\square$

Corollary   *If U is strictly convex, the time optimal control is unique.*

The proof is analogous to that of Corollary 1 of Theorem 7.3.3.

7.4.2 The case when a bounded inverse of $S(t, s)$ exists

We assume that $S(t, s)$ has a bounded inverse for each t, s in $0 \leqslant s \leqslant t \leqslant T$ and that $S(t, s)^{-1}$ is also strongly continuous in t and s. For example, we consider the case in which (7.23) can be solved for the past as well as for the future.

Theorem 7.4.3 (Bang-Bang Principle) *If $\bar{u}$ is a time optimal control with respect to $\{x_0, x_1\}$ and τ_0 is its optimal time, then $\bar{u}(t) \in \partial U$ for almost all $t \in [0, \tau_0]$.*

Proof Suppose to the contrary that the conclusion of the theorem does not hold. Then, there exists an $e \subset [0, \tau]$ with $|e| > 0$ and $0 < \tau < \tau_0$ and a $\delta > 0$ such that $\text{dist}\,(\bar{u}(t), \partial U) \geqslant \delta > 0$ for each $t \in e$. We write $\bar{x}(t) = x(t; \bar{u})$. Let χ be the defining function of e and $\tau \leqslant \sigma < \tau_0$. Putting

$$v(t) = |e|^{-1} \chi(t) S(\sigma, t)^{-1} (\bar{x}(\tau_0) - \bar{x}(\sigma)),$$

we have

$$x_1 = \bar{x}(\tau_0) = \bar{x}(\sigma) + (\bar{x}(\tau_0) - \bar{x}(\sigma))$$

$$= S(\sigma, 0) x_0 + \int_0^\sigma S(\sigma, s)(\bar{u}(s) + v(s))\,\mathrm{d}s.$$

If σ is sufficiently close to $\tau_0, \|v(t)\| \leqslant \delta$ for each t, so that $\bar{u} + v$ is an admissible control, which contradicts the fact that τ_0 is the optimal time. $\square$

Theorem 7.4.4 (Maximum Principle) *Let $\bar{u}$ and τ_0 be the same as in the preceding theorem. Then, there exists a non-zero $f \in X^*$ such that*

$$\max_{v \in U} (S^*(\tau_0, t)f)(v) = (S^*(\tau_0, t)f)(\bar{u}(t)). \tag{7.32}$$

for almost all $t \in [0, \tau_0]$.

Proof Define

$$\Omega = \left\{ y : y = \int_0^{\tau_0} S(\tau_0, s)v(s)\,\mathrm{d}s, v \text{ is an admissible control} \right\}$$

$$\hat{\Omega} = \left\{ y : y = S(\tau_0, 0)x_0 + \int_0^{\tau_0} S(\tau_0, s)v(s)\,\mathrm{d}s\,(= x(\tau_0; v)), \right. \tag{7.33}$$

$$\left. v \text{ is an admissible control} \right\}.$$

Let $U \supset \{y : \|y\| < \rho\}$ and $\|S(t, s)^{-1}\| \leqslant M$. If y is an arbitrary element of X satisfying $\|y\| < \tau_0 \rho / M$, then $v(t) = \tau_0^{-1} S(\tau_0, t)^{-1} y$ satisfies $\|v(t)\| < \rho$, so that

v is an admissible control. Furthermore, since

$$y = \int_0^{\tau_0} S(\tau_0, s)v(s)\,ds \in \Omega,$$

0 is an interior point of Ω. Therefore, $\hat{\Omega}$ also contains an interior point. Clearly, $x_1 \in \hat{\Omega}$, and next we want to show that $x_1 \in \partial\hat{\Omega}$. If $S(\tau_0, 0)x_0 = x_1$, then $v \equiv 0$ becomes an admissible control, which contradicts Theorem 7.4.3. Therefore, we must have $z = x_1 - S(\tau_0, 0)x_0 \neq 0$. Let $\varepsilon > 0$ and assume that $x_1 + \varepsilon z \in \hat{\Omega}$. Then, there exists an admissible control v such that

$$x_1 + \varepsilon z = S(\tau_0, 0)x_0 + \int_0^{\tau_0} S(\tau_0, s)v(s)\,ds,$$

which implies that

$$x_1 = S(\tau_0, 0)x_0 + \int_0^{\tau_0} S(\tau_0, s)\frac{v(s)}{1+\varepsilon}\,ds,$$

so that $(1+\varepsilon)^{-1}v(s)$ is a time optimal control, and, furthermore, as in the proof of (7.31) for $t \in e$, we obtain dist $((1+\varepsilon)^{-1}v(s), \partial U) \geq \varepsilon\rho(1+\varepsilon)^{-1}$, which contradicts Theorem 7.4.3. Thus, we have obtained $x_1 \in \partial\hat{\Omega}$. Since $\hat{\Omega}$ is clearly a convex set, on account of Theorem 1.1.9, there exists a non-zero $f \in X^*$ such that $f(y) \leq f(x)$ for all $y \in \hat{\Omega}$. By rewriting this, we have

$$f\left(\int_0^{\tau_0} S(\tau_0, s)v(s)\,ds\right) \leq f\left(\int_0^{\tau_0} S(\tau_0, s)\bar{u}(s)\,ds\right)$$

for all admissible controls v. Similarly to the proof of (7.17) or (7.19), let t be a Lebesgue point of u and v be an element of U and $t < t + \varepsilon < \tau_0$; if we take $v(s) = v$ for $t < s < t + \varepsilon$ and $v(s) = u(s)$ elsewhere, then it can be shown that (7.32) holds for each Lebesgue point of u. $\square$

Corollary 1   *If U is strictly convex, then the time optimal control is unique.*

The proof is similar to that of Corollary 1 of Theorem 7.3.3.

Corollary 2   *If X is a Hilbert space and U is the unit ball, then*

$$\bar{u}(t) = \frac{S^*(\tau_0, f)f}{\|S^*(\tau_0, t)f\|} \tag{7.34}$$

for almost all $t \in [0, \tau_0]$.

Proof   Since $S^*(\tau_0, t)$ has an inverse by assumption, $S^*(\tau_0, t)f \neq 0$. From this and (7.32), the result (7.34) follows immediately. $\square$

Remark 7.4.1 If we put $w(t) = S^*(\tau_0, t)f$, then w is a solution in some sense of

$$dw(t)/dt = -A^*(t)w(t), \qquad w(\tau_0) = f.$$

7.4.3. Problem of reaching a target set

So far we have considered the case in which we enforce arrival at one point x_1; we will now consider the problem of reaching a *target set* W in the earliest time. Assume that W is a convex closed set which does not contain x_0 and that its interior int W is not empty. Assume also that for some $\tau > 0$ there exists an admissible control u satisfying $x(\tau; u) \in W$. If W has no interior points, the case may arise in which all controls, not necessarily admissible, are unable to transfer the system to W, so that int $W \neq \varnothing$ is considered as a natural assumption. On this topic, the reader is referred to Fattorini [60], Sakawa [151] and so on. The *optimal time* is defined by the lower limit τ_0 of τ, such that $x(\tau; u) \in W$ for some admissible control u. The existence of an admissible control satisfying $x(\tau_0, u) \in W$, that is, of a *time optimal control* u with respect to $\{x_0, W\}$, can be proved similarly to Theorem 7.4.1; it is seen by noting that W is weakly closed by Corollary 2 of Theorem 1.1.8.

Theorem 7.4.5 (Friedman [64]) *If $\bar{u}$ is a time optimal control with respect to $\{x_0, W\}$, then there exists a non-zero $f \in X^*$ such that*

$$\max_{v \in U} (S^*(\tau_0, t)f)(v) = (S^*(\tau_0, t)f)(\bar{u}(t)) \tag{7.35}$$

for almost all t.

Proof Put $\bar{x}(t) = x(t; \bar{u})$. Let $\hat{\Omega}$ be a set defined by (7.33). Suppose $y \in (\text{int } W) \cap \hat{\Omega}$. Then, there exists an admissible control v such that $y = x(\tau_0; v) \in \text{int } W$. Since $x(t; v)$ is continuous in t, we can find a $\tau_1 < \tau_0$ satisfying $x(\tau_1; v) \in W$, which contradicts the fact that τ_0 is the optimal time. Hence, $(\text{int } W) \cap \hat{\Omega}$ is empty. Consequently, by the next lemma, there exists a non-zero $f \in X^*$ such that the inequality

$$\sup_{y \in \hat{\Omega}} f(y) \leqslant \inf_{y \in \text{int } W} f(y) \tag{7.36}$$

holds. $\square$

Lemma 7.4.5 *Let C and K be two convex subsets of X satisfying $C \cap K = \varnothing$ and int $C \neq \varnothing$. Then, there exists a non-zero $f \in X^*$ such that $f(x) \leqslant f(y)$ for every $x \in K$ and every $y \in C$.*

For the proof of <u>the lemma</u>, see Dunford and Schwartz [4], p. 417. Since the result $W = \text{int } \overline{W}$ is seen easily from Theorem 1.1.8, we obtain

$$\sup_{y \in \Omega} f(y) \leqslant \inf_{y \in W} f(y) \leqslant f(x(\tau_0)) \tag{7.37}$$

by (7.36). The rest of the proof is analogous to that of Theorem 7.4.4.

Remark 7.4.2 In the proof above, we have not used the condition that int U is not empty.

From the proof of the corollary of Theorem 7.3.3, we obtain the following result.

Corollary *If the adjoint equation of (7.23) has a unique weak solution for the past, then $u(t) \in \partial U$ for almost all $t \in [0, \tau_0]$. Therefore, if U is strictly convex, the time optimal control with respect to $\{x_0, W\}$ is unique. In particular, if X is a Hilbert space and U is the unit ball, then*

$$\bar{u}(t) = \frac{S^*(\tau_0, t)f}{\|S^*(\tau_0, t)f\|}$$

for almost all $t \in [0, \tau_0]$.

The material used in Chapters 6 and 7 was mainly adapted from Friedman [62–64].

Bibliography

1 General monographs

1 Agmon, S. *Lectures on Elliptic Boundary Value Problems*, D. Van Nostrand Company, Princeton, 1965.

2 Brezis, H. *Opérateurs maximaux monotones et semi-groupes de contractions dans les espaces de Hilbert*, North-Holland Publishing Company, Amsterdam, London, 1973.

3 Carroll, R. W. *Abstract Methods in Partial Differential Equations*, Harper & Row, New York, Evanston, London, 1969.

4 Dunford, N. & Schwartz, J. T. *Linear Operators*, Interscience Publishers, New York, Part I, 1966; Part II, 1963.

5 Friedman, A. *Generalized Functions and Partial Differential Equations*, Prentice-Hall, Engelwood Cliffs, N.J., 1963.

6 Friedman, A. *Partial Differential Equations*, Holt, Reinhart & Winston, New York, 1969.

7 Hille, E. & Phillips, R. S. *Functional Analysis and Semigroups*, American Mathematical Society Colloquium Publication, Vol. 31, Providence, R.I., 1957.

8 Kato, T. *Perturbation Theory for Linear Operators*, Springer-Verlag, Berlin, Heidelberg, New York, 1966.

9 Kreĭn, S. G. *Linear Differential Equations in Banach Space*, American Mathematical Society Translations of Mathematical Monographs, Vol. 29, Providence, R.I., 1971.

10 Ladas, G. E. & Lakshmikantham, V. *Differential Equations in Abstract Spaces*, Academic Press. New York, 1972.

11 Lions, J. L. *Équations différentielles opérationnelles et problèmes aux limites*, Springer-Verlag, Berlin, Göttingen, Heidelberg, 1961.

12 Lions, J. L. *Contrôle optimal de systèmes gouvernés par des équations aux dérivées partielles*, Dunod, Gauthier-Villars, Paris, 1968.

13 Lions, J. L. *Quelques méthodes de résolution des problèmes aux limites non linéaires*, Dunod, Gauthier-Villars, Paris, 1969.

14 Lions, J. L. *Perturbations singulières dans les problèmes aux limites et en contrôle optimal*, Springer-Verlag, Berlin, Heidelberg, New York, 1973.

15 Lions, J. L. & Magenes, E. *Problèmes aux limites non homogènes et applications*, Dunod, Paris, Vols 1, 2, 1968; Vol. 3, 1970.

16 Mandelbrojt, S. *Séries de Fourier et classes quasi-analytiques de fonctions*, Gauthier-Villars, Paris, 1935.

17 Mizohata, S. *The Theory of Partial Differential Equations*, Cambridge University Press, Cambridge, 1973.

18 Yosida, K. *Functional Analysis*, 2nd ed., Springer-Verlag, Berlin, Heidelberg, New York, 1968.

2 Literature

19 Agmon, S. On the eigenfunctions and on the eigenvalues of general elliptic boundary value problems, *Comm. Pure Appl. Math.*, **15,** 119–147, 1962.

20 Agmon, S. & Nirenberg, L. Properties of solutions of ordinary differential equations in Banach space, *Comm. Pure Appl. Math.*, **16,** 121–239, 1963.

21 Agmon, S. & Nirenberg, L. Lower bounds and uniqueness theorems for solutions of differential equations in a Hilbert space, *Comm. Pure Appl. Math.*, **20,** 207–229, 1967.

22 Agmon, S., Douglis, A. & Nirenberg, L. Estimates near the boundary for solutions of elliptic partial differential equations satisfying general boundary conditions I, *Comm. Pure Appl. Math.*, **12,** 623–727, 1959; II, **17,** 35–92, 1964.

23 Aizawa, S. A semigroup treatment of the Hamilton–Jacobi equation in one space variable, *Hiroshima Math. J.*, **3,** 367–386, 1973.

24 Asplund, E. Averaged norms, *Israel J. Math.*, **5,** 227–233, 1967.

25 Bardos, C. A regularity theorem for parabolic equations, *J. Func. Anal.*, **7,** 311–322, 1971.

26 Bardos, C. & Brezis, H. Sur une classe de problèmes d'évolution non linéaires, *J. Differential Equations*, **6,** 345–394, 1969.

27 Bardos, C. & Cooper, J. M. A nonlinear wave equation in a time dependent domain, *J. Math. Anal. Appl.*, **42,** 29–40, 1973.

28 Bardos, C. & Tartar, L. Sur l'unicité rétrogarde des équations d'évolution, *C.R. Acad. Sci., Paris, Sér. A-B*, **273,** A1239–A1241, 1971.

29 Bobisud, L. E. & Calvert, J. Energy bounds and virial theorems for abstract wave equations, *Pacific J. Math.*, **47,** 27–37, 1973.

30 Brezis, H. Équations et inéquations non linéaires dans les espaces vectoriels en dualité, *Ann. Inst. Fourier,* **18,** 115–175, 1968.

31 Brezis, H. On some degenerate nonlinear parabolic equations, in *Nonlinear Functional Analysis, Proceedings in Pure Mathematics,* Vol. 18 (Part 1) (F. Browder, ed.), American Mathematical Society, Providence, R.I. 1970, pp. 28–38.

32 Brezis, H. Perturbations non linéaires d'opérateurs maximaux monotones, *C.R. Acad. Sci. Sér. A,* **269,** 566–569, 1969.

33 Brezis, H. Propriétés régularisantes de certains semi groupes non linéaires, *Israel J. Math.,* **9,** 513–534, 1971.

34 Brezis, H. Problèmes unilatéraux, *J. Math. Pures Appl.,* **51,** 1–168, 1972.

35 Brezis, H. & Stampacchia, G. Sur la régularité de la solution d'inéquations elliptiques, *Bull. Soc. Math. France,* **96,** 153–180, 1968.

36 Browder, F. E. On the spectral theory of elliptic differential operators I, *Math. Ann.,* **142,** 22–130, 1961.

37 Browder, F. E. On nonlinear wave equations, *Math. Z.,* **80,** 249–264, 1962.

38 Browder, F. E. Nonlinear monotone operators and convex sets in Banach spaces, *Bull. Amer. Math. Soc.,* **71,** 780–785, 1965.

39 Browder, F. E. Nonlinear maximal monotone operators in Banach space, *Math. Ann.,* **175,** 89–113, 1968.

40 Browder, F. E. Pseudo-monotone operators and the direct method of the calculus of variations, *Archiv Rat. Mech. Anal.,* **38,** 268–277, 1970.

41 Browder, F. E. Recent results in nonlinear functional analysis and applications to partial differential equations, *Actes, Congrès intern. Math.,* **2,** 821–829, 1970.

42 Burak, T. On semigroups generated by restrictions of elliptic operators to invariant subspaces, *Israel J. Math.,* **12,** 79–93, 1972.

43 Burak, T. Two point problems and analyticity of solutions of abstract parabolic equations, *Israel J. Math.,* **16,** 404–417, 1973.

44 Burak, T. Regularity properties of solutions of some abstract parabolic equations, *Israel J. Math.,* **16,** 418–445, 1973.

45 Calderon, A. P. Commutators of singular integral operators, *Proc. Nat. Acad. Sci. U.S.A.,* **53,** 1092–1099, 1965.

46 Calderon, A. P. & Zygmund, A. Singular integral operators and differential equations, *Amer. J. Math.,* **79,** 901–921, 1957.

47 Calvert, B. Nonlinear evolution equations in Banach lattices, *Bull. Amer. Math. Soc.,* **76,** 845–850, 1970.

48 Carroll, R. W. and Cooper, J. M. Remarks on some variable domain problems in abstract evolution equations, *Math. Ann.,* **188,** 143–164, 1970.

49 Carroll, R. W. & Mazumdar, T. Solutions of some possibly noncoercive evolution problems with regular data, *Applicable Anal*, **1**, 381–395, 1972.

50 Carroll, R. W. & State, E. Existence theorems for some abstract variable domain hyperbolic problems, *Canad. J. Math.* **23,** 611–626, 1971.

51 Cooper, J. M. Evolution equations in Banach space with variable domain, *J. Math. Anal. Appl.*, **36,** 151–171, 1971.

52 Cooper, J. M. Two point problems for abstract evolution equations, *J. Differential Equations*, **9,** 453–495, 1971.

53 Crandall, M. G. The semigroup approach to first order quasi-linear equations in several space variables, *Israel J. Math.*, **12,** 108–132, 1972.

54 Crandall, M. G. & Liggett, T. M. Generation of semigroups of nonlinear transformations on general Banach spaces, *Amer. J. Math.*, **93,** 265–298, 1971.

55 Crandall, M. G. & Liggett, T. M. A theorem and a counter example in the theory of semigroups on nonlinear transformations, *Trans. Amer. Math. Soc.*, **160,** 263–278, 1971.

56 Crandall, M. G. & Pazy, A. Nonlinear equations in Banach spaces, *Israel J. Math.*, **11,** 57–94, 1972.

57 Daleckiĭ, Ju. L. On a problem in fractional powers of self-adjoint operators, *Voronez. Gos. Univ. Trudy Sem. Funkcional Anal.*, No. 6, 44–48, 1958 (in Russian).

58 Duffin, R. J. Equipartition of energy in wave motion, *J. Math. Anal. Appl.*, **32,** 386–391, 1970.

59 Fattorini, H. O. Time optimal control of solutions of operational differential equations, *J. SIAM Control, Ser. A*, **2,** 54–59, 1964.

60 Fattorini, H. O. On complete controllability of linear systems, *J. Differential Equations*, **3,** 391–402, 1967.

61 Fattorini, H. O. An observation on a paper of A. Friedman, *J. Math Anal. Appl.*, **22,** 382–384, 1968.

62 Friedman, A. Optimal control in Banach spaces, *J. Math. Anal. Appl.*, **18,** 35–55, 1967.

63 Friedman, A. Optimal control for parabolic equations, *J. Math. Anal. Appl.*, **18,** 479–491, 1967.

64 Friedman, A. Optimal control in Banach space with fixed end point, *J. Math. Anal. Appl.*, **24,** 161–181, 1968.

65 Friedman, A. & Schuss, Z. Degenerate evolution equations in Hilbert space, *Trans. Amer. Math. Soc.*, **161,** 401–427, 1971.

66 Fujie, Y. & Tanabe, H. On some parabolic equations of evolution in Hilbert space, *Osaka J. Math.*, **10,** 115–130, 1973.

67 Glassy, R. T. On the asymptotic behavior of nonlinear wave equations, *Trans. Amer. Math. Soc.*, **182**, 187–200, 1973.
68 Goldstein, J. A. Time dependent hyperbolic equations, *J. Func.
   Anal.*, **4**, 50–70, 1969.
69 Goldstein, J. A. An asymptotic property of solutions of wave equations, *Proc. Amer. Math. Soc.*, **23**, 359–363, 1969; II, *J. Math. Anal.
   Appl.*, **32**, 392–399, 1970.
70 Goldstein, J. A. On the growth of solutions of inhomogeneous
   abstract wave equations, *J. Math. Anal. Appl.*, **37**, 650–654, 1972.
71 Heinz, E. Beitrage zur Störungstheorie der Spektralzerlegung, *Math.
   Ann.*, **123**, 415–438, 1951.
72 Heinz, E. & von Wahl, W. Zu einem Satz von F. E. Browder über
   nichtlineare Wellengleichungen, *Math. Z.*, **141**, 33–45, 1975.
73 Inoue, A. Sur $\Box u + u^3 = f$ dans un domaine noncylindrique, *J. Math.
   Anal. Appl.*, **46**, 777–819, 1974.
74 Jörgens, K. Das Anfangswertproblem im Großen für eine Klasse
   nichtlinearer Wellengleichungen, *Math. Z.*, **77**, 295–308, 1961.
75 Jörgens, K. Über die nichtlinearer Wellengleichungen der
   mathematischen Physik, *Math. Ann.*, **138**, 179–202, 1959.
76 Kato, T. Integration of the equation of evolution in a Banach space,
   J. Math. Soc. Japan, **5**, 208–234, 1953.
77 Kato, T. On linear differential equations in Banach spaces, *Comm.
   Pure Appl. Math.*, **9**, 479–486, 1956.
78 Kato, T. Note on fractional powers of linear operators, *Proc. Japan
   Acad.*, **36**, 94–96, 1960.
79 Kato, T. Abstract evolution equations of parabolic type in Banach
   and Hilbert spaces, *Nagoya Math. J.*, **5**, 93–125, 1961.
80 Kato, T. Fractional powers of dissipative operators, *J. Math. Soc.
   Japan*, **13**, 246–274, 1961.
81 Kato, T. Fractional powers of dissipative operators, II, *J. Math. Soc.
   Japan*, **14**, 242–248, 1962.
82 Kato, T. A generalization of the Heinz inequality, *Proc. Japan
   Acad.*, **37**, 305–308, 1961.
83 Kato, T. Nonlinear evolution equations in Banach spaces, *Proc.
   Symp. Appl. Math.*, **17**, 50–67, 1964.
84 Kato, T. Nonlinear semigroups and evolution equations, *J. Math.
   Soc. Japan*, **19**, 508–520, 1967.
85 Kato, T. Demicontinuity, hemicontinuity and monotonicity, *Bull.
   Amer. Math. Soc.*, **70**, 548–550, 1964; **73**, 886–889, 1967.
86 Kato, T. Accretive operators and nonlinear evolution equations in
   Banach spaces, in *Proceedings of the Symposium on Nonlinear Functional Analysis, Chicago,* American Mathematical Society, Providence, R.I., 1968, pp. 138–161.

87 Kato, T. Linear evolution equations of "hyperbolic" type, *J. Fac. Sci. Univ. Tokyo, Sec. I*, **17**, 241–258, 1970.

88 Kato, T. Linear evolution equations of "hyperbolic" type, II, *J. Math. Soc. Japan*, **25**, 648–666, 1973.

89 Kato, T. The Cauchy problem for quasi-linear symmetric hyperbolic systems, *Archiv Rat. Mech. Anal.*, **58**, 181–205, 1975.

90 Kato, T. & Tanabe, H. On the abstract evolution equation, *Osaka Math. J.*, **14**, 107–133, 1962.

91 Kato, T. & Tanabe, H. On the analyticity of solution of evolution equations, *Osaka J. Math.*, **4**, 1–4, 1967.

92 Kenmochi, N. Existence theorems for certain nonlinear equations, *Hiroshima Math. J.*, **1**, 435–443, 1971.

93 Kenmochi, N. Nonlinear operators of monotone type in reflexive Banach spaces and nonlinear perturbations, *Hiroshima Math. J.*, **4**, 229–263, 1974.

94 Kielhöfer, H. Halbgruppen und semilineare Anfangs-Randwertprobleme, *Manuscripta Math.*, **12**, 121–152, 1974.

95 Komatsu, H. Abstract analyticity in time and unique continuation property of solutions of a parabolic equation, *J. Fac. Sci. Univ. Tokyo, Sec. I*, **9**, 1–11, 1961.

96 Komura, Y. Nonlinear semigroups in Hilbert space, *J. Math. Soc. Japan*, **19**, 493–507, 1967.

97 Komura, Y. On nonlinear semigroups, *Sûgaku*, **25**, 148–160, 1973 (in Japanese).

98 Konishi, Y. Nonlinear semigroups in Banach lattices, *Proc. Japan Acad.*, **47**, 24–28, 1971.

99 Konishi, Y. A remark on perturbation of *m*-accretive operators in Banach space, *Proc. Japan Acad.*, **47**, 452–455, 1971.

100 Konishi, Y. A remark on semi-groups of local Lipschitzians in Banach space, *Proc. Japan Acad.*, **47**, 970–973, 1971.

101 Konishi, Y. Une méthode de résolution d'une équation d'évolution non linéaire dégénéré, *J. Fac. Sci. Univ. Tokyo, Ser. IA*, **19**, 353–361, 1972.

102 Konishi, Y. Sur un système dégénéré des équations paraboliques semi-linéaire avec les conditions aux limite non linéaire, *J. Fac. Sci. Univ. Tokyo, Ser. IA*, **19**, 353–361, 1972.

103 Konishi, Y. Some examples of nonlinear semigroups in Banach lattices, *J. Fac. Sci. Univ. Tokyo, Ser. IA*, **18**, 537–543, 1972.

104 Konishi, Y. On the uniform convergence of a finite difference scheme for a nonlinear heat equation, *Proc. Japan Acad.*, **48**, 62–66, 1972.

105 Konishi, Y. Une remarque sur la perturbation d'opérateurs *m*-accrétifs dans un espace de Banach, *Proc. Japan Acad.*, **48**, 157–160, 1972.

106 Konishi, Y. Sur la compacité des semi-groupes non linéaires dans les espaces de Hilbert, *Proc. Japan Acad.*, **48,** 278–280, 1972.

107 Konishi, Y. On $u_t = u_{xx} - F(u_x)$ and the differentiability of the non-linear semigroup associated with it, *Proc. Japan Acad.*, **48,** 281–286, 1972.

108 Konishi, Y. A remark on fluid flows through porous media, *Proc. Japan Acad.*, **49,** 20–23, 1973.

109 Konishi, Y. Semi-linear Poisson's equations, *Proc. Japan Acad.*, **49,** 100–105, 1973.

110 Konishi, Y. Compacité des résolvantes des opérateurs maximaux cycliquement monotones, *Proc. Japan Acad.*, **49,** 303–305, 1973.

111 Kreĭn, S. G. & Laptev, G. I. An abstract scheme for the examination of parabolic problems in noncylindrical regions, *Differencial'nye Uravnenija*, **5,** 1158–1169, 1969, (in Russian).

112 Kurtz, T. G. Convergence of sequences of semigroups of nonlinear operators with an application to gas kinetics, *Trans. Amer. Math. Soc.*, **186,** 259–272, 1973.

113 Lagnes, J. On equations of evolution and parabolic equation of higher order in t, *J. Math. Anal. Appl.*, **32,** 15–37, 1970.

114 Lions, J. L. Sur les semi-groupes distributions, *Portugal Math.*, **19,** 141–164, 1960.

115 Lions, J. L. Espaces d'interpolation et domaines de puissances fractionaires d'opérateurs, *J. Math. Soc. Japan*, **14,** 233–241, 1962.

116 Lions, J. L. Remarks on evolution inequalities, *J. Math. Soc. Japan*, **18,** 331–342, 1966.

117 Lions, J. L. Inéquations variationelles d'évolution, *Actes, Congrès intern. Math.*, **2,** 841–851, 1970.

118 Lions, J. L. & Magenes, E. Espaces de fonctions et distributions du type de Gevrey et problèmes aux limites paraboliques, *Ann. Mat. Pura Appl.*, **68,** 341–418, 1965.

119 Lions, J. L. & Magenes, E. Espaces du type de Gevrey et problèmes aux limites pour diverses classes d'équations d'évolution, *Ann. Mat. Pura Appl.*, **72,** 343–394, 1966.

120 Lions, J. L. & Malgrange, B. Sur l'unicité rétrograde, *Math. Scand.*, **8,** 277–286, 1960.

121 Lions, J. L. & Stampacchia, G. Variational inequalities, *Comm. Pure Appl. Math.*, **20,** 493–519, 1967.

122 Lions, J. L. & Strauss, W. A. Some non-linear evolution equations, *Bull. Soc. Math. France*, **93,** 43–96, 1965.

123 Lumer, G. & Phillips, R. S. Dissipative operators in a Banach space, *Pacific J. Math.*, **11,** 679–698, 1961.

124 Maruo, K. Integral equation associated with some non-linear evolution equation, *J. Math. Soc. Japan*, **26,** 433–439, 1974.

125 Maruo, K. & Yamada, N. A remark on integral equation in a Banach space, *Proc. Japan Acad.*, **49,** 13–16, 1973.

126 Masuda, K. On the holomorphic evolution operators, *J. Math. Anal. Appl.*, **39,** 706–711, 1972.

127 Matsuzawa, S. Sur une classes d'équations paraboliques dégénérées, *Ann. Sci. École Norm. Sup.* (4), **4,** 1–19, 1971.

128 Matsuzawa, S. Sur les équations $-d^2u/dt^2 + t^\alpha \Lambda u = f$, $\alpha \geqslant 0$, *Proc. Japan Acad.*, **46,** 609–613, 1970.

129 Medeiros, L. A. Non-linear wave equations in domains with variable boundary, *Archiv Rat. Mech. Anal.*, **47,** 47–58, 1972.

130 Minty, G. On the maximal domain of a monotone function, *Michigan Math. J.*, **8,** 135–137, 1961.

131 Minty, G. Monotone (nonlinear) operators in a Hilbert space, *Duke Math. J.*, **29,** 341–346, 1962.

132 Minty, G. On a monotonicity method for the solution of nonlinear equations in Banach spaces, *Proc. Nat. Acad. Sci. U.S.A.*, **50,** 1038–1041, 1963.

133 Minty, G. On the monotonicity of the gradient of a convex function, *Pacific J. Math.*, **14,** 243–247, 1964.

134 Minty, G. A theorem on maximal monotone sets in Hilbert space, *J. Math. Anal. Appl.*, **11,** 434–439, 1965.

135 Minty, G. Monotone operators and certain systems of nonlinear ordinary differential equations, in *Proceedings of a Symposium on System Theory, Polytechnic Institute of Brooklyn, 1965*, pp. 39–55.

136 Minty, G. On a generalization of the direct method of the calculus of variations, *Bull. Amer. Math. Soc.*, **73,** 315–321, 1967.

137 Miyadera, I. Some remarks on semi-groups of nonlinear operators, *Tôhuku Math. J.*, **23,** 245–258, 1971.

138 Miyadera, I., Oharu, S. & Okazawa, N. Generation theorems of semigroups of linear operators, *Publ. RIMS Kyoto Univ.*, **8,** 509–555, 1973.

139 Mizohata, S. Le problème de Cauchy pour le passé pour quelque équations paraboliques, *Proc. Japan Acad.*, **34,** 693–696, 1958.

140 Moreau, J. J. Proximité et dualité dans un espaces hilbertien, *Bull. Soc. Math. France*, **93,** 273–299, 1965.

141 Oharu, S. On the generation of semigroups of nonlinear contractions, *J. Math. Soc. Japan*, **22,** 526–550, 1970.

142 Oharu, S. & Takahashi, T. A convergence theorem of nonlinear semigroups and its application to first order quasi-linear equations, *J. Math. Soc. Japan*, **26,** 124–160, 1974.

143 Okazawa, N. Two perturbation theorems for contraction semigroups in a Hilbert space, *Proc. Japan Acad.*, **45,** 850–853, 1969.

144 Okazawa, N. A perturbation theorem for linear contraction semigroups on reflexive Banach spaces, *Proc. Japan Acad.*, **47**, 947–949, 1971.

145 Okazawa, N. Operator semigroups of class (D_n), *Math. Japonica*, **18**, 33–51, 1973.

146 Ouchi, S. On the analyticity in time of solutions of initial boundary value problems for semi-linear parabolic differential equations with monotone non-linearity, *J. Fac. Sci. Univ. Tokyo, Ser. IA*, **21**, 19–41, 1974,

147 Pazy, A. Asymptotic behavior of the solution of an abstract evolution equation and some applications, *J. Differential Equations*, **4**, 493–509, 1968.

148 Phillips, R. S. Dissipative operators and hyperbolic systems of partial differential equations, *Trans. Amer. Math. Soc.*, **90**, 193–254, 1959.

149 Phillips, R. S. Dissipative operators and parabolic partial differential equations, *Comm. Pure Appl. Math.*, **12**, 249–276, 1959.

150 Pogorelenko, V. A. & Sobolevskiǐ, P. E. Hyperbolic equations in a Hilbert space, *Sibirsk. Mat. Ž.*, **8**, 123–145, 1967 (in Russian)

151 Sakawa, Y. Controllability for partial differential equations of parabolic type, *SIAM J. Control*, **12**, 389–400, 1974.

152 Sakawa, Y. Observability and related problems for partial differential equations of parabolic type, *SIAM J. Control*, **13**, 14–28, 1975.

153 Schechter, M. Integral inequalities for partial differential operators and functions satisfying general boundary conditions, *Comm. Pure Appl. Math.*, **12**, 37–66, 1959.

154 Schechter, M. General boundary value problems for elliptic partial differential equations, *Comm. Pure Appl. Math.*, **12**, 457–486, 1959.

155 Schechter, M. Remarks on elliptic boundary value problems, *Comm. Pure Appl. Math.*, **12**, 561–578, 1959.

156 Seeley, R. The resolvent of an elliptic boundary problem, *Amer. J. Math.*, **91**, 889–920, 1969.

157 Seeley, R. Norms and domains of the complex power A_B^z, *Amer. J. Math.*, **93**, 299–309, 1971.

158 Sobolevskiǐ, P. E. Equations of parabolic type in a Banach space, *Trudy Moskov. Mat. Obšč.*, **10**, 297–350, 1961 (in Russian); English translation, *Amer. Math. Soc. Transl. (2)*, **49**, 1–62, 1965.

159 Sobolevskiǐ, P. E. First-order differential equations in Hilbert space with a variable positive definite self-adjoint operator, a fractional power of which has a constant domain of definition, *Dokl. Akad. Nauk SSSR*, **123**, 984–987, 1958 (in Russian).

160 Sobolevskiǐ, P. E. Parabolic equations in a Banach space with an unbounded variable operator, a fractional power of which has a

constant domain of definition, *Dokl. Akad. Nauk SSSR*, **138**, 59–62, 1961 (in Russian); English translation, *Soviet Math. Dokl.*, **2**, 545–548, 1961.

161 Sobolevskiĭ, P. E. Degenerate parabolic operators, *Dokl. Akad. Nauk SSSR*, **196**, 302–304, 1971 (in Russian); English translation, *Soviet Math. Dokl.*, **12**, 129–132, 1971.

162 Stampacchia, G. Variational inequalities, *Actes, Congrès intern. Math.*, **2**, 877–883, 1970.

163 Strauss, W. A. Decay and asymtotics for $\square u = F(u)$, *J. Func. Anal.*, **2**, 409–457, 1968.

164 Strauss, W. A. On weak solutions of semi-linear hyperbolic equations, *Anais Acad. Brazil. Ciências*, **42**, 645–651, 1970.

165 Suryanarayana, P. The higher order differentiability of solutions of abstract evolution equations, *Pacific J. Math.*, **22**, 543–561, 1967.

166 Tanabe, H. On the equations of evolution in a Banach space, *Osaka Math. J.*, **12**, 363–376, 1960.

167 Tanabe, H. Convergence to a stationary state of the solution of some kind of differential equations in a Banach space, *Proc. Japan Acad.*, **37**, 127–130, 1961.

168 Tanabe, H. Note on singular perturbation for abstract differential equations, *Osaka J. Math.*, **1**, 239–252, 1964.

169 Tanabe, H. On regularity of solutions of abstract differential equations of parabolic type in Banach space, *J. Math. Soc. Japan*, **19**, 521–542, 1967.

170 Tanabe, H. & Watanabe, M. Note on perturbation and degeneration of abstract differential equations in Banach space, *Funkcialaj Ekvacioj*, **9**, 163–170, 1966.

171 Trotter, H. F. Approximation of semi-groups of operators, *Pacific J. Math.*, **8**, 887–919, 1958.

172 Trotter, H. F. On the product of semi-groups of operators, *Proc. Amer. Math. Soc.*, **10**, 545–551, 1959.

173 Ushijima, T. Some properties of regular distribution semigroups, *Proc. Japan Acad.*, **45**, 224–227, 1969.

174 Ushijima, T. On the strong continuity of distribution semigroups, *J. Fac. Sci. Univ. Tokyo, Ser. I*, **17**, 363–372, 1970.

175 von Wahl, W. Gebrochene Potenzen eines elliptischen Operators und parabolische Differentialgleichungen in Raumen hölderstetiger Funktionen, *Nachr. Akad. Wiss. Göttingen II, Math. Phys. Klasse Jahrgang*, Nr. 11, 231–258, 1972.

176 Watanabe, J. On certain nonlinear evolution equations, *J. Math. Soc. Japan*, **25**, 446–463, 1973.

177 Webb, G. Continuous non-linear perturbations of accretive operators in Banach spaces, *J. Func. Anal.*, **10**, 191–203, 1972.

178 Yagi, A. On the abstract evolution equations in Banach spaces, *J. Math. Soc. Japan*, **28,** 290–303, 1976.

179 Yosida, K. A perturbation theorem for semi-groups of linear operators, *Proc. Japan Acad.*, **41,** 645–647, 1965.

Index